Graphene

Graphene

A New Paradigm in Condensed Matter and Device Physics

E. L. Wolf

Professor of Physics, Polytechnic Institute of New York University

OXFORD

UNIVERSITY PRESS

OXFORD
UNIVERSITY PRESS

Great Clarendon Street, Oxford, OX2 6DP,
United Kingdom

Oxford University Press is a department of the University of Oxford.
It furthers the University's objective of excellence in research, scholarship,
and education by publishing worldwide. Oxford is a registered trade mark of
Oxford University Press in the UK and in certain other countries

First Edition published in 2014
Impression: 1

Published in the United States of America by Oxford University Press
198 Madison Avenue, New York, NY 10016, United States of America

British Library Cataloguing in Publication Data
Data available

Library of Congress Control Number: 2013940539

ISBN 978–0–19–964586–2

Printed and bound by
CPI Group (UK) Ltd, Croydon, CR0 4YY

Dedicated To
P. R. Wallace
G. W. Semenoff
A. K. Geim
K. S. Novoselov

Preface

Graphene, discovered in 2004 as a new phase of crystalline matter one atom thick, exhibits electronic conduction distinct from and superior to conventional metals and semiconductors and thus opens new opportunities for device design and fabrication. The single plane of graphite is now known to represent a new class of two-dimensional materials, one to several atoms in thickness that are conventionally crystalline in lateral dimensions micrometers to centimeters. The electrons in graphene move rapidly in a way that resembles massless photons and nearly massless neutrinos. They indeed exhibit Klein tunneling, a quantum phenomenon of unit probability specular tunneling through a high potential barrier that was originally conceived as a property of electrons and positrons in vacuum. These remarkable properties, endowed by the hexagonal "honeycomb" carbon-atom array onto ordinary electrons, fortunately are well explained by the methods of condensed matter physics, but that "explaining" has several initially puzzling aspects that we address.

This book is intended as such an explication: to introduce and simply explain what is so remarkably different about graphene. It describes the unusual physics of the material, that it offers linear rather than parabolic energy bands. The Dirac-like electron energy bands lead to high constant carrier speed, similar to light photons. The lattice symmetry further implies a two-component wave-function, which has a practical effect of cancelling direct backscattering of carriers. The resulting high carrier mobility allows observation of the quantum Hall effect at room temperature, unique to graphene. The material is two-dimensional, and in sizes micrometers to nearly meters displays great tensile strength but vanishing resistance to bending. We are intent as well to summarize the progress toward better samples and the prospects for important applications, mostly in electronic devices. The book is aimed at researchers and advanced undergraduate and beginning graduate students as well as interested professionals. This book is intended not as a text but a comprehensive summary and resource on a scientific and technological area of rapid advance and promise. The hope is to span the range between the painstaking small-science extraction from graphite of high quality graphene layers (that are, of course, part of every pencil lead) and high flying physics topics, including an anomalous integer quantum Hall effect at room temperature, bipolar transmission of Cooper pairs in a superconducting proximity effect, light-like charged particles only explainable by the Dirac equation, evidence for a unit-probability tunneling behavior (Klein tunneling) heretofore predicted, but never before observed. This book is also intended to suggest possibilities for new families of electron devices in a post-Moore's Law version of nanoelectronics. Benzene rings, whose "radii" are about 0.190 nm, are excellent conductors (it is estimated that a screening current of $\sim$3.9 nanoamperes/Tesla is estimated to flow around a benzene ring at room temperature) and might be viewed as basic units in graphene electronics.

Silicon and metal crystals lose their conductivity in small scale structures but the basic unit of graphene, essentially a benzene ring, still conducts well. A form of "chemistry" appears in the arrangement of broken bonds at the edges of graphene ribbons (e.g., terminations in zig-zag vs. armchair edges). A premium now is placed on experimental methods for epitaxial growths, from which a new "semiconductor technology" might arise. Device technologies that will necessarily depend on fabrication and patterning schemes for graphene layers are in a rapid state of development.

This book is dedicated to four physicists, two of them theorists and two experimentalists. P. R. Wallace first understood the unusual linear bandstructure in graphene (conceived as an approximation to graphite). G. W. Semenoff first understood the unusual two-sublattice origin of the chiral carriers, avoiding conventional backscattering and improving the mobility. A. K. Geim and K. S. Novoselov, two brilliant, resourceful and persistent experimentalists, showed how to isolate the individual planes and convincingly demonstrated their unique properties, indeed as representative of a new class of two-dimensional crystals.

The author is grateful to Sönke Adlung and Jessica White at Oxford University Press for invaluable help in conceiving and completing this project. It is a pleasure to acknowledge assistance from the Department of Applied Physics at NYU Poly, particularly from Prof Lorcan Folan and Ms. DeShane Lyew. Ankita Shah, Harsh Bhosale, Vijit Jain, Kiran Koduru and Manasa Medikonda, who have assisted with aspects of preparing the manuscript and clearing the way to publication. My wife Carol has helped in many ways and has been a constant source of support and encouragement.

E. L. Wolf, Brooklyn, New York, September, 2013

Contents

1
Introduction

"Graphene" is the name given to a single-layer hexagonal lattice of carbon atoms, an extended two-dimensional lattice of benzene rings, devoid of hydrogen atoms. This one-atom-thick material has recently been found to be robust, if not completely planar, in samples tens of micrometers up to 30 inches in extent, on a supporting substrate. Graphene is a contender in the new information technology (and other) applications, beyond being a scientific breakthrough and curiosity. As we will see, electrons in graphene display properties similar to photons and neutrinos, never before observed in a condensed-matter environment. The new electron properties arise in a straightforward way from the symmetry of the atomic positions and the resulting cone-shaped, rather than parabolic, regions in the electron energy surfaces. It is reassuring to see that all the new effects are well described by the Schrödinger-equation-based methods of condensed matter physics that have served well in understanding solids from semiconductors to superconductors. Dirac-equation-like electron behavior in graphene is obtained directly from appropriate simplification of the Schrödinger theory of atoms, molecules and solids. Beyond this, graphene is the first example of a new class of two-dimensional crystals, a new phase of matter. This is a surprise in many ways that offers new opportunities, especially, in electronics.

The discovery of graphene extends, beyond some theoretical predictions, what useful forms matter can take. It is truly a new paradigm.

1.1 "Crystals" one atom thick: a new paradigm

A crystal is an ordered array of identical repeating units. We can think of the unit, in graphene, as the hexagonal benzene ring, whose diameter (between opposite carbon atoms, say those numbered 1 and 4) is $2a = 284$ pm,[1] where a is the carbon-carbon spacing (the 1–2 distance) $a = 142$ pm. Benzene, C_6H_6, has one electron per atom binding a hydrogen atom at each ring location 1, 2, . . . to 6. In graphene, H is absent and that one electron per atom is delocalized over the whole crystal. The resulting perfectly ordered honeycomb lattice, for a 10 μm sheet, is thus 35 211 benzene ring diameters (at 284 pm/ring) in linear size, certainly showing long-range order.

---

[1] One picometer (pm) $= 10^{-12}$ m. Common units on the atomic scale include Ångströms (10^{-10} m) and nanometers 1nm $= 10^{-9}$ m. The Bohr radius of the hydrogen atom, also taken as the base unit of length on the atomic scale, is $a_0 = 0.0529$ nm.

(Actually the lattice repeat distance, the "cell constant," is the 1–3 distance in the ring, namely 246 pm.) And for the 30-inch sample the number of benzene ring diameters is 2.68 billion! (The corresponding two-dimensional honeycomb crystalline array will then certainly have defects, grain boundaries and dislocations, as are well known in conventional crystals.)

The honeycomb array in graphene is dictated by the facile three-fold planar bonding, via Schrödinger's equation, of the quantum states of the carbon atom called 2s and 2p (discussed in Chapter 3). (One possibly might ask how honeybees chose the honeycomb lattice, composed of hexagons? Perhaps in the evolution of honeybees, the 3-fold lattice (that would put centers in all the hexagons) did not leave enough room for honey, and the cubic 4-fold lattice might collapse flat, like a cardboard box without the ends, squeezing the honey out.)[2]

In fact, the most economical description of the honeycomb lattice is that generated by fundamental translations of the *basis atoms* 1 and 2 (called A and B by physicists). This two-atom unit, when translated by ± multiples of the translation vectors : **1→3**, **1→5** gives the honeycomb lattice.[3]

The angle between these vectors is 60°, and we see that atoms 1, 3, 5 and 2, 4, 6 form triangles (they lie on the A and B sublattices, respectively), and that the two sublattices are separated by the interatomic vector **1→2**. So the honeycomb lattice is fundamentally two interpenetrating triangular lattices, known as A and B. Thus *nearest-neighbor* atoms lie on *opposite* sublattices, with profound consequences in the unusual electronic bandstructure, as first recognized by the American physicist Wallace in 1947.

But the achieved 30-inch, one-atom-thick graphene sample certainly will be so floppy that it will have to be supported on some surface. This is the real question as to whether it is a crystal. If we imagine the honeycomb sheet as unsupported, we realize it is very susceptible to being bent out of its flat planar condition. The chemical bonds ("pi-bonds" = "π-bonds" between two $2p_z$ electrons) will tend to return it to a flat planar condition, but this restoration force is weak. The large graphene sheet is very strong in tension, but weak against flexing motion. It is like a bedsheet, in being flexible but inextensible, but, unlike a bedsheet, it retains a weak restoring force toward a perfectly planar condition. We may italicize the word "crystal," because inherent in two dimensions (2D) (embedded in three dimensional space) are long-wavelength flexural phonons that allow large root-mean-square (rms) fluctuational displacements, much larger than a lattice constant. How floppy the sheet will be depends on its size, as we will see shortly. It may be a matter of semantics whether a slightly bent crystal is still crystalline. From a familiar example: on a diving board, the deflections imposed by the diver's weight exceed the cell dimension, but obviously do not suggest

[2]Why the honeybee evolution avoided 5-fold rings or tilings, all having unequal angles that do not permit an infinite crystal (but of course a honeycomb is finite), may have to do with eyes and brains better able to generate 120° angles, than the several angles in any 5-fold tiling.)

[3]A slightly different definition of the basis vectors as **1→3**, **5→1** is given in Fig. 1.2b. In that choice, the angle between the basis vectors is 120°. Figure 4.1 shows the same choice of basis vectors as our present text.

collapse of the material supporting the diver. By formal definition, long-range order does not occur, but in practice the local distortions can be small, so that it is still useful to consider the sample as a crystal, if slightly distorted. For graphene in practice, the out-of-plane deflections are the main concern as to whether the system is crystalline.

But there is more, fortunately not of much practical importance, to the story of crystallinity in two dimensions. In addition, there are more subtle points, really only of academic interest that lead theorists to say that any 2D array, *even if arbitrarily kept absolutely planar*, cannot have long-range (infinite) order except at $T = 0$. (The planarity would have to be imposed without transverse pinning; the closest system of this type may be electron crystals on the surface of liquid helium.) We will discuss these points in Chapter 2, including a proof that an infinitely large 2D array would exhibit, at any finite temperature, large absolute in-plane motions (but without sensibly affecting local inter-atom distances). This might have a real effect, for example, in smearing the electron- or x-ray- diffraction spots on a sufficiently large sample, unless that sample was in effect pinned to be stationary at the measurement site. But since the phonon wavelengths (now in 2D) involved are large, local regions move *intact* so that local order is not disrupted. For example, the cohesive energy of the system is not reduced and this has nothing to do with the melting point of the system (that we connect with local order). In the words of Das Sarma (2011) "There is nothing mysterious or remarkable about having finite 2D crystals with quasi-long-range positional order at finite temperatures, which is what we have in 2D graphene flakes." We return to this subject in Chapter 2, but simply comment here that the academic points in the literature do not in any way detract from the important potential uses of graphene in electronics and nano-electromechanical systems, as examples.

While there had been earlier suggestions that the single planes of graphite might be extracted for individual study (contrary to a theoretical literature that suggested that crystals in two dimensions should not be stable), Novoselov *et al.* (2004, 2005) were the first to demonstrate that such samples were viable, and indeed represented a new class of 2D materials with useful properties and potential applications. (Hints toward isolating single layers had earlier been given by Boehm *et al.* (1962), Van Bommel *et al.* (1975), Forbeaux *et al.* (1998) and Oshima *et al.* (2000), among others. And, as we will see in Section 5.1.2, chemists, since 1859, with notable work in 1898, have developed bulk processes to "exfoliate" graphite, as extracted from the ground, into "expanded," typically oxidized, forms exposing, to a greater or lesser degree, the individual planes now called graphene.)

On small size scales, perhaps 10 nm to 10 μm, the graphene array of carbon atoms is "crystalline," and has sufficient local order to provide electronic behavior as predicted by calculations based on an infinite 2D array. Micrometer-size samples of graphene show some of the best electron mobility values ever measured. In microscopy, on scales 10 nm to 1 μm, it sometimes may appear that the atoms are not entirely planar, but undulate slightly out of the plane. While it has been suggested that such "waves" are intrinsic (Morozov *et al.*, 2006), it is quite likely, on the contrary that they actually originate as the classical response of the thin membrane to inevitable stress from its mounting, or as a result of adsorbed molecules, since in graphene every carbon atom is exposed. Monolayer graphene is strong and continuous, but, because of its small thickness,[4] $t \sim 0.34$ nm, all but the shortest samples are extremely "soft" in the sense of easily bending with a small transverse force. This can be understood from the classical "spring constant K"[5] for deflection x of a cantilever of width w, thickness t and length L (with Young's modulus Y) under a transverse force F: $F = -Kx$. Since $K \sim Ywt^3/L^3$ (discussed in Section 7.4), with t near a single atom size, one sees that graphene, in spite of a large value of Young's modulus, $Y \sim 1$ TPa, is the softest possible material against transverse deflection.

As we will see in Chapter 7, graphene rectangles, length L, width w and thickness t, quantitatively bend and vibrate as predicted by classical engineering formulas. For example, the spring constant K defined for deflection and applied force at the center of a rectangle clamped on two sides depends strongly on the dimensions as $K = 32Yw\,t^3/L^3$. A square of graphene, of size $L = w = 10$ nm, from the above formula, gives $K = 12.6$ N/m, while a square of size 10 μm has $K = 12.6 \times 10^{-6}$ N/m. If the sample is short, approaching atomic dimensions, the spring constant is large and the object appears to be rigid. For example, the spring constant of a graphene square ten benzene molecules on a side against bending can be estimated as ~156 N/m, using the formula, while the spring constant of a carbon monoxide (CO) molecule (in extension), deduced from its measured vibration at 64.3 THz, is known to be 1860 N/m. A further quantity in the graphene literature is Yt, a 2D rigidity that has a value of about 330 N/m. But for graphene longer than a few micrometers, with the spring constant K of a square falling off as $1/L^2$, the material is exceedingly soft.

Accordingly, graphene, on micrometer-size scales, conforms to any surface under the influence of attractive van der Waals forces. In an electron micrograph, graphene on a substrate appears adherent, more like a wet dishrag or "membrane" than a playing card, quite unlike a 10-inch diameter wafer of silicon. These 2D "crystals" cannot, at present, be grown from a melt, as is silicon and as were graphite and diamond in the depths of the earth at high temperature. Graphene crystals can only be obtained (see Chapter 5) by extraction from an existing crystal of graphite, or by being grown epitaxially on a suitable surface such as SiC or catalytically on Cu or Ni from a carbon-bearing gas such as methane.

---

[4]The space per layer in graphite is 0.34 nm that is widely quoted as the nominal thickness of the graphene layer. An equivalent elastic thickness of graphene, closer to the actual atomic thickness, is about 0.1 nm, see Section 2.7.

[5]The spring constant K is a macroscopic dimension-related engineering quantity quoted in SI units as N/m. It is related to the "bending rigidity" or "rigidity" $\kappa = Yt^3$, a microscopic property usually quoted in eV that is about 1 eV for graphene. (The Young's modulus Y, an engineering quantity, is defined as pressure/(relative strain) $= P/(\delta x/x)$ and is about 10^{12} N/m^2 = 1 TPa for graphene, but see Section 2.7.1) The rigidity κ has units of energy, as force times distance. One sees that the rigidity κ of graphene, by virtue of the minimal atomic value of thickness t, is the lowest of any possible material. In connection with extension of a chemical bond, the spring constant K relates to the bond energy E as $K = d^2E/dx^2$.

Notably, graphene is an excellent electronic conductor, somewhat like a semimetal, but with conical rather than parabolic electron energy bands near the Fermi energy with a characteristic *linear* dependence of energy on crystal momentum, $k = p/\hbar$: i.e., $E = \text{``}pc\text{''} = c^*\hbar k$. These electrons move like photons, at speed $c^* \approx 10^6 \text{m/s}$ and with vanishing effective mass. There is nothing magic about this; it simply results, in band theory, from the particular crystal lattice. This aspect also presents a new paradigm in the realm of condensed matter physics. Not only is Graphene nature's closest approach to a two-dimensional (2D) self-supporting material, it also has charge carriers moving in a different way, as if their mass were zero. The physics of the situation also confirms that "back-scattering" is "forbidden" leading to measurably larger carrier mobility.

In the real world of atoms, no material can be mathematically two dimensional: the probability distribution $P(x,y,z)$ must extend in the z-direction by at least one Bohr radius. There are well-known examples of 2D subsystems of particles, notably electrons on the surface of liquid helium and the "2-DEG" two-dimensional electron gases engineered into leading semiconductor devices. The latter useful electron systems are supported by quantum well heterostructures. The remarkable difference, in graphene, is that there is no external supporting system, the layer of carbon atoms is the mechanical support, as well as the medium exhibiting light-like propagation of electrons. How is this possible?

The answer was not entirely clear before the discoveries of Geim and Novoselov: indeed the existence of free-standing graphene layers with novel electronic properties was a surprise, worthy of a Nobel Prize in Physics. Other one-layer materials include BN $(BN)_n(C_2)_m$, with n, m, integers; MoS_2, TaS_2, $NbSe_2$ and the superconductor $Bi_2Sr_2CaCu_2O_x$, although the last is seven atoms thick (Novoselov *et al.*, 2005). So the Nobelists, in fact, confirmed the practical reality of a new class of 2D locally crystalline materials.

The binding energy of a crystal, an extended periodic array of atoms, for temperatures below a melting temperature, T_M, is a subject of solid state physics. The methods of this discipline do not always predict binding of an infinite 2D crystal. Indeed, thin layers of many substances are found to break up into "islands" as their thickness is reduced, especially if the attraction of atom to substrate exceeds the attraction atom-to-atom. This island breakup definitely does not occur with graphene: on the contrary, graphene is found to be among the strongest known materials under tension. Tenth-millimeter scale sheets of one-atom-thick graphene have been studied as elastic beams and sheets, whose vibrational frequencies have been measured consistent with a Young's modulus ~ 1 TPa. At a lattice constant of 0.246 nm, a 20 μm graphene sheet (80 000 unit cells) looks flat, if suspended across a trench, but may bend in response to van der Waals forces from the mounting. In some cases, 10 nm-scale "waves" or "ripples" of ~ 1 nm amplitude have been inferred from transmission electron microscope measurements, with a likely origin in a combination of molecular surface adsorbates and mounting strain. Subtle physics is involved here, but experiments trump the situation; these "crystals" are large enough to be useful in many circumstances.

Molecules have vibrations: in an extended crystal these are called phonons. The vibrational motions of molecules are 3D in nature and any real 2D crystal[6] will have vibrational motion in the z-direction, termed flexural. In an extended real 2D sample the flexural motion extends to low frequency and large amplitude, at any finite temperature T. Even when restricted completely to planar motion, the methods of solid state physics have predicted that thermal vibrations at any finite temperature will lead to excessive transverse motion and destroy long-range (as distinct from short-range) order (Mermin and Wagner, 1966; Mermin, 1968). These theorems, it seems, do not prohibit (Das Sarma *et al.*, 2011) the observed finite size samples of graphene, although the theoretical predictions probably had a deterring effect on experimenters prior to the pioneers, Geim and Novoselov.

The benzene ring is planar, with restoring forces against bending. The two- and three-ring compounds naphthalene and anthracene, respectively, are also planar. We saw above that an effective spring constant $K \approx Ywt^3/L^3$ provides a useful estimate of the rigidity, depending on the length L. If graphene is an extended polymer of benzene rings, with the same 2s and 2p electrons (4 per atom) supporting the structure, it should similarly resist bending, tending to remain planar. Yet the resistance to bending is not so strong as to prevent rolled forms of graphene, including carbon nanotubes and scrolls (Braga *et al.*, 2004). An energy in favor of the 3D conformation is the sum of bonding energies of otherwise dangling or weakly satisfied bonds and of course, the van der Waals attraction that binds graphene layers into graphite. More stable states of carbon will clearly occur in 3D, where roughly there will be 6–8 nearest neighbors vs. 3 or 4 in a 2D configuration. A typical conformation of an extended micron-scale sheet of graphene laid onto a substrate is shown in Fig. 1.1. The behavior is that of a limp but nearly inextensible sheet, with wrinkles and conical cusps where there are singular elastic strains. It resembles a wrinkled sheet of paper, except that when removed from the substrate the wrinkles and conical cusps will disappear. The classical physics of this situation is described in Section 2.7.3 "Isometric distortions of a soft inextensible membrane".

On size scales of tens to hundreds of micrometers, adequate for electronic devices, ambient temperature graphene is planar, exhibiting high carrier mobility. We will see later (Section 7.1.5) that removing adsorbates by heating to a modest temperature ~400°C can greatly improve the carrier mobility and electrical conductivity. There is some danger that unrecognized adsorbates have led to false impressions of "intrinsic" rippled behavior in graphene. We find that most small-scale non-planarity in the form of "corrugations" with lateral size scales of 10 to 100 nm with ~1 nm vertical amplitude, are the response of the nearly inextensible sheet to boundary conditions that inevitably introduce strain. The system will minimize the energy cost of the waves by maximizing their wavelength.

---

[6] A useful notation for a "real 2D crystal" is "2D-3" meaning that motion into the third dimension is available. A "pure" 2D system is one, like electrons on the surface of liquid helium, where no motion into the third dimension is allowed, the z-motion is represented by a single quantum state. We take the electron system inside graphene to be "pure 2D" as confined by the graphene lattice, even though that lattice may slightly undulate or flex into the third dimension.

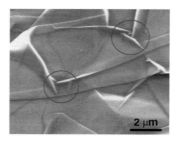

Fig. 1.1 Micrometer-scale transmission electron microscope (TEM) image of graphene sheet draped onto a substrate. It can be seen that the sheet is continuous but has little resistance to bending and folding. On a micrometer scale the single atomic layer, showing ripples with cusps or corners at their terminations, conforms to its supporting surface under the influence of van der Waals forces. Under tension it is very strong, with Young's modulus on the order of 1 TPa. Only on a much smaller size scale can it be considered to be stiff or rigid against bending. (From Pereira *et al.*, 2010. © 2010 by the American Physical Society).

1.2 Roles of symmetry and topology

The interesting observed electronic properties of graphene are usefully predicted by theoretical solid state physics. The hexagonal honeycomb lattice, formally a triangular lattice with two atoms per unit cell (in 2D), leads to electronic bands in the usual way. In fact the electron bands, if not the subtle implications of their symmetry, were correctly predicted starting in 1947 (Wallace, 1947; McClure, 1956). The Brillouin zone is shown in Figs. 1.2a and 1.3b; the unit cell in Fig. 1.2b. A view of the lattice is again shown in Fig. 1.3a.

It is of extreme importance that A and B sublattices interpenetrate to form the honeycomb and the two sublattices, generated by the two atoms per unit cell, represent two separate groups of allowed states or bands. A carrier can be described by a two-component wavefunction, conveniently as if it had an "iso-spin" one-half that is not related to the physical spin of the electron.

1.2.1 Linear bands, "massless Dirac" particles

From an electronic band structure point of view, the essential novel feature is a set of linear electron bands: twin collinear intersecting vertical cones with apices at inequivalent corners **K, K′** of the hexagonal Brillouin zone. The Fermi energy lies near these degenerate intersection points in pure material, but E_{F} can be pulled up into the upper cones, making an n-type metal, or depressed into the lower cones, making a p-type metal, by an electric field (or by chemical doping). While the pure crystal at $T = 0$ has zero carriers, the material is typically observed to have finite conductivity, on the order of e^2/h. (With further work, this aspect has turned out to be an experimental artifact and a really pure and ordered sample is insulating at $T = 0$.) In

(a)

(b)

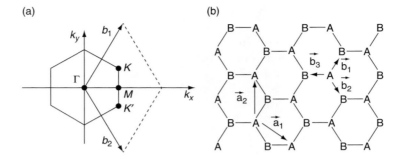

Fig. 1.2 (a) Hexagonal Brillouin zone of honeycomb lattice, resulting from A and B inter-penetrating triangular lattices. If the nearest-neighbor distance is $a = 142$ pm, the lattice constant is $3^{1/2}a$ and the zone boundary M (half the reciprocal lattice vectors $\mathbf{b_1}$, $\mathbf{b_2}$), is $2\pi/3a$. The coordinates of the corner point K are $(2\pi/3a, \pi/3\sqrt{3}a)$ so that the distance from the origin to point K is $4\pi/(3\sqrt{3}a)$. Since the conduction and valence bands touch precisely at K, we have $k_\mathrm{F} = |\mathrm{K}|$ and the Fermi wavelength $\lambda_\mathrm{F} = 2\pi/k_\mathrm{F} = 3\sqrt{3}a/2 = 369$ pm. (b) Honeycomb lattice, resulting from A and B interpenetrating triangular lattices. If the nearest-neighbor distance is $a = 142$ pm, the lattice constant is $3^{1/2}a = 246$ pm. The basis vectors of the triangular lattice are $\mathbf{a_1} = (\sqrt{3}/2, -1/2)a$, $\mathbf{a_2} = (0,1)a$, and the sublattices are connected by $\mathbf{b_1} = (1/2\sqrt{3}, 1/2)a$, $\mathbf{b_2} = (1/2\sqrt{3}, -1/2)a$, $\mathbf{b_3} = (-1/\sqrt{3}, 0)a$. (From Semenoff, 1984. © 1984 by the American Physical Society).

(a)

(b)

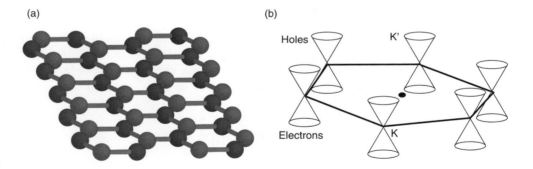

Fig. 1.3 (a) Honeycomb lattice emphasizing its composition as two interlocking triangular lattices. One may imagine the light shaded balls as the A lattice, the dark shaded balls as the B lattice. (b) Hexagonal Brillouin zone of honeycomb lattice, showing intersecting conical electron bands. The linear bands and the two "valleys" at K and K′ give unusual electronic properties to graphene. In a pure sample the lower cones are filled with electrons and the upper cones are empty.

experimental practice, the resistivity is typically less than $h/e^2 \sim 25$ kΩ (Ohms per square [$\Omega/\square$] in 2D).

This gives an aspect of a zero-band-gap semiconductor, except for the linear dispersion (Wallace, 1947). (This is in contrast to the usual case, as in Si and GaAs, where the electron bands have parabolic minima, leading to effective masses given by $m^* = \hbar^2/\mathrm{d}^2 E/\mathrm{d}k^2$, with the result that carrier speeds vary linearly with crystal momentum, $k = p/\hbar$.) In this linear region for graphene, the particle velocity, $c^* = v_\mathrm{F} = 3ta/(2\hbar)$ (Wallace, 1947; Katsnelson and Novoselov, 2007), is constant, entering light-like relations, $E = $ "pc" $= c^*\hbar|\mathbf{k} - \mathbf{K}|$, valid near each "Dirac point". In this expression t is a nearest neighbor hopping energy, about 2.8 eV and a is the nearest-neighbor distance, 142 pm.

While the linear dispersion near the Fermi energy is novel for electrons, the calculation, including the speed c^*, is straightforward in the conventional, Schrödinger-equation-based, "tight-binding" method, using a nearest- neighbor hopping interaction t and bond length a. Taking t as 2.8 eV and $a = 142$ pm, one finds $c^* = 0.91 \times 10^6$ m/s, or about $c/300$. The connection between the linear dispersion and term "massless fermion" comes from the total energy formula of special relativity, $E = [(pc)^2 + (mc^2)^2]^{1/2}$ where the observed linear dependence $E = pc$ arises in the limit $m = 0$. The conical band structure, as mentioned, predicts zero conductivity for neutral graphene with the Fermi energy at the Dirac point where the density of states vanishes. Clearly, because (at zero temperature) "undoped graphene has no free electrons, an infinite sample cannot conduct electricity," as stated by Snyman and Beenakker (2007).

Experiments (Novoselov *et al.*, 2004) showed that the electrical conductivity as a function of charge density of graphene rises symmetrically on either side of a minimum at the neutrality point. In the experiment, a gate applied electric fields to induce charge of either sign, much as in a field-effect transistor. The sharp peak in resistivity at the crossing point, and other features, are confirmed by Zhang *et al.* (2005), as shown in Fig. 1.4.

As indicated in Fig. 1.4, the mobility in graphene rises as the carrier concentration falls, and values as high as 200 000 cm^2/Vs = 20 m^2/Vs have been obtained in suspended samples (Bolotin *et al.*, 2008).

While the mobility is clearly sample dependent, the maximum resistivity values near $h/4e^2 \approx 6.45$ kΩ, were initially suggestive of a quantum condition, catalogued by Novoselov *et al.* (2005) as shown in Fig. 1.5, a bit larger than the ~ 4 kΩ shown in Fig. 1.4. The exact formula for the quantum limit was still discussed until 2011. The leading theoretical value, $h/4\pi e^2$, seems small by a factor of π, but is in disagreement with a basic paper (Abrahams *et al.*, 1979) that predicts an infinitely large resistivity at zero temperature.

It is now believed that the "minimum conductivity," a practical matter in dealing with graphene, is actually an artifact of extrinsic electron and hole "puddles," for which similar theoretical conductivity estimates can be found, on the basis of percolation and tunneling between adjacent puddles. An early indication of this non-metallic behavior was offered by Bolotin *et al.* (2008) who described "a nonuniversal conductivity that

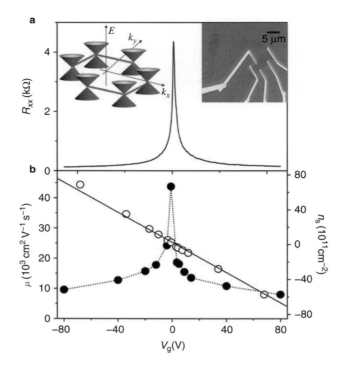

Fig. 1.4 Electric field effect in single-layer graphene obtained by variation of voltage V_g on gate underlying the sample (see the upper right inset). (a) A resistance maximum about 4 kΩ is seen here at 1.7 K at gate voltage corresponding to the neutrality point, where Fermi level drops to the apices shown in upper left inset. (b) The carrier density n, shown as open circles, from a Hall measurement, and the mobility are plotted. The mobility, on the order of 4 m^2/Vs at peak, implies a long mean free path $\lambda \sim 1$ μm. The line in lower panel, well matching the measured carrier density points, comes from an estimate of the charge induced by the gate voltage, $n = C_g V_g / e$, where $C_g = 115$ $aF/(\mu m)^2$ is obtained from the geometry. (From Zhang *et al.*, 2005, by permission from Macmillan Publishers Ltd., © 2005).

decreases with decreasing T," after applying a heating procedure to their 4-point-probed graphene sample to remove adsorbed impurities. The simple cleaning procedure was found to increase the conductivity by about a factor of 10. Surface scattering is thus a factor in much of the literature before 2008, for example, in the earlier work in the Kim group, Tan *et al.* (2007).

A recent experiment on screened graphene which succeeded in nearly removing the "puddles," reveals what appears to be a conventional Mott–Anderson transition to an insulating state at low temperature, shown in Fig. 1.6.

These new results are revealed by Ponomarenko *et al.* (2011) in a paper entitled "Tunable metal-insulator transition in double-layer graphene heterostructures". The

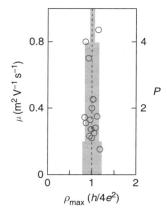

Fig. 1.5 Maximum values of resistivity ρ_{max} (circles) exhibited by devices with mobilities (left y axis). Histogram in darker shade shows the number of devices P exhibiting ρ_{max} within 10% intervals around the average value $\sim h/4e^2$. Bolotin *et al.* (2008) found that mobilities about a factor of 10 larger were easily available by gently heating the samples to release adsorbed gases that evidently strongly scatter charge carriers. (From Novoselov *et al.*, 2005, by permission from Macmillan Publishers Ltd., © 2005).

tuning is of the carrier concentration in the graphene layer intervening between the (puddle-inducing) substrate, containing charged impurities, and the measured graphene layer. The two graphene layers are separated by $d = 12$ nm and the concentration in the screening layer is 3×10^{11} cm^{-2}. The 12 nm spacing is large enough that the measured graphene layer does not exchange carriers with the screening layer: more details are contained in Section 7.5. (A tunneling FET configuration is attainable at smaller spacings d, as will be described in Chapter 9.)

These features are in accord with the band diagram in Fig. 1.3 and the upper left inset to Fig. 1.4 and are further discussed in Section 7.5. But there are further unusual aspects of graphene, subtle but important, due to the dual sublattices arising from the two atoms per unit cell that we now discuss.

1.2.2 "Pseudo-spins" from dual sublattices and helicity

The need for a two-component or "pseudo-spin" electron wavefunction near these "Dirac points" was more recently realized (DiVincenzo and Mele, 1984; Semenoff, 1984; Novoselov, *et al.*, 2004, 2005, 2006, 2007). These workers showed that the double-sublattice origin of the states in the cones **K, K′** requires such a treatment, based on a "pseudo-spin" of lattice origin. Semenoff described the graphene Brillouin zone as having two "right- and left-handed degeneracy points" (where valence and conduction bands meet). In each hexagonal ring, three atoms are in the A lattice and three atoms are in the B lattice, as we have seen. (We will return to the band structure of graphene

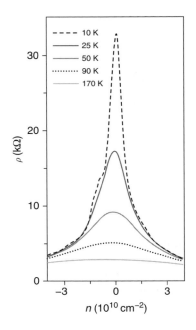

Fig. 1.6 Maximum value of resistivity more recently measured is in fact strongly temperature dependent, and reaches 33 kΩ (about 5.12 $h/4e^2$, well above the Anderson localization value) at 10 K. This is now regarded as intrinsic behavior, attained by screening away extrinsic potential fluctuations that supported "puddles" of locally high and conducting electron and hole concentrations, now recognized as subtle artifacts. The authors describe the curves in Fig. 1.6 as exhibiting an "insulating temperature dependence". This is a device of the type shown in Fig. 7.31a with BN spacer 12 nm and screening layer concentration 3×10^{11} cm^{-2}. (From Ponomarenko *et al.*, 2011, by permission from Macmillan Publishers Ltd., © 2011).

in Chapter 4.) The tight-binding Hamiltonian used is of the simplest form, allowing only nearest-neighbor interactions $t(a_i^* b_j + b_i^* a_j)$ (on opposite sublattices) plus second-nearest-neighbor interactions (same sublattice): $t'(a_i^* a_j + b_i^* b_j)$.[7]

By the symmetry of the lattice, the Schrödinger theory expression simplifies into a Dirac-like form giving conical bands.

In this range of energies, $|E| <\approx 1$ eV, the Hamiltonian for the single-layer electron system, making use of the "tight binding" and "k • p" approximations, widely used in semiconductor physics (Yu and Cardona, 2010; Semenoff 1984), reduces to the matrix form

$$\hat{H} = \hbar c^* \begin{pmatrix} 0 & k_x - i k_y \\ k_x + i k_y & 0 \end{pmatrix} = \hbar c^* \sigma \bullet \mathbf{k} \qquad (1.1)$$

---

[7]In such expressions, $a_i^*(a_i)$ represent operators that create (destroy) an electron on site i.

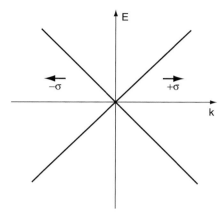

Fig. 1.7 Model energy dispersion $E = \hbar c^* |\mathbf{k}|$ at each Dirac point. The ascending line arises from one sublattice, the descending line from the other, suggesting that transitions between branches are "spin-forbidden". Back-scattering is thus reduced, making the quantum Hall effect observable even at room temperature.

where $\mathbf{k}$ is the quasiparticle momentum and $\boldsymbol{\sigma}$ the two-dimensional Pauli matrix. This reduction of the Hamiltonian into a form similar to the Dirac equation for massless particles comes from the crystal symmetry and the two equivalent sublattices A and B. The cosine-like energy bands (each C atom has one participating $2p_z$ valence electron that form π^* anti-bonding bands at positive energy and π bonding bands at negative energy) intersect in the cones at the six corners of the Brillouin zone. That the electronic states arise from two (A and B) triangular sublattices leads, as mentioned above, to a two-component wavefunction (Semenoff, 1984) mathematically treated as spin $^1/_2$. The Pauli matrix $\boldsymbol{\sigma}$ here refers to the *pseudo-spin*, not to the actual electron spin that is neglected at this point. These features only appear near the degeneracy (Dirac) points, where their importance is clear from measurements on graphene.

The wavefunction that is needed near the Dirac point K' can be written (Kane and Mele, 1997; Castro Neto *et al.*, 2009) as

$$\psi_{\pm,\mathbf{K}'}(\mathbf{k}) = \frac{1}{\sqrt{2}} \begin{pmatrix} e^{i\theta_{\mathbf{k}}/2} \\ \pm e^{-i\theta_{\mathbf{k}}/2} \end{pmatrix} \tag{1.2}$$

$$\theta_{\mathbf{q}} = \arctan\left(\frac{q_x}{q_y}\right) \tag{1.3}$$

for π^* states above $(+)$ and π states $(-)$ below the Fermi energy. Here $\mathbf{q}$ is the momentum measured from the Dirac point $\mathbf{K}'$, and the angle θ is measured around that point. Similar equations apply for point $\mathbf{K}$, with the opposite choice of $\pm$ sign in eqn (1.2), to reverse the helicity. In both cases, the two components of the wavefunction indicate the separate contributions from the separate sublattices, A and B. Thus one may say that each π-electron carries, in addition to its physical spin and momentum, an internal pseudo-spin index, labeling the sublattice state, and a further "pseudo-spin" index, labeling the two independent Dirac spectra derived from the **K** and **K'** points in the Brillouin zone.

The *pseudo-spin* of the particle, in relation to its motion, gives rise to helicity or chirality. Helicity is the projection of spin $\boldsymbol{\sigma}$ onto the direction of motion **k** and is positive (negative) for electrons (holes) defined as:

$$h = {}^1\!/_2\, \boldsymbol{\sigma} \bullet \mathbf{p}/|\mathbf{p}| \tag{1.4}$$

The concept of helicity is valid near the Fermi energy and to the extent that second-nearest-neighbor interactions are negligible. Its relevance is confirmed by observation of anomalous quantum Hall effects in graphene. Further, the long, micrometer-scale, mean free paths seen in metallic carbon nanotubes (McEuen *et al.*, 1999) have been related to cancellation of backscattering of chiral electrons within a given valley (see Section 8.4). The reversal of k_x to $-k_x$, because of the chirality, involves a rotation of the pseudo-spin (that always points in the direction of motion). But the reversal of the pseudo-spin is forbidden because the electronic wavefunctions of the A and B sublattice contributions are orthogonal. We will return to this topic in Chapter 8.

As an alternative narrative, the "spin" part of the wavefunction eqn (1.2) has half-angles, so that, if the particle executes a closed path, with angle θ gaining 2π, as it might in returning from a scattering center, the wavefunction phase advances by $\theta/2$, an angle π, leading to a minus sign and cancellation of backscattering.

The electron states, finally, have 4-fold degeneracy, including the valley (K, K') degeneracy and the electron spin degeneracy. The density of states per unit cell is $g(E) = 2\,A_{\mathrm{C}}\,|E|\,/(\pi\hbar^2 v_{\mathrm{F}}^2)$, where $v_{\mathrm{F}} = c^*$ and $A_{\mathrm{C}} = 3\sqrt{3}$, $a^2/2$ is the cell area, with a the nearest-neighbor distance. (See also eqn (4.13), and related text.)

1.3 Analogies to relativistic physics backed by experiment

A 2D electron system in a perpendicular magnetic field gives rise to Landau levels (LL) whose energies conventionally are $E = (n + {}^1\!/_2)\,(eB\hbar/m)$, with $n = 0,\ 1,\ 2,\ \ldots$ and m the electron mass. The minimum energy in this set is ${}^1\!/_2\,\hbar\omega_{\mathrm{c}}$, where the cyclotron frequency is $\omega_{\mathrm{c}} = eB/m$. (We use SI units, note that much of the theoretical literature on graphene and quantum Hall effect uses cgs (centimeter-gram-second) units, where the cyclotron frequency would be written as eB/mc.) For a system of area A, the total number of orbital states at the LL energy is $N = AB/\varphi_0$, where $\varphi_0 = h/e$ is the (one electron) magnetic flux quantum. The quantum Hall effect occurs when N is similar to the total number of mobile electrons in the system so that "all electrons are in fully quantized states". This effect is only present in two-dimensional 2D systems, as will be discussed in Chapter 2. (The quantum Hall effect is observable in graphene even at room temperature, as we will see in Section 8.3, because of the exceptionally large mean free

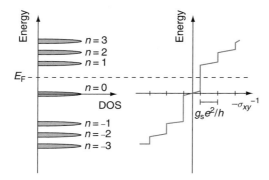

Fig. 1.8 Anomalous quantum Hall effect (QHE) in monolayer graphene shows a Landau level at zero energy [corresponding to central peak in density of states (DOS)]. This anomalous observation is the signature of the massless Dirac Fermion. The conventional QHE has a gap at zero energy. The measured quantity on the right, contrary to the label, is $-\sigma_{xy} \approx R_{xy}^{-1}$ that conventionally rises by $g_s\, e^2/h$, with g_s the spin degeneracy, as a Landau level crosses the Fermi energy. (From Zhang *et al.*, 2005, by permission from Macmillan Publishers Ltd., © 2005).

path.) The graphene LL spectrum is anomalous, including a prominent level at zero energy that supports the pseudo-spin wavefunction. The anomalous observed levels can be written (Novoselov, 2011) as

$$E_{\mathrm{n}} = \pm v_{\mathrm{F}}[2e\hbar B(n + {}^1\!/_2 \pm {}^1\!/_2)]^{1/2} \qquad \text{where } n = 0,\ 1,\ 2,\ \ldots \qquad (1.5)$$

In this expression, describing a half-integer quantum Hall effect, the $\pm {}^1\!/_2$ term is related to the chirality of the quasiparticles and ensures the existence of two energy levels (one electron-like and one hole-like) at exactly zero energy, each with degeneracy half that of all other Landau levels (McClure, 1956, 1960).

The data of Zhang *et al.* (2005) are shown in Fig. 1.8 (very similar data were reported on p. 197 of the same journal by Novoselov *et al.* [2005]). We will return to this topic later (in Section 8.3), in the book.

The spacing of the Landau levels shown in Fig. 1.8 is similar to that predicted for massless Dirac electrons (as measured by a scanning single electron transistor) shown in Fig. 1.9a and directly observed by Miller *et al.* (2009) in scanning tunneling spectroscopy, in Fig. 1.9b.

The data in Fig. 1.9a are due to Martin *et al.* (2009), using a scanning single electron transistor (SSET, described in Chapter 6) to locally measure what is called the "inverse compressibility" of the electron system.[8]

---

[8]The measured quantity is $d\mu/dn$ where μ is the chemical potential and n is the density of carriers.

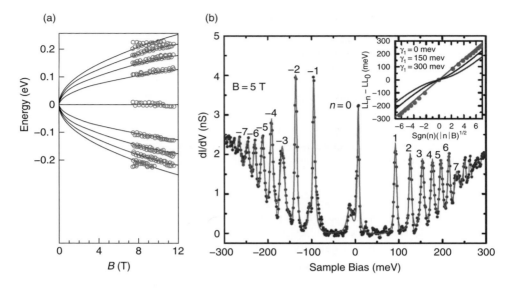

Fig. 1.9 (a) The observed anomalous $B^{1/2}$ energy dependence of Landau levels, contrary to the usual linear dependence, is strong evidence that the electron carriers in graphene behave as "massless Dirac fermions," because of the unusual lattice symmetry. The conventional formula $\omega = eB/m$ is untenable for m essentially zero. Measurements by method of scanning Single Electron Transistor (SET) by Martin *et al.* (2009). (This measurement method will be mentioned in Chapter 6.) (Martin *et al.*, 2009, Fig. 1). (From Martin *et al.*, 2009, by permission from Macmillan Publishers Ltd., © 2009) (b) The observed anomalous $B^{1/2}$ energy dependence of Landau levels, as well as the zero-energy Landau level, are clearly seen in Scanning Tunneling Spectroscopy. (From Miller *et al.*, 2009, with permission from AAAS).

The data are obtained at 11.7 T with the scanning SET tip hovering "a few tens of nanometers" above the monolayer graphene, supported on the conventional oxidized silicon wafer. The indexing used by Martin *et al.* (2009) for the Landau levels is

$$E_i = \text{sgn}(i)\, [2e\hbar v_{\rm F}^2 |i| B]^{1/2} \qquad (1.6)$$

with integer i running from $i = -4$ to $i = +4$. Positive i values indicate electron-like Landau levels, and negative i values indicate hole-like Landau levels. The solid lines are calculated on this formula using the best fit value $v_{\rm F} = 1.1 \times 10^6$ m/s that the authors say is ten percent higher than the theoretical estimate. The anomalous zero-energy Landau level is prominent.

To summarize, the Landau level spectrum of graphene is doubly anomalous, first in the presence of the zero-energy peak and second in the uneven spacing of the levels, represented by the square-root factor in eqn (1.5).

1.4   Possibility of carbon ring electronics

Transistors can be made from single-layer carbon films, a saving in material costs for a start. One does not have to grow a large crystal and saw it into wafers, although it is true that an epitaxial substrate such as SiC may still be needed. The mobility of the carriers is exceptionally high, translating into the chance for a ballistic device working at room temperature. Further, the mobility remains high even for high carrier concentrations. The valley degeneracy and spinor property prohibiting backscattering offer new avenues for making devices. The drawback that one sees at the outset is that graphene does not have a high resistivity state, so On/Off ratios in a conventional FET configuration are limited. Patterning graphene into nanoribbons, with space quantization of levels across the ribbon, offers the chance to introduce energy barriers, even though in some situations anomalous Klein tunneling of carriers is expected. Patterning is itself a major, if not impossible, challenge, on the size scale of a benzene ring, as is potentially available. The details of the patterned edge, whether it is "armchair" or "zigzag" in the notation of carbon nanotubes, significantly affect the electrical properties. How the unpaired electrons at the "broken bonds" are passivated may also be an important aspect. The graphene sheet can, in principle, be patterned to act as a single electron transistor (SET) as well as a field-effect transistor, and indeed on a molecular scale. As we will see in Chapter 9, activity in these directions is occurring at leading firms. There is no question that graphene elements will play important supplemental roles in the silicon chip industry. A more realistic and detailed account of graphene field-effect FET transistors, including a possible family of graphene tunneling T-FET switching devices is given in Chapter 9, especially Section 9.8.

1.5   Nobel Prize in Physics in 2010 to Andre K. Geim and Konstantin S. Novoselov

The 2010 Nobel Prize in Physics was shared by Andre K. Geim and Konstantin S. Novoselov for their work on graphene. They had isolated and named the single layers in 2004. The method that they used, basically using Scotch tape to pull the graphite apart, was simple and accessible, and provided samples still the best available. This caused a rush of new experiments. Graphene, as they showed, was really only one example of a class of two-dimensional crystals that they had discovered. Their experimental work revealed the remarkable electronic properties of graphene, and they tied these to earlier work in relativistic particle physics. The material itself has unique and superior properties, both mechanically and electrically, and has many possible applications. The explosion of publications related to graphene has been truly remarkable. A very recent review of the rapidly expanding literature and assessment of applications is given by Novoselov *et al.* (2012).

The Nobel Lectures are: Andre K. Geim (2011), "Random Walk to Graphene" and K. S. Novoselov (2011), "Graphene: Materials in the Flatland". These are excellent sources of information including references up to 2010. The citation for the Nobel Prize mentions "groundbreaking experiments regarding the two-dimensional material graphene." An added source of information is "Scientific Background on the

Nobel Prize in Physics 2010: Graphene" compiled by the Class for Physics of the Royal Swedish Academy of Sciences. The two physicists have been knighted by Queen Elizabeth II in the UK.

1.6 Perspective, scope and organization

The discovery of 2004 has led to an explosion of literature on graphene and to a lesser extent on the class of 2D crystalline systems.

The brilliant experimental work has revealed a vast amount of new information. Experimental work on graphene is very difficult because every carbon atom is exposed to the environment that contains contaminating atoms and molecules, and also stray electric fields that have the effect of inducing charge carriers in the graphene. The new field has gone through phases where effects were seen repeatedly and thought to be fundamental when, in light of improving methods and sample quality, the effects were later realized as consequences of contaminants, mounting strains or stray electric fields. The field has also appeared more mysterious than it really is by the uncertain relevance of the theoretical literature on the limitations of crystalline order in two dimensions. The other mysterious behaviors, analogies between the electrons in graphene and photons and neutrinos in high energy physics are confirmed and explained by the brilliant experiments including observation of the Klein tunneling effect, to make graphene indeed an interesting material.

This book undertakes to systematically review all of the work, taking seriously even those aspects that now appear less urgently relevant. We cover in Chapter 2 all of the literature on 2D systems, theoretical and experimental. The theoretical literature has two branches, one devoted to ideal 2D arrays with no motion allowed into the third dimension, where excessive in-plane thermal motion, described by logarithmic divergences in atomic excursions from lattice sites, are correctly predicted at finite temperature, but have not been seen in experiment. Purely 2D systems, before discovery of the free electron subsystem contained in graphene, were realized experimentally as electrons on the surface of liquid helium and electrons in the quantum Hall effect. We review the quantum Hall effect, as it was a key to understanding 2D electron behavior in graphene. In the second branch of 2D that we can refer to as 2D-3, a planar system, like graphene, can distort into the third dimension. A large literature here comes from polymer work, and the central question has often been whether or how the 2D system ("membrane") may "crumple" at high temperature, in such a way that an initially flat system of area L^2 eventually fills a volume of size up to L^3. Mechanical engineering also understands 2D-3 systems in the context of bending beams and plates, and it appears that these descriptions are actually more applicable to graphene than are the polymer-related treatments. In the end it appears that graphene does not in fact crumple, but disintegrates near 4900 K. Graphene, suggested as unstable, certainly seems to exist at least to 3900 K, the measured sublimation temperature of graphite.

In Chapters 3, 4 and 5 we deal, respectively, with properties of carbon as atoms, molecules (with attention to benzene rings), and in its solid forms; with the electron bandstructure of mono-layer and bi-layer graphene; and with the sources and types of graphene, ribbons and bilayers. Chapter 6 introduces several of the less-familiar experimental methods that have been helpful in understanding graphene, including angle-resolved photoemission spectroscopy (ARPES), electron scanning tunneling microscopy and spectroscopy (STM, STS) and the scanning single-electron transistor (SSET). Chapter 7 is a review of the physical properties of graphene, with attention to the lattice stability of the material as affected by the flexural modes of vibration well studied by neutron diffraction and other methods. The question of undulations and waves is examined, from the view of experiment and the view of theory, leading to the suggestion that the effects are subtle responses of the atomically thin system to typical environments including gaseous contamination and strain introduced by a mounting structure. Chapter 8 describes several areas of graphene behavior that can be regarded as anomalous, including the predicted disintegration at 4900 K primarily into linear chain fragments of carbon moving away into space. This chapter also covers the Klein tunneling discovery, recent results in the quantum Hall effect, discovery of non-local behavior and work suggesting a nematic phase transition of the electron system. Chapter 9 is devoted to applications of graphene. The emphasis is on electronic devices with particular attention to transistors, including radio-frequency (100 GHz) transistors, flash memory elements, optical devices and the interesting questions relating to a possible new class of switching transistors that can be miniaturized beyond the limits of Moore's Law as it applies to silicon. Finally, emphasis is on types of graphene transistors that are potentially manufacturable in existing technology and that will operate as switches in spite of the essentially semi-metallic nature of graphene. Chapter 10 is a summary and assessment, with attention to key questions of expanding methods to obtain high quality samples for electronic applications at reasonable cost. Briefly it is suggested that the primary advantage of graphene in device applications is the continuity and high conductivity available literally down to a thickness of one atom.

2
Physics in two dimensions (2D)

2.1 Introduction

In a world of three dimensions, no physical object can exist in only two dimensions (2D). However the 2D rules apply to 2D subsystems such as spins and to several cases where motion in the third dimension is constrained to a single quantum state. We will describe in Section 2.2 below the behavior of electrons on a liquid helium surface and also their behavior in semiconductor layers where the quantum Hall effect was discovered. The majority of these systems are supported by—or embedded in—3D systems, for example a planar array of spins existing in a 3D crystal. The new paradigm uncovered by Geim and Novoselov (Novoselov *et al.*, 2004, 2005) is a class of self-supported 2D crystals that are large enough to be useful. Yet, as we shall see, the importance of the graphene system for applications is not so much that it is self-supporting in small sizes, as that it remains conductive at one atom thickness and has a new type of band structure with massless carriers exhibiting inhibited scattering. In supported samples this electron system can extend for distances as large as 30 inches at present.

A well-studied 2D subsystem is the 2D arrangement of spins, treated by the Ising model (eqn 2.1). This should be regarded as a 2D subsystem of a crystal, as the spins are not self-supporting. The energy is expressed as

$$E = -J_{ij} \sum \sigma_i \, \sigma_j, \tag{2.1}$$

representing nearest-neighbor interaction of spins $\pm 1/2$. Expressions for the magnetization $M = (1/N)\Sigma \, \sigma_i$ and transition temperature

$$T_C = 2J/[\log(1 + \sqrt{2})], \tag{2.2}$$

where Boltzmann's constant is understood, for an infinite 2D array were given by Onsager (1949), while it had earlier been shown that, in dimensions less than two, no phase transition would occur.

The mathematics of diffusion and random walks in two dimensions are also well known, as additional examples. See for example Polya (1921). In two dimensions the probability of reaching any point in a random walk is exactly unity, while it smaller than one in higher dimensions.

There are formal limitations on the size of crystals in 2D at finite temperature (however, providing no known limitations in practice) that we will describe in Section 2.4.1. Before doing so, we will describe the observed properties of important 2D electron systems on the surface of liquid helium and at semiconductors hetero-junctions.

2.2 2D electrons on liquid helium and in semiconductors

Crystallization in 2D of ultra-high-mobility electrons deposited onto the surface of liquid helium was observed (Grimes and Adams, 1979) near 0.5 K. The mobile electrons, contained laterally by a ring electrode, are observed to freeze from a 2D liquid to a triangular crystal, and the relevant theory is well understood (Fisher *et al.*, 1979). The experimental cell approximates a 5 cm diameter capacitor, with one plate about 2 mm above the helium level, and containing a tiny metal point used to spray electrons to charge the surface. The equilibrium electron density is $N_S = \varepsilon \varepsilon_0 V_T / ed$, where V_T is the upper plate voltage, d the distance between the helium surface and the lower electrode, about 1 mm, and ε the permittivity of liquid helium, about 1.06. The electron density is modulated by an ac voltage on a confining ring electrode and lock-in detection used for the impedance measurement. The critical temperature is seen to be about 0.45 K at an electron density N_S near 4.6×10^8 e/cm^2, in Fig. 2.1. This is a large spacing, on the order of 5 µm, and the system is classical, not quantum, in the nature of its lateral Coulomb interactions. An electron at the helium–vacuum interface has long-range attraction toward the surface by its image charge in the dielectric medium and short-range Pauli-principle repulsion from the helium atom electrons, against entering the surface layer. The electron has a quantum bound state at a characteristic height above the helium surface ≈ 4.1 nm with a binding energy 0.7 meV. The vertically bound electron is free in transverse motion.

Below the 2D freezing point of the electron system, longitudinal and transverse vibrations (phonons) of the 2D electron solid appear, and these interact with the

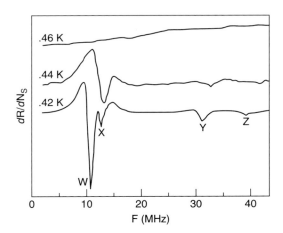

Fig. 2.1 Measurement of real part R of impedance vs. frequency F in MHz of capacitor-like cell containing electron-decorated superfluid helium surface. The electron system of surface density N_S crystallizes near 0.45 K. Characteristic and well-understood excitations of the 2D electron solid are prominent in these data. (From Grimes and Adams, 1979. © 1979 by the American Physical Society).

characteristic capillary waves of the superfluid helium surface, called "ripplons." The ripplons have a wavelength, at 1 K, ~0.13 μm, a frequency ≈80 MHz, and energy less than 0.004 $k_B T$. The resonances in the data occur at frequencies where the coupled-mode "phonon" wavelengths match the size of the lateral confinement, about 3 cm in the experimental geometry. The theory of Fisher *et al.* (1979) fits the data well. The data are consistent with a triangular lattice of electrons, but not with a square lattice. The parameter that controls the state of the electrons is the ratio

$$\acute{\Gamma} = e^2 \left(\pi N_S\right)^{1/2}/4\pi\varepsilon\varepsilon_0 k_B T \tag{2.3}$$

of the Coulomb potential energy to the kinetic energy. Crystallization in 2D is observed to occur near $\acute{\Gamma} = 127$. The lattice spacing is on the order of 4.6 μm, and the electrons are completely free to move laterally on the helium surface, influenced only by the Coulomb repulsion between electron charges. The mobility of these electrons is extremely high, as shown (Lea *et al.*, 1997) in Fig. 2.2, up to 2090 m²/Vs at 0.4 K (not shown in this plot).

A computer device application of electrons on liquid helium has been recently reported by Bradbury *et al.* (2011). Their system requires a temperature around 0.5 K that would limit application to a large facility, perhaps as a part of a quantum computer. In silicon technology computing there is an increasing problem with power consumption, leaving open the eventual possibility of a switch to a more energy-efficient large-scale technology. The cost of refrigeration in a large system could well be offset by power consumption lower than that of present silicon devices that also entails expensive cooling and air conditioning equipment.

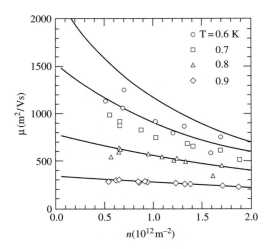

Fig. 2.2 Mobility of 2D electrons on the surface of liquid helium. These are the largest mobilities of electrons known. (From Lea *et al.*, 1997. © 1997 by the American Physical Society).

The most common physical 2D examples are those where an electron subsystem is supported on a surface, as we have seen for electrons on liquid helium, or trapped between two planar walls, as in a semiconductor heterostructure.

An exemplary semiconductor 2D electron subsystem is the GaAlAs/GaAs/GaAlAs tri-layer, although a more important 2D electron subsystem in practice is the accumulation layer in the forward-biased field-effect transistor (FET), sketched in Fig. 2.3, with the AlGaAs/GaAs heterostructure version sketched in Fig. 2.4. The 2D electron gas in the FET allowed the discovery of the quantum Hall effect, a purely 2D phenomenon. The FET case leaves a narrow triangular potential well V(z) as sketched. This supports a ground state electronic wavefunction

$$\psi \sim e^{ikx}e^{iky}\, f(z) \tag{2.4}$$

with electronic energies

$$E \sim (\hbar^2/2m^*)(k_x^2 + k_y^2) + E_0. \tag{2.5}$$

An approximate formula for E_0 is

$$E_0 \sim h^2/8m^*t^2, \tag{2.6}$$

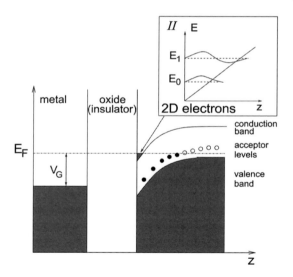

Fig. 2.3 Sketch of bands in metal oxide semiconductor transistor, with gate electrode (on the left) positively biased at V_G to accumulate electrons at the front of the p-type semiconductor. This can be called an N-FET, an N-channel field-effect transistor. The accumulating electrons are 2D when only the states marked E_0 are populated. On the right, the dark circles represent negatively ionized acceptor impurities, whose electric charges tend to scatter the electrons in the accumulation layer. (From Goerbig, 2011).

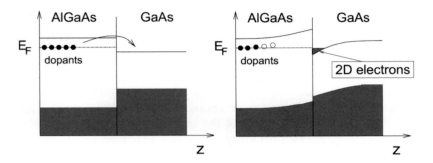

Fig. 2.4 Sketch of bands in AlGaAs/GaAs heterostructure, leading to electrons of mobility 100 m^2/Vs. (From Goerbig, 2011).

where t is an equivalent thickness and m^* is the electron's effective mass (in kg). An estimate for the energy E_1, the bottom of the second sub-band, is $4E_0$ at 1.7 K at the interface, whose surface density is $N_S = 4.5 \times 10^{11}$/cm^2. This electron gas is characterized by a mean free path of 11 μm, Fermi energy 16 meV and Fermi wavelength 37 nm. In 2D, the Fermi energy formula is

$$E_F = (\pi \hbar^2 / m^*)\, N_S \tag{2.7}$$

[rather than $(h^2/8m^*)(3n/\pi)^{2/3}$ for 3D, with n in particles/m^3]. Taking N_S as 4.5×10^{11}/cm^2 and $E_F = 16$ meV, we find $m^* = 6.09 \times 10^{-32}$ kg $= 0.067\, m_e$, an accepted value for electrons in GaAs. By using the 3D formula and the effective mass we find an average 3D electron density $n = 1.61 \times 10^{23}$ m^{-3} and an effective width $t = 28$ nm. This gives an estimate of the energy E_0 as about 6.7 meV.

Having all the electrons in the first band is the defining quality of the 2D state, and we may check how accurate this is by confirming that the occupancy of higher energy states is negligible. The bottom of the second band will be near $4E_0$ or 26.8 meV. The chance of an excitation into this state (out of the ground state) is roughly

$$P = \exp(-\Delta E / k_B T). \tag{2.8}$$

This is extremely small for $\Delta E = 3E_0 - E_F = 4.1$ meV and $k_B T = 0.147$ meV for 1.7K: thus $P = \exp(-4.1/0.147) = 7.7 \times 10^{-13}$. So the system qualifies as 2D on this criterion. (A detailed study of a similar system is offered by Topinka *et al.*, 2000.)

2.3 The quantum Hall effect, unique to 2D

The quantum Hall effect was discovered by von Klitzing *et al.* (1980) in a structure similar to that shown in Fig. 2.3. The most unexpected part of the experimental discovery was the width in parameter values over which the accurate quantization of the Hall resistance persisted. The quantum Hall effect can only occur in a fully

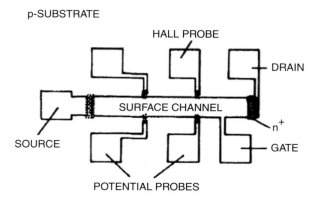

Fig. 2.5 Configuration of MOS transistor in Hall experiment of von Klitzing *et al.* (1980). Taking x along the surface channel, y in the vertical direction and z out of the page for the B field, the Hall voltage V_H is measured across opposing vertical terminals and the Hall resistance is V_H/I. (From Von Klitzing, 1980. © 1980 by the American Physical Society).

quantized system of electrons and is thus a unique indicator of a 2D electron system. The geometry of the Hall device of von Klitzing *et al.* (1980) is shown in Fig. 2.5. We take x, y and z, respectively, as the horizontal (channel) direction, the vertical and out of the page directions.

2.3.1 Hall effect at low magnetic field

In such a geometry, the low field Hall effect is a conventional measure of the carrier type and density in the channel, of dimensions L, w and thickness t. With B in the z direction, the Lorentz force F, with ν the carrier velocity,

$$F = q\nu B \qquad (2.9)$$

is vertical, and leads to a compensating Hall field,

$$E_H = \nu B = jB/nq, \qquad (2.10)$$

if the 3D density of carriers of charge q is n, with current density

$$j = nq\nu. \qquad (2.11)$$

The conventional Hall coefficient, R_H, defined by

$$E_H = R_H jB, \qquad (2.12)$$

is thus

$$R_H = 1/nq. \tag{2.13}$$

In the Hall-bar geometry, the measured Hall resistance

$$R_{xy} = wE_H/I = wjB/nqjwt = B/qn_{2d}, \tag{2.14}$$

where $n_{2d} = nt$, with t the thickness. This recovers the formula for the Hall resistance $R_{xy} = -B/n_{2d}e$. The connection with the quantum Hall formula,

$$R_H = (h/e^2)(1/i), \quad \text{with } i \text{ an integer}, \tag{2.15}$$

is that in a high field with a filled Landau level system,

$$n_{2d} = i(e/h)B = iB/\varphi_0, \ i = 1, 2, 3, \ldots \tag{2.16}$$

The corresponding longitudinal resistance $R_{xx} = R_{pp}$, for potential probe spacing L, is

$$R_{pp} = \rho L/tw = (L/w)(1/n_{2d}\mu) \quad \text{where } \mu = e\tau/m. \tag{2.17}$$

At high B-field the ratio of the vertical and horizontal voltages is

$$R_{xy}/R_{xx} = (w/L)\omega_c\tau, \quad \text{where } \omega_c = eB/m, \tag{2.18}$$

that can be large. The situation for high magnetic field is suggested in Figs. 2.6 and 2.7.

In the usual Hall-bar geometry, where W and L are the width and length of the channel between the voltage probes, the relation between resistance quantities and the conductivity σ_{xy} is

$$\sigma_{xy} = -R_{xy}/(R_{xy}^2 + W^2 R_{xy}^2/L^2) \tag{2.19}$$

The quantized electron levels in a perpendicular magnetic field B are

$$E = (n + 1/2)(eB\hbar/m), \quad \text{with } n = 0, 1, 2, \ldots \tag{2.20}$$

The minimum energy in this set is $1/2\,\hbar\omega_c$, where the cyclotron frequency is $\omega_c = eB/m$. For a system of area A, the total number of orbital states at the LL energy is

$$N = AB/\varphi_0, \tag{2.21}$$

where $\varphi_0 = h/e$ is the (one-electron) magnetic flux quantum. (Each orbital state has a further degeneracy because of the electron spin and the fact that the conduction band in Si has equivalent valleys.) The quantum Hall effect occurs when N is similar to the total number of mobile electrons in the system. When all Landau levels up to n are fully occupied (Fermi level lies between Landau levels), the "Hall resistance" is determined solely by fundamental constants,

$$R_{\mathrm{H}} = V_{\mathrm{H}}/I = R_{xy} = n^{-1}(h/e^2) = 25{,}813 \; \Omega/n, \quad n = 1, 2, \ldots \qquad (2.22)$$

The value does not depend on the sample width, nor on the ratio of its length to width. In addition, the longitudinal resistance R_{xx}

$$R_{xx} = 0 \qquad (2.23)$$

is zero when the Fermi level is between filled Landau levels. Observation of this effect means the electron system is 2D in the sense of our discussion. It also requires a field B larger than B_{c}, the field such that $\omega_{\mathrm{c}} \tau = eB\tau/m > 1$.

2.3.2 High field effects

The first change that appears above this field, $B > B_{\mathrm{c}}$ is shown in Fig. 2.6. Namely, an oscillation occurs in longitudinal resistance R_{xx} (this is the Shubnikov–de Haas effect), while R_{xy} increases linearly.

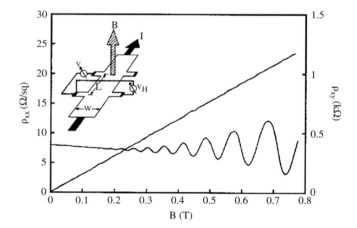

Fig. 2.6 Transport coefficients ρ_{xx} and ρ_{xy} of GaAs/Al$_x$Ga$_{1-x}$As device sketched at 0.35 K. The inset shows the measurement geometry. The voltages V and V_{H} are respectively measured along and perpendicular to the current I. $\rho_{xx} = (V/L)/(I/W)$ is the resistance across a square, independent of square size and $\rho_{xy} = V_{\mathrm{H}}/I$ is the Hall resistance, independent of sample width. For B, beyond $B = B_{\mathrm{c}} \sim 0.3 \, T$, such that $\omega_{\mathrm{c}} \tau = eB\tau/m > 1$, R_{xx} begins to oscillate (Shubnikov-de Haas effect) while R_{xy} increases linearly. The oscillations reflect changes in the density of states at the Fermi level, as Landau levels pass through that energy. (From Tsui, 1999).

The origin of the oscillations in longitudinal resistance is the oscillation in the density of states at the Fermi energy, as succeeding Landau levels cross. This can be seen in the formula for the longitudinal conductivity,

$$\sigma_{xx} = e^2 D g(E_{\mathrm{F}}), \tag{2.24}$$

where D is the electron diffusion constant and $g(E)$ the density of electronic states.

The measured quantities are explained in the caption to Fig. 2.6. In 2D the resistivity is measured directly in Ohms per square. The theoretically natural transport coefficients σ_{xx} and σ_{xy} are obtained from the measured coefficients from the relations

$$\sigma_{xx} = \rho_{xx}/(\rho_{xx}^2 + \rho_{xy}^2). \tag{2.25}$$

and

$$\sigma_{xy} = -\rho_{xy}/(\rho_{xx}^2 + \rho_{xy}^2). \tag{2.26}$$

The first relation shows that σ_{xx} is zero if ρ_{xx} is zero.

2.3.3   von Klitzing's discovery of the quantized Hall effect

The quantum Hall effect was discovered in measurements shown in Fig. 2.7 (Klitzing *et al.*, 1980).

The carriers are electrons at variable density controlled by the gate voltage in an inversion layer on p-type Si. Plateaus in the Hall voltage are seen, apparently at gate voltages such that E_{F} is in the gap between Landau levels, and here the longitudinal voltage has a zero or a minimum. At the left of the figure, in the region marked "n = 0," the minima due to the lifting of the lifting of the electron spin and the (twofold) valley degeneracy are also apparent. The authors measured several devices and found that the number values for the Hall resistances were identical, exactly matching $R_{\mathrm{H}} = (h/e^2)n^{-1}$, even with devices have different ratios of L to W. So the effect occurs when all the mobile carriers are in fully quantized states: in z by the confinement and in x, y by the magnetic field. The extent in gate-voltage of the plateaus was unexpected and made the effect more useful. The width of plateaus is related to localization of electrons away from the peaks of the Landau level spectrum.

The mobile-electron wavefunctions are changed by the magnetic field from

$$\psi \sim \exp(ik_x x + ik_y y) \tag{2.27}$$

to

$$\psi \sim L^{-1/2} \exp(ik_x x)\varphi(y), \tag{2.28}$$

where $\varphi(y)$ is a wavefunction for a (1D) harmonic oscillator centered at $y_0 = l^2 k_y$. In terms of the fundamental magnetic length (cyclotron orbit radius),

$$l = (\hbar/eB)^{1/2} \tag{2.29}$$

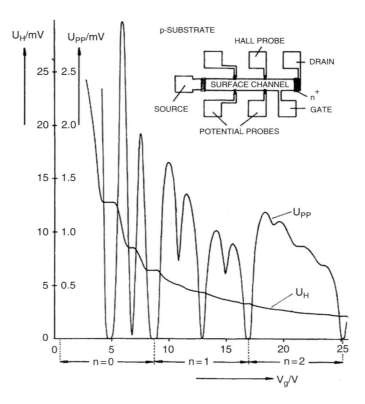

Fig. 2.7 Original observation of quantum Hall effect, plots of Hall (vertical) and longitudinal voltages, U_H and U_{PP}, at 1 μA, 1.5 K and 18 T. Plateaus in the Hall voltage are accompanied by zeroes or minima in the longitudinal voltage. The abscissa is the gate voltage that increases the inversion carrier density and thus causes the Fermi energy to rise through the set of Landau levels established at 18 T. (From Von Klitzing, 1980. © 1980 by the American Physical Society).

(e.g., 6 nm at 18T) this is

$$\Phi_{n,k}(y) = H_n[(y - y_0)/l] \exp[-(y - y_0)^2/4l^2], \qquad (2.30)$$

where H_n is the Hermite polynomial, with $n = 0, 1, 2, \ldots$ The delocalized wavefunction centered at y_0 is then

$$\psi_{n,k}(x, y) = L^{-1/2} \exp(ik_x x) \, H_n[(y - y_0)/l] \, \exp[-(y - y_0)^2/4l^2]. \qquad (2.31)$$

The spacing between positions y_0, neighboring quantum states in y, assuming length L and width W, is $2\pi \, l^2/L$ with l the magnetic length, so that each state

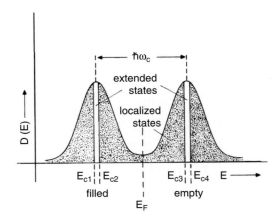

Fig. 2.8 A simplified sketch of the density of localized and delocalized states and Fermi level for observation of the integer quantum Hall effect plateaus and zero longitudinal resistance. (From Von Klitzing, 1986).

occupies an area $2\pi\, l^2$. For a given Landau level (n) the number of states per unit area is $1/(2\pi l^2) = B/(h/e)$, namely the magnetic field measured in quanta per unit area. So the number of quantum states in a given Landau level is the number of flux quanta cutting the surface of area $A = LW$.

Scanning tunneling spectroscopy (Niimi *et al.*, 2006; Hashimoto *et al.*, 2008) supports the existence of delocalized states near the Landau level peaks and localized states between those peaks.

Because the numbering of the Landau levels starts at $n = 0$, for n filled Landau levels, the index of the last one filled is $n - 1$. It can be argued as long as at least one orbit at the center of a Landau level peak is delocalized, the transverse Hall conductance e^2/h will appear. It was argued, by Ando (1974), that electrons in impurity bands, arising from short range scatterers, do not contribute to the Hall current; whereas the electrons in the Landau level give rise to the same Hall current as that obtained when all the electrons are in the level and can move freely.

When the Fermi level is in a range of localized states, the longitudinal conductivity σ_{xx} must be zero that requires, via the relation $\sigma_{xx} = \rho_{xx}/(\rho_{xx}^2 + \rho_{xy}^2)$ (eqn 2.25) that R_{xx} is zero, as is seen in the plateau regions. These two features will not be changed as the Fermi level is shifted by adding more localized states, since the localized states have no effect on the measured quantities. These arguments seem to be confirmed by the extremely wide plateaus seen in Fig. 2.9. The Nobel Lecture of Laughlin includes a more general derivation of the quantization rule $\sigma_{xy} = ne^2/h$ that allows for localized states. It is also argued that for a perfect system that would be invariant along the x direction, the rule would actually be $\sigma_{xy} = \mathrm{P}c/B$. Here P (capital rho) is the charge density that certainly is not quantized. Thus it is argued that disorder is

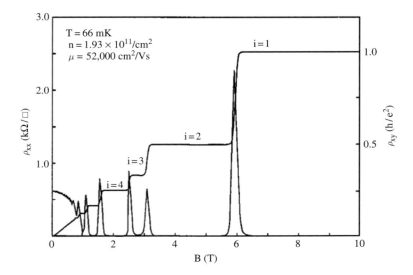

Fig. 2.9 Quantum Hall effect showing broad plateaus in Hall resistance, accompanied by broad regions of zero longitudinal resistance. The interpretation is that the plateaus correspond to Fermi energy lying in range of localized states (shown hatched in Fig. 2.8). (From Tsui, 1999).

needed to break the translational invariance, in order for the Hall quantization to be observed.

This discussion describes one of the essentially 2D phenomena. In these experiments the needed fully-quantized electrons are possible since the extent of their probability cloud in the z-direction is invariant and short, with $f(z)$ a bound state in a potential well narrower than the electron de Broglie wavelength.

The observed effects require at least one delocalized electron state connecting across the width W of the sample. In a disordered 2D metal the theory of Abrahams *et al.* (1979) predicts that all electron states be localized, in apparent contradiction, unless the prediction is inaccurate at high magnetic field. The conductance of a disordered 2D metal sample is predicted at $T = 0$, $B = 0$ to decrease exponentially with length L, falling below the value of the conductance quantum e^2/h.

The work of Levine *et al.* (1983) finds that a magnetic field, as in the quantum Hall experiments, creates at least a few delocalized states, removing a discrepancy that was commented on by von Klitzing (1986).

To summarize, we have seen that electrons on liquid helium crystallize to form a well-defined lattice with a sharp transition to a liquid phase. Electrons in confined semiconductor layers at high magnetic field exhibit a new phase of matter, the quantum Hall state, that exists only in 2D. Since the quantum Hall effect is observed in graphene, one can be sure that the electrons in graphene are in the 2D category.

2.4   Formal theorems on 2D long-range order

We return to the broader questions of 2D systems; stability, crystallinity and conductivity, as the size L is expanded. For the electronic conductivity of a 2D system at zero temperature, there is prediction, at zero magnetic field, of no true metallic behavior, but rather an exponentially falling zero temperature conductance with increasing sample size (Abrahams *et al.*, 1979). While early results on graphene suggested a minimum conductivity, more recent work (Ponomarenko *et al.*, 2011) does find the insulating state. Graphene is generally an excellent electrical conductor. We will see in Chapter 9 that field-effect transistors of graphene are capable of current densities in the order of milliamperes per micron channel width, with equivalent 3D current densities 3×10^{12} A/m^2.

The recently discovered two-dimensional crystals, self-supporting in small but useful lateral size scales, are relatively few, and graphene is unique among these. A broader class of self-supporting 2D systems are membranes floating in solution that are weakly cohesive but not ordered in two dimensions. As in the case of linear polymers, tending to coagulate by the larger cohesive energy afforded by a larger number of near neighbors, planar forms of matter are basically unstable in comparison to 3D arrangements of atoms. Crumpling and folding are the first steps in this direction, from an initially planar arrangement.

2.4.1   Absolute vs. relative thermal motions in 2D

We have given examples of 2D electrons supported in semiconductors and on liquid helium, exhibiting long-range order as far as one can tell. We have also discussed graphene sheets containing many millions of unit cells in a crystalline array, that are unique among 2D conductors in that they are self-supporting. Yet there are fundamental theorems that limit, except at zero temperature, positional order of crystals in 2D. In particular, for $T > 0$, the mean square deviation of an atom from its equilibrium position increases logarithmically with the size L of the system, as $|\mathbf{u_i}|^2 \sim T \ln (L/a)$. At the same time the *relative* local fluctuation $|\mathbf{u_i} - \mathbf{u_j}|^2$ of near neighbor atoms/particles i and j remains finite (Jancovici, 1967), and may be used to define a melting point (Bedanov *et al.*, 1985). If atoms i and j are widely separated by distance L, then $|\mathbf{u_i} - \mathbf{u_j}|^2$ diverges as $T \ln (L/a)$.

In many cases it seems that the limiting theorems in fact permit large systems, the word "astronomical" has been used in this context. There is no experimental evidence for the absolute divergence $|\mathbf{u_i}|^2 \sim T \ln(L/a)$, in part it may be that an observation would pin the local portion of the system, moving the possible divergence to the edges. It is also possible that candidate systems are so robust as to be effectively in the $T = 0$ limit. The electron on helium system is not available in large lateral dimensions L.

2.4.1.1   Analysis of Landau and Lifshitz.   We follow a discussion by Landau and Lifshitz (1986) that clarifies these topics. Going back to the 3D case, the meaning of crystalline order is that every unit cell in the arbitrarily large system can be reached by a vector

$$\mathbf{R} = n\mathbf{a} + m\mathbf{b} + l\mathbf{c} \qquad (2.32)$$

with suitable integer choices of n, m and l. This is a very high standard, indeed, and cannot be met in any system capable of flexing. But it is not hard to show that the root mean square deviation in position **u** of a small region of the hypothetical infinite 3D sample remains finite at finite temperature. Landau and Lifshitz (1986) start with the Fourier decomposition of the fluctuational displacement

$$\mathbf{u} = \sum_{\mathbf{k}} \mathbf{u_k} e^{i\,\mathbf{k} \bullet \mathbf{r}}. \tag{2.33}$$

If the small region has size d, the phonon waves of interest to move this region intact have small wavevectors, $k \leq 1/d$.

The probability of a fluctuation will be weighted by its free energy cost, ΔF, according to

$$P \propto \exp(-\Delta F/k_\mathrm{B}T). \tag{2.34}$$

The potential energy cost comes from quadratic product terms, collected as $\varphi_{i,l}(k_x, k_y, k_z)$, among the components of the displacement:

$$\Delta F = V/2 \sum_{\mathbf{k}il} \mathbf{u_{ik}}\mathbf{u_{lk}}\,\varphi_{i,l}(k_x, k_y, k_z). \tag{2.35}$$

This is described as a quadratic functional of the displacements. These are terms in the nature of $1/2\,Kx^2$, with K a spring constant and the change is of the total free energy of the entire body of volume V as a result of the displacement **u**.

Landau and Lifshitz find that this leads to an expectation value, with T the temperature,

$$<\mathrm{u_{ik}u_{lk}^*}> = T/V \sum_{il} A_{il}(\mathbf{n})/k^2 \tag{2.36}$$

where the A_{il} depend only on the direction **n** of the vector **k** ($\mathbf{n} = \mathbf{k}/k$). The values of $<\mathrm{u_i\,u_i^*}>$ are found by summing over wavevector k. Changing from sum to integral, one finds for the mean-square displacement vector

$$<\mathbf{u^2}> = T \int [d^3k/(2\pi)^3]A_{ll}(\mathbf{n})/k^2 \tag{2.37a}$$

$$= T \int [dk\, d\sigma/(2\pi)^3]A_{ll}(\mathbf{n}), \tag{2.37b}$$

with σ an area element.

This integral converges at $k = 0$, where the integrand is linear in k. (The integrand is written only for k not too large, so the upper limit is assumed to converge.) This means that the mean square fluctuation displacement is, as it should be, a finite quantity independent of the size of the body, QED.

Briefly, to turn this analysis to the 2D case, Landau and Lifshitz (1986) find in the same way

$$<u_{ik}u_{lk}^*> = T \int [dk_x \, dk_y/(2\pi)^2]A_{il}(\mathbf{n})/k^2. \qquad (2.38)$$

This integral diverges logarithmically at $k = 0$, since $dk_x dk_y = 2\pi k dk$, and the integral is const. $\times \ln (k)$ with limits on the integration variable k on the order of $1/L$ and $1/a$, if a is the lattice constant. So the result for a system of size L is on the order of

$$<\mathbf{u^2}> = \text{const.} \times T \ln (L/a). \qquad (2.39)$$

This is the divergent result: the absolute thermal motion of a point in a 2D system of size L becomes indefinitely large proportional to $T \ln (L/a)$, with a the lattice constant and T the temperature. (There is no problem at $T = 0$.) If L is one meter and a 0.34 nm, then the offending ln term would be 21.8.[1]

This result, strictly speaking (Landau and Lifshitz, 1986), means only that the fluctuational displacement becomes infinite when the size L of the two-dimensional system increases without limit, so that the wave number $k \sim 1/L$ may be arbitrarily small. But, because of the slow (logarithmic) divergence of the integral, the size of the film (two-dimensional system), where the displacements remain small, may still be very great. In such cases, according to Landau and Lifshitz (1986), a film of large finite size might have practically "solid state" properties and be approximately describable as a two-dimensional lattice. It can be shown that the "solid state" properties of the system become more marked at low temperatures.

To show this, Landau and Lifshitz (1986) consider the correlations between fluctuations at different locations of the 2D system.

To start, it is clear that at $T = 0$ there is no problem:[2] a crystal in 2D of any size can occur.

The problem in the 2D case is with thermal fluctuations at finite T. Let $\rho_0(\mathbf{r})$ represent the density distribution at $T = 0$. At low T, if the displacement $\mathbf{u}(\mathbf{r})$ of atoms at locations $\mathbf{r}$ can be considered as slowly varying over distances on the order of the lattice constant a, then the change in density at each point may be regarded as simply the result of a shift of the lattice by an amount equal to the local value of the displacement vector. Thus the density can be expressed as

$$\rho(\mathbf{r}) = \rho_0[\mathbf{r} - \mathbf{u}(\mathbf{r})]. \qquad (2.40)$$

---

[1] Such enhanced thermal movements in 2D have never been observed to the author's knowledge. In such an experiment, it might be that the measurement apparatus would pin the system locally and the remote edges at ±0.5 m would be vibrating laterally. Then the measurement would be limited to the relative motions that, as we will see shortly, show no peculiar property. This whole analysis ignores any excursion into the perpendicular direction. Further, one hardly ever encounters a crystal as large as one meter, even in 3D.

[2] In practice the effective "zero temperature" of a material may be gauged by the Debye temperature that is high for light, rigid materials like graphene. We will see in Section 7.2 that the Debye temperature for in-plane motions in graphite, similar to graphene, is 2500 K.

The correlation between fluctuations at different locations is given by the correlation function

$$<\rho(\mathbf{r_1})\rho(\mathbf{r_2})> \; = \; <\rho_0[\mathbf{r_1} - \mathbf{u}(\mathbf{r_1})]\rho_0[\mathbf{r_2} - \mathbf{u}(\mathbf{r_2})]>. \qquad (2.41)$$

The periodic function $\rho_0(\mathbf{r})$ can be expanded as a Fourier series

$$\rho_0(\mathbf{r}) = \rho_{av} + \sum_{\mathbf{b}\neq 0} \rho_{\mathbf{b}} e^{i\mathbf{b}\bullet\mathbf{r}} \qquad (2.42)$$

where the vectors $\mathbf{b}$ are in the reciprocal lattice of the 2D system. When this is inserted into the correlation function, surviving products include

$$|\rho_{\mathbf{b}}|^2 \exp[i\mathbf{b}\bullet(\mathbf{r_1} - \mathbf{r_2})] < \exp[-i\mathbf{b}\bullet(\mathbf{u_1} - \mathbf{u_2})] >, \qquad (2.43)$$

where $\mathbf{u_1} = \mathbf{u}(\mathbf{r_1})$, etc.

It is shown by Landau and Lifshitz (1986) that an important result toward constructing the correlation function is

$$<\exp[-i\mathbf{b}\bullet(\mathbf{u_1} - \mathbf{u_2})]> = \exp(-1/2 \; b_i b_l \chi_{il}) \qquad (2.44)$$

where

$$\chi_{il}(\mathbf{r}) = T \int [dk_x \; dk_y/(2\pi)^2] \; (1 - \cos \mathbf{k}\bullet\mathbf{r}) A_{il}(\mathbf{n})/k^2. \qquad (2.45)$$

The last integral converges for small k because $(1 - \cos \mathbf{k}\bullet\mathbf{r}) \propto k^2$ when k is near zero. For large k the integral diverges logarithmically, but this is an artifact of approximation. Identifying a maximum $k_{max} = T/(\hbar c)$, with c the velocity of sound, and dropping the rapidly oscillating factor $\cos \mathbf{k}\bullet\mathbf{r}$, one finds

$$\chi_{il}(\mathbf{r}) = T/\pi \; A_{il \; av} \; \ln \; (k_{max}r). \qquad (2.46)$$

This term appears in the exponent, leading to the central result for the correlation function, expressed with a fictitious temperature T'

$$<\rho(\mathbf{r_1})\rho(\mathbf{r_2})> - \rho_{av}^2 = \cos(\mathbf{b}\bullet\mathbf{r})/r^{T/T'}, \qquad (2.47)$$

where $1/T' = b_i b_l A_{il \; av}/2\pi$, with $\mathbf{b}$ chosen as the reciprocal lattice vector corresponding to the minimum value of $b_i b_l A_{il \; av}/2\pi$.

Thus, in the two-dimensional lattice, the correlation of motion at locations 1 and 2 does tend to zero as r goes to infinity, but quite slowly, especially at low temperature. For comparison, in a liquid, the correlation function decays exponentially with distance r, and in a 3D lattice it approaches a constant. For nearest-neighbor sites the correlation function is well-defined and can be used to define the melting temperature, as we will see. The motion of near-neighbors is highly correlated, much as in a 3D crystal, but anomalous behavior exists in the relative motion of distant neighbors.

A similar result was given earlier by Jankovici (1967), who identified a temperature T':

$$T' = 4\pi mc^2/(b^2 k_B) \qquad (2.48)$$

(not a thermodynamic quantity), where m is the mass of the atom in the 2D lattice, b is a reciprocal lattice vector, c is the speed of sound and k_B is Boltzmann's constant.

2.4.1.2 Melting criterion in 2D. The important result is that the nearest-neighbor correlations are finite for any size L, at finite T. Their magnitude, in comparison with the lattice constant a, was used by Bedanov *et al.* (1985) to find the melting temperature of the 2D lattice. As they point out, the conventional Lindemann criterion for melting $\gamma_M = <|u|^2>/a^2$ (where $a = 1/(\pi n)^{1/2}$ with n the number of atoms per unit area), cannot be used in 2D because the quantity diverges according to the Landau–Peierls analysis, given above. However, using a differential criterion

$$\gamma_M = <|\mathbf{u}(\mathbf{r}+\mathbf{a}) - \mathbf{u}(\mathbf{r})|^2>/a^2 \qquad (2.49)$$

Bedanov *et al.* (1985) showed that the adapted method works as well as the conventional form has for three-dimensional systems. The values of γ_M are typically in the range 0.1 to 0.12 at the melting point.

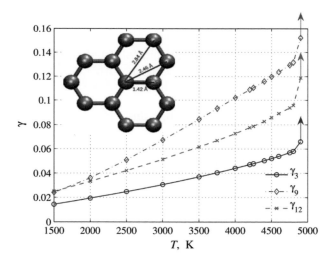

Fig. 2.10 Simulated temperature dependences of Lindemann differential vibrational amplitude melting criterion for graphene, all showing melting indication at 4900 K. The lowest curve (circles) is based on 3 nearest-neighbor distances, the intermediate curve (crosses) based on 12 neighbors, and upper curve (diamonds) based on 9 nearest neighbors. (Reprinted with permission from Zakharchenko *et al.*, 2011. © IOP Publishing 2011).

Recently a modeling of the local melting of graphene has been carried out by Zakharchenko *et al.* (2011), yielding the plots in Fig. 2.10, consistent with melting at 4900 K. Further discussion of this is given in Chapter 8.

2.4.1.3 Is an oscillating crystal in 1D or 2D no longer crystalline? We now consider a 2D sheet embedded in 3D space (2D-3), so it can move in the perpendicular direction, much as the xylophone key sketched in Fig. 2.11. If the sample is long, the flexing motion leads to large displacements, but no loss of local order occurs. An extreme statement about this situation might be: "thermal fluctuations should destroy long-range order, resulting in melting of a 2D lattice at any finite temperature." The first part is correct by the definition, of displacements exceeding a lattice constant, but the second statement is not reasonable, since we insist that melting means loss of local order. When one talks of graphene (graphite) that sublimates at 3900 K, loss of local order much below 3900 K seems hardly a realistic possibility. Annealing above 2300 K of graphene layers in transmission electron microscopy (TEM) experiments (Huang *et al.*, 2009; Jia *et al.*, 2009) smooths the edges as atoms fly off, but melting is not observed.

Simulations run up to 3500 K on graphene (Fasolino *et al.*, 2007) did not find lattice defects, but later work by Zakharchenko *et al.* (2011) predicts melting at 4900 K initiated by clustering of defects and formation of octagons leading to carbon chains.

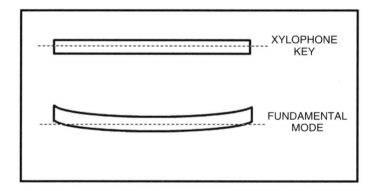

Fig. 2.11 Sketch of xylophone key, say of dimensions L, w and t and its fundamental distortion. The frequency calculation is the same as for a cantilever of half the length. The analysis for the fundamental frequency will be the same for a square of side L, but the estimate of the classical rms thermal fluctuation will differ. At any temperature the amplitude of the fundamental mode scales as a positive power of the length, making the movement at the end arbitrarily large for arbitrarily large L. Perhaps the motion, when it exceeds the lattice constant at any point in the sample, invalidates crystallinity, or perhaps it is more reasonable to say it remains a crystal in an oscillating state. But it certainly does not melt. (Courtesy Ankita Shah).

The implication of a sample *melting* as a consequence of a larger size (where the motion would exceed a lattice constant) is counter-intuitive. More likely (but an estimate involving dislocation formation as a step to crumpling will be discussed in Chapter 8) the breakdown of order that will occur in graphene is the bending of the atomic lattice so that unit cells some distance away, due to slow thermal fluctuations of large amplitude, are no longer predictably on the lattice positions stated by $\mathbf{R} = n\mathbf{a} + m\mathbf{b} + l\mathbf{c}$. In a large scale system, viewed globally, the result might resemble melting, but locally order is retained. One can make an elementary prediction that a one meter square of graphene clamped on opposite edges will have a slowest mode of thermal oscillation at 300 K with root mean square (rms) amplitude on the order of 2 mm and a period around 270 hours. It is no longer crystalline on the definition, because its unit cells wander over many lattice constants (but it retains most of its useful crystalline properties). Compare this to a one meter cube of silicon, where no unit cell will move at all appreciably. (One might consider a one meter cube of diamond, bringing to mind that crystals in the real world are of very limited extent.) The difference is simply the dimensionality. The hypothetical unsupported one meter sheet is not unstable, except as we have described, and it is not impossible that it will behave similarly nearly to 3900 K with an rms amplitude around 7 mm. The temperature where a drastic change in the local order of the graphene may occur is difficult to establish, but this temperature was recently estimated (See Fig. 2.10) to be 4900 K (to be discussed in Chapter 8).

2.4.2   The Hohenberg–Landau–Mermin–Peierls–Wagner Theorem

The Hohenberg–Landau–Mermin–Peierls–Wagner (HLMPW) theorem of statistical mechanics was developed from earlier work of Peierls and Landau, see Peierls (1923), Landau (1937) and Hohenberg (1967). In more general terms than offered in the original papers, this theorem asserts that a continuous symmetry, like rotational symmetry, cannot be spontaneously broken in 2D (or lower dimensions) to create a (crystalline, infinite) phase at any finite temperature. (*Discrete* symmetry in dimensions ≤ 2 *can* be broken, as illustrated by ferromagnetism.) The formal theorem does not consider any excursion of the system from strict planarity.

An example of a broken symmetry in 3D is the growth of a diamond or graphite crystal in the earth, where a particular orientation is spontaneously chosen by the crystal, out of the 4π directions equally available in the melt. The resulting crystal, at any temperature below its melting point, can be grown stably to a size L not limited by statistical mechanics. The HLMPW theorem says that the same is not true of graphene, assuming it is 2D and maintained flat, or a carbon nanotube, assuming it is 1D: limits exist on the sizes L of such crystals. In practice the size L that is available (at 300 K), for micromechanically cleaved or epitaxially grown and released graphene (in 3D space), seems to be so large as to make the formal theorem irrelevant. Indeed, in his 1968 paper, Mermin commented that his treatment "may... allow two-dimensional systems of *less than astronomical size* to display crystalline order" (italics added).

The importance in practice of this limit is thus almost zero, because the largest crystal in the universe may well be the one meter sample of silicon mentioned above, apart from some chance of a neutron star, whose diameter might be 20 km (Haskell *et al.*, 2007).

Useful crystals in technology are more commonly a centimeter in size. While it has not been possible, thus far, to grow graphene crystals in the usual sense, it has certainly been possible to grow very long carbon nanotubes, 1D, by using a catalyst in a chemical vapor deposition (CVD) apparatus. Zheng *et al.* (2004) show, in detail, a single-wall carbon nanotube 4 cm (about 200 million unit cells) in length, grown by CVD on a Si substrate with an Fe catalyst, at growth rate estimated as 11 μm/s. This nanotube, resting on the silicon substrate, is likely stable at any reasonable temperature. However, if it were free, it might well be unstable against bending (thermal crumpling) as discussed by Schelling and Keblinski (2003), to provide displacements of magnitude comparable to its length. Flexural modes have low frequencies and large amplitudes as the size scale increases. It might be regarded as "crumpled" if its thermal distortions have the same scale as its length.

The theorem, not addressing flexure, would say that there is a fundamental limit on how long such nanotubes can be without having in-line thermal motions exceeding the interatomic distance.

2.4.3   2D vs. 2D embedded in 3D

The HLMPW theorem assumes a strictly 2D system, while a related subsequent, and more practical, literature deals with 2D systems embedded in 3D space, to allow for distortion, flexure, out of the plane. Such systems may be referred to as membranes, and it is said that distortion into the third-dimension can stabilize the membrane. On such a basis it has been predicted (Peliti and Leibler, 1985) that the lateral coherence length is

$$\xi = a \, \exp(4\pi\kappa/3k_{\mathrm{B}}T). \tag{2.50}$$

Taking a reasonable value for the rigidity $\kappa = Yt^3 = 1\mathrm{eV}$, a as the lattice constant and $T = 300$ K, the allowed size (before the membrane crumples) is on the order of $a \exp(161) = 0.2$ nm $\times 9.29 \times 10^{69} = 1.86 \times 10^{60}$ m. (Crumpling implies displacement of unit cells by distances similar to the size of the system, as in crumpling a sheet of paper to a small ball.) A somewhat similar formula, but dealing with strictly planar motions, was published by Thompson-Flagg *et al.* (2009). In their view, the Hohenberg–Landau–Mermin–Peierls–Wagner theorem predicts a breakdown of long-range order in a strictly planar crystal, by virtue of low energy rotations when the sample size L is large. The idea is that, for large sample size L, long-range rotational fluctuations occur because they have small energy cost, are favored at finite temperature by increasing the entropy. Thompson-Flagg *et al.* estimate, without giving details, that the limiting length L is

$$L = a \, \exp(Ga^2/k_{\mathrm{B}}T), \tag{2.51}$$

where a is the lattice constant and G the shear modulus. Adopting the bulk shear modulus of graphite $\mu = G/a = 440$ GPa, and taking a as 0.1 nm, they find L for

300 K to be larger than 10^{30} m, about 7×10^{18} times the earth–sun distance. Equation (2.51) addresses the same strictly planar situation as do eqns (2.33–39), and it is not obvious that the two estimates are compatible. Apart from the separate assumptions on planarity, we note that the two estimates above for the limiting length L (based on eqn [2.50] vs. eqn [2.51]) differ by 30 orders of magnitude at 300 K. A further confusion, on the size scale of graphene to remain flat, is the estimate of a related parameter $L_T = 1$ Ångström at 300 K offered by Castro Neto *et al.* (2009) [see their eqn (139)], with the implication "that free-floating graphene should always crumple at room temperature due to thermal fluctuations associated with flexural phonons." Physically, L_T is the "thermal wavelength of flexural (phonon) modes," and it appears that the quoted statement is incorrect in associating L_T in a direct way with the size L of a sample that will be stable, that is clearly much larger than 1 Ångström. We will return to this question in Chapter 7, see eqns 7.9–7.10.

It seems probable that the observed corrugation effects are either familiar classical results of shear or tensile strain imposed by boundary conditions on a soft but inextensible sheet, or arise from randomly adsorbed chemical impurities such as OH that distort nearby C–C bonds, pulling some atoms out of the plane. Oxygen chemisorbed to graphene has been recently studied by Hossain *et al.* (2012). In Fig. 2.12, nine chemisorbed oxygen atoms on epitaxial graphene are shown, in a high resolution, three-dimensionally rendered STM image. The image was taken with sample bias 2.4 V and tunneling current 50 pA. A chemically uniform state of oxygen on graphene is presented in this image. It is found that the oxygen is released thermally at temperatures as low as 260 K. This will be discussed again in Chapter 8, including modeling of the bonding and resultant deformation of the graphene, distorting the STM image as shown in Fig. 2.12.

Graphene is entirely surface and hard to keep clean at the required level. Large changes in the resistance of graphene have been shown after current-annealing to modest temperatures, releasing weakly-bound adsorbates. Such annealing was not available to early workers. These topics will be further explored in Chapter 7.

2.4.4   Soft membrane, crumpling instability

Melting, in the strict 2D situation of the electron crystal on liquid helium, was successfully modeled by a theory (Fisher *et al.*, 1979) based on dislocations. The observed electron solid is stabilized by in-plane Coulomb repulsion between electrons. A dislocation core is composed of paired disclinations, illustrated for the triangular lattice in Fig. 2.13. In the left panel, the central atom has five nearest neighbors, while in the right panel the central atom has seven nearest neighbors. Graphene has the same triangular lattice but its instability is likely to involve motion into the third direction. Melting based on such defects appears to play a role in the disintegration of the 2D graphene crystal (Zakharchenko *et al.*, 2011). The work of Nelson (1982) relates to a strict 2D system. It is conceivable that the 2D defects in the 2D melting of Nelson may first appear in regions of graphene that fluctuate out of plane, in the real world situation.

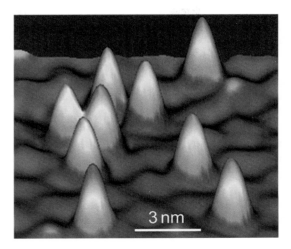

Fig. 2.12 STM evidence for O atoms as nucleating localized distortion of stressed graphene. The large observed heights imply a buckling of the graphene layer surrounding the adsorbate. (From Hossain *et al.*, 2012, by permission from Macmillan Publishers Ltd., © 2012).

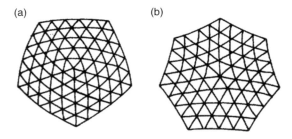

Fig. 2.13 Five-fold (a) and seven-fold (b) disclinations embedded in a triangular 2D lattice. Such defects are believed (Fisher *et al.*, 1979) to play a role in the melting of the electron solid on liquid helium. (From Nelson, 1982. © 1982 by the American Physical Society).

A large literature centered on polymers is based on the concept of a "tethered" surface, where fixed connectivity (e.g., three neighbors) is enforced, but elastic bonds are not assumed. That this is a reasonable model is more easily seen when one realizes that, in conjunction with a hard-core repulsive potential, this represents a square-well potential V connecting nearest-neighbor atoms. Only a small portion of this literature assumes a crystalline layer with elastic bonds.

The model of Kantor *et al.* (1987) is one of hard spheres connected by inextensible strings. A modeled crumpled surface is shown in Figure 3 of their paper. This figure illustrates their work in a three-dimensional rendering of bonds and atom positions, in a triangular tethered surface with $L = 6$ in Monte Carlo simulation. Increasing entropy is the sole consideration, in predicting system behavior, starting from an assumed planar initial condition.

This model has no role for crystallinity. A flat phase here is a stretched, locally-disordered array at low temperature and crumpling is the high temperature limiting case driven by increasing entropy.

If weak attractive interactions are added (Abraham and Kardar, 1991) modeling suggests folding rather than "crumpling," with singly- and doubly-folded phases leading to a "compact" phase of bulk 3D density. This literature envisions crumpling as occurring at a critical temperature, but the simulations of Abraham and Kardar in fact do not find crumpling, but rather folding as a precursor to a compact phase. Advance in this field occurred when the concept of bending energy was included.

What exactly is meant by crumpling has been elusive. In practical terms, perhaps closer to the situation with graphene, an experimental approach to "crumpling" was reported by Kantor *et al.* (1987), summarized in their Figure 12. These authors literally crumpled sheets of paper and of aluminum foil, measuring carefully the resulting diameter. The starting sheets ranged from 1 inch in linear size up to about 30 inches in linear size, while the crumpled products ranged in diameter from around 2 mm up to around 80 mm. A log-log plot of resulting diameter vs. starting linear size gave straight lines. The authors find that the diameters are consistent with slope $\nu = 0.8$, in the relation (R_G is the radius of gyration):

$$\text{Diameter} \propto R_G \propto L^\nu.$$

The experimental data of Kantor *et al.* (1987), in their Figure 12, show that the crumpled result is not compact, as that would require $\nu = 2/d = 0.667$. That value would correspond to crumpling the sheet of volume $L^2 h$, with h its thickness, to a sphere of the same volume $L^2 h = (4/3)\pi r^3$ so $r \propto L^{2/3}$. Here the radius of gyration R_G is defined as the mean square distance of particles from the center of mass. The exponent $\nu = 0.8$, found experimentally, is predicted by the authors' extension of Flory theory (de Gennes, 1979) for a sheet embedded in dimension $d = 3$, namely $\nu = 4/(d+2)$. The authors comment "If the surface is elastic then we will find that simple folding requires energy which increases faster than the surface area." Monte Carlo modeling of surfaces such as the $L = 6$ surface as illustrated in Figure 3 of their paper, support this value for ν for a self-avoiding system. If self-avoidance is not required then $R_G \propto [\ln(L)]^{1/2}$.

A rather different picture had already been introduced for the case when the membrane is endowed with a bending modulus $\kappa = Yt^3$. According to Peliti and Leibler (1985), mentioned above, for membranes of dimension larger than two, or if there are long-range forces, a crumpling transition separates a rigid low T phase from a crumpled high T phase. They found a size scale, above which the membrane appears flexible and crumpled. They used a perturbation theory renormalization to find that the effective bending modulus and spontaneous curvature were changed. They suggested that the surface of the 2D membrane becomes so corrugated that it tends to fill up the whole space. This means for a sample of lateral dimension L, one will have transverse displacements also of size L. It appears that this would apply to the membrane with long-range forces above the crumpling transition. On such a basis it has been predicted (Peliti and Leibler, 1985) (see eqn [2.50] above) that the coherence length is $\xi = a \exp(4\pi\kappa/3k_{\mathrm{B}}T)$. Taking a reasonable value for the rigidity $\kappa = Yt^3 = 1\mathrm{eV}$, and a as the lattice constant, the allowed size (before the membrane crumples) is on the order of $\xi = a \exp(161) = 0.2$ nm $\times\ 9.29\ \times 10^{69} = 1.86\ \times 10^{60}$ m at 300 K. While this number is "astronomically large" at 300 K, it is drastically reduced, to 48 μm, at 3900 K and to 4.0 μm at 4900 K. This formula was derived by solving a renormalization equation and differs only slightly from that presented earlier by de Gennes and Taupin (1982):

$$\xi = a \ \exp(2\pi\kappa/k_{\mathrm{B}}T), \tag{2.52}$$

so that it appears to be on a firm footing. In the words of de Gennes and Taupin (above T_{c}) the interface can become extremely wrinkled and the resulting entropy gain exceeds the loss of energy due to the departure from a periodic array.

The estimate of eqn (2.50) is endorsed by Radzihovsky (2004) who comments "Crumpling, however, only takes place beyond an exponentially long persistence length $\xi_{\mathrm{p}} = a \exp(4\pi\kappa/3k_{\mathrm{B}}T)$, that, in a typical liquid membrane at room temperature far exceeds its size and therefore even a liquid membrane appears not crumpled.". We can take a as the lattice constant and $\kappa = Yt^3$, the bending rigidity, as 1 eV [Kim and Castro Neto (2008), who reference the rigidity measurement by Bunch *et al.* (2007) and theory by Lu *et al.* (2009a)].

Confirming the above, at 300K we find $\xi_{\mathrm{p}} = a \exp(4\pi\kappa/3k_{\mathrm{B}}T) = a \exp(161) = 0.2$ nm $\times\ 9.29\ \times 10^{69} = 1.86\ \times 10^{60}$ m.

Explicit incorporation of crystalline order into this literature came in the papers of Nelson and Peliti (1987) and Paczuski *et al.* (1988). (However, it appears that crystallinity per se adds little, once the bending rigidity κ is assumed.) In the view of Nelson and Peliti, if the connectivity of the crystalline state is preserved, the membrane remains uncrumpled at low temperature. When dislocations are allowed, screening of elastic stresses by buckling reduces dislocation energies and promotes dislocation unbinding, leading to a "hexatic" phase. This is a theoretical construct in 2D, intermediate between liquid and solid, a stable phase that retains six-fold orientational order but loses positional order. The stiffness associated with orientation can lead to logarithmic enhancement of bending rigidity that counteracts the thermal softening found in fluid surfaces. A finite-temperature crumpling transition is predicted for crystalline membranes and possibly for hexatic membranes as well.

Paczuski *et al.* (1988) in their Figure 1 show a plot of the radius of gyration R_{G}, defined as the mean square distance of particles from the center of mass, of a polymerized membrane with linear dimension L as a function of temperature. The 2D membrane is embedded in 3D space, presumably applicable to graphene. The R_{G} curve shows a large, smoothed, downward step, identified as the crumpling transition, at temperature T_{C}. At low temperature, $R_{\mathrm{G}} \approx \zeta L$, where ζ is an order parameter that measures the shrinkage of the membrane in the flat phase due to undulations. (For a disk, e.g., $R_{\mathrm{G}} = R/\sqrt{2}$ so $\zeta = 0.707$.) A second interpretation of ζ is shrinkage due to small scale fluctuations (Kantor and Nelson, 1987). The prediction $R_{\mathrm{G}} \propto L^{0.8}$

appears at high temperature, as a crumpled phase becomes compact. This situation was anticipated by Peliti and Leibler (1985).

Some reality to the concept of the crumpling transition is provided in simulations of Kantor and Nelson (1987), who found a peak in the specific heat at T_c. The elastic energy related to undulations was modeled as suggested in their Figure 1, a depiction of normals, used to define the bending energy, for two elements in a triangular lattice. The bonds in their Figure 1 indicate the pairs of atoms connected via a square well potential V. Kantor and Nelson (1987) establish that when $K_0 = 4a^2\mu(\mu + \lambda)/(2\mu + \lambda)$ greatly exceeds κ_{av}, coupling between modes is sufficient to suppress transverse fluctuations. These workers calculated the specific heat per atom in samples $L = 3$, to $L = 19$ (indicating atoms per side in simulated samples) and with Monte Carlo simulations extending to $N = 271$.

Figure 13 of Kantor and Nelson (1987) plots specific heat C per atom as a function of bending constant κ for Monte Carlo-simulated samples of several sizes L (in atoms per side). The simulated specific heat peak becomes more pronounced for large sizes L.

The calculated quantity is

$$C = (k_B \kappa^2/N) \, | < \sum (\mathbf{n}_\alpha \cdot \mathbf{n}_\beta)^2 > - < \sum \mathbf{n}_\alpha \cdot \mathbf{n}_\beta >^2 \, | + 3/2 \, k_B \qquad (2.53)$$

(the last term is omitted in the plot). The sums are over adjacent pairs as indicated in their Figure. The simulation supports the idea of a transition from a corrugated state to a flat state, as the bending energy is increased (temperature lowered). The size of the effect suggests the transition is in second order. The analysis has not included self-avoidance, but the authors suggest that this is a minor point, perhaps slightly shifting the transition.

2.5   Predictions against growth of 2D crystals

Fluctuations described above at finite temperature in 2D place fundamental restrictions on the existence and synthesis of 1D and 2D crystals. Growth or synthesis generally requires elevated temperatures where only crystallites of very limited size can be stable and remain flat. Following Novoselov (2011), because of the usually low bending rigidity of these crystallites they would generally crumple and fold easily and form 3D structures (that might also help in reducing the energy of unsaturated bonds at the perimeter). In fact the largest flat molecule synthesized to date is C_{222} (Simpson *et al.*, 2002). The method is room temperature cyclo-dehydrogenation of a 3D precursor molecule. Counter to this discussion, however, is the growth of 4 cm-long carbon nanotubes, based on catalysts but without a supporting epitaxial surface (Zheng *et al.*, 2004, mentioned earlier). By analogy it might seem possible to grow a sheet of graphene from a line of catalysts without support. Growth of large flat sheets of graphene on copper or nickel foil at 1000°C may seem close to this, not even requiring a line of catalysts, but this growth can also be regarded as being enabled only in 3D by the rigid substrate. The 2D crystal is later removed at much lower temperature.

2.6 "Artificial" methods for creating 2D crystals

"Artificial" methods, in the sense of a strategy or artifice, fortunately have allowed isolation (distinct from creation or growth) from 3D precursors. Excellent 2D crystals of several layered crystalline compounds (as we will discuss in Section 5.1.1) have been isolated by "peeling" from the parent 3D crystal. Growth of the parent 3D crystal is at high temperature and the fluctuations of the subsequently isolated 2D layer are much reduced by doing the isolation at room temperature. As we will see, chemical vapor deposition at high temperature onto a suitable substrate allows actual growth of large 2D crystals, up to 30 inches in the case of graphene, and the isolation of the grown 2D crystal again is performed at room temperature, where thermal fluctuations are reduced.

2.7 Elastic behavior of thin plates and ribbons

Vibrations and waves on thin plates of elastic material have long been studied in the engineering literature and much of this is applicable to graphene. An elastic solid is characterized by an innate Young's modulus Y and a shear modulus G. If the thickness of the plate or sheet is t, then a bending modulus $\kappa = Yt^3$ comes into play. Elastic deformation waves representing displacement y into the perpendicular direction are well known. For a review of the physics we follow Witten (2007).

Each set $\mathbf{x} \equiv (x_1, x_2, x_3, \ldots)$ identifies a particular bit of matter. Physical materials have three material co-ordinates. This material is embedded in space: to each set of points $\mathbf{x} \equiv (x_1, x_2, x_3, \ldots)$, we identify a set of points $\mathbf{r} = (r_1, r_2, r_3, \ldots)$ in space where the corresponding bit of matter is positioned. If the material is a solid, there is an energy functional $E(\mathbf{r})$ that gives the elastic energy of the particular embedding $\mathbf{r}(\mathbf{x})$. This energy has a minimum for some particular $\mathbf{r_0}(\mathbf{x})$ and its rigid-body translations and rotations, and we take this state as the zero of energy.

For a wide range of deformations, interactions between material points are local and $E[\mathbf{r}]$ is a local functional in material co-ordinates. It depends only on the distances between nearby pairs of points; thus it can be expressed as a function of spatial derivatives such as $\partial r_3 / \partial x_2 \equiv \partial_2 r_3$ and $\partial^2 r_1 / \partial x_1 \partial x_2 \equiv \partial_1 \partial_2 r_1$.

In the resting state of the solid, the distance ds between nearby points is given by

$$ds^2 = \sum_i dx_i^2. \tag{2.54}$$

However, embedding the membrane in space can alter this distance, so that ds^2 takes the form $g_{ij} \, dx_i \, dx_j$ where the metric tensor is defined as

$$g_{ij} = d\mathbf{r}/dx_i \cdot d\mathbf{r}/dx_j = \delta_{ij} + \gamma_{ij}. \tag{2.55}$$

The (dimensionless) strain tensor $\boldsymbol{\gamma}$ with coefficients γ_{ij} may be taken as symmetric, and evidently in the undistorted state this quantity is zero. An inextensible sheet is said to be isometric. An illustration of a change in metric tensor g_{ij} is shown in Fig. 2.14.

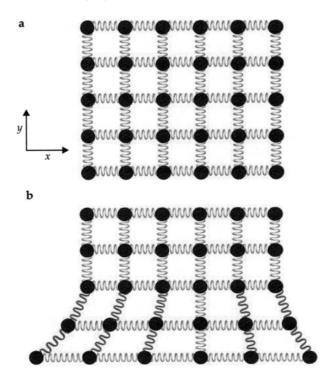

Fig. 2.14 Illustration of change in metric tensor. (a) Elastic network in equilibrium with all masses in reference state. (b) The equilibrium lengths of horizontal springs in successive rows are increased, but vertical springs are not changed. In the lower panel, the metric tensor has been altered only in its x component, g_{xx} that has become a function of coordinate y: $g_{xx}(y)$. In the bottom row the horizontal springs are 50% longer, and in the second row from the bottom, 40% longer. With increasing horizontal spacing, vertical springs in the bottom two rows are seen to be under tension (excepting the vertical column third from the right). All masses are here shown in the same x–y plane, but the response of the system to the high energy distortion would be to buckle in the bottom two rows out of the plane. In this case the bottom row might be described in its position z (out of the page) as $z = z_0 \cos(2\pi y/\lambda)$. Flared edges are often observed along a tear in a material where the local thickness is reduced. (Reprinted with permission from Marder *et al.*, 2007. © 2007 AIP).

According to a theorem of Gauss, the change in metric illustrated in Fig. 2.14 implies a curvature $K(y)$ satisfying the relation

$$K(y) = (1/\sqrt{g_{xx}})\,d^2(\sqrt{g_{xx}})/dy^2. \tag{2.56}$$

Any small departure of γ_{ij} from zero brings an energy cost quadratic in strain γ. There are two independent isotropic scalars that are quadratic in any symmetric tensor; they may be taken as $(\mathrm{Tr}\,\boldsymbol{\gamma})^2$ and $\mathrm{Tr}(\boldsymbol{\gamma}^2)$.

2.7.1   Strain Nomenclature and Energies

It can be shown (Witten, 2007) that the strain energy can be written as

$$E[\bar{r}] = \int_x \frac{1}{2} \lambda (\mathrm{Tr}\,\boldsymbol{\gamma})^2 + \mu \mathrm{Tr}\,\boldsymbol{\gamma}^2. \tag{2.57}$$

The Lame coefficients λ and μ are properties of the material.

When the material is distorted uniformly so that some of the g_{ij} are nonzero, the work per unit strain $\partial E/\partial \gamma_{ij}$ is the force per unit area required to maintain the distorted state. This derivative is called the stress and denoted σ_{ij}. Since the energy is quadratic in γ, each component σ_{ij} is proportional to components of the strain tensor $\boldsymbol{\gamma}$. An important case is when the solid is pulled in one direction (eg, direction 1), without other stress. Then the proportionality of that stress to the corresponding strain is called the Young's modulus Y. In terms of the Lame coefficients,

$$Y = \sigma_{ij}/\gamma_{11} = \mu\,(3\lambda + 2\mu)/(\lambda + \mu). \tag{2.58}$$

The solid then contracts in the orthogonal directions, e.g., the 2 direction, by a factor $\gamma_{22} = -\nu\,\gamma_{11}$. The Poisson ratio ν is then given by

$$\nu = \lambda/2(\lambda + \mu). \tag{2.59}$$

Poisson's ratio, ν, is 0.165 for the basal plane of graphite (Blakslee *et al.*, 1970). If we consider a material in the form of a sheet of thickness t, the description is reduced to two coordinates x_1, x_2 in the plane plus displacements $\mathbf{r} \cdot \mathbf{n}$ normal to the sheet. The curvature tensor C_{ij} may be defined by

$$C_{ij} \equiv \mathbf{n} \cdot \partial^2 \mathbf{r}/(\partial x_i \partial x_j). \tag{2.60}$$

This is a symmetric tensor with dimensions of inverse length.

In the absence of curvature, the stretching energy can be expressed in terms of the transverse or in-surface components 1, 2 of the strain tensor. For ordinary two-dimensional surfaces energy may be expressed (Landau and Lifshitz, 1986) in terms of the in-surface strain tensor $\boldsymbol{\gamma}$ and the curvature tensor $\mathbf{C}$. To lowest order in these tensors, the energy takes the form:

$$\begin{aligned} E(\mathbf{r}) = S(\boldsymbol{\gamma}) + B(\mathbf{C}) = &\int \mathrm{d}x_1\,\mathrm{d}x_2 [\lambda/2\,(\mathrm{Tr}\,\boldsymbol{\gamma})^2 + \mu\,\mathrm{Tr}\,\boldsymbol{\gamma}^2)] \\ &+ \int \mathrm{d}x_1\,\mathrm{d}x_2 [\kappa/2\,\{\mathrm{Tr}\,\mathbf{C})^2 + \kappa_G/2[(\mathrm{Tr}\,\mathbf{C})^2 - \mathrm{Tr}(\mathbf{C})^2]\} \end{aligned} \tag{2.61}$$

The energy is the sum of a part S involving only the in-surface strain and another part B involving only the curvature. There are no cross-terms linear in $\mathbf{C}$ and $\boldsymbol{\gamma}$. Such terms would imply an asymmetry between one side of the surface and the other, assumed absent. It is natural to denote S as the stretching energy and B as the bending energy. Again the derivatives of these energies give the local forces on a small element of surface. Specifically, $\partial S/\partial\gamma_{ij} = \Sigma_{ij}$ is the membrane stress or force per unit length in direction j in the surface acting across a line in direction i. Likewise, $\partial B/\partial C_{ij} = M_{ij}$ is the torque per unit length in the j direction, acting across a line in the i direction. A given set of stresses and torques acting on an element of surface also implies certain normal stresses, denoted Q_i, acting across a line in the i direction (Landau and Lifshitz, 1986).

The two-dimensional (2D quantities in italics) Young's modulus $Y = hY$ (with h the layer thickness, so this quantity has units N/m) relating uniaxial membrane stress to strain is

$$Y = 4\mu\,(\lambda + \mu)/(\lambda + 2\mu). \tag{2.62}$$

The membrane's Poisson ratio is

$$\nu = \lambda/(\lambda + 2\mu). \tag{2.63}$$

The ratio of bending coefficients κ and κ_G to stretching coefficients λ and μ gives the relative importance of bending to stretching energy for a given deformation. The bending coefficients have dimensions of energy (typically 1 eV for graphene) while the stretching coefficients have dimensions of energy per unit length squared (typically 20 eV/A^2 for graphene). Thus their ratios define characteristic lengths for graphene. To be specific, if $\kappa = Yh^3 = 1$ eV and $Y = Yh$ (see eqn [2.67] below) is quoted as $Y = 20.2$ eV/A$^2 = 323$ N/m, then by taking the ratio of these expressions we find h = 0.449 nm = 4.5A. This is on the order of the nominal layer thickness, $h = t = 0.34$ nm. This nominal value t is more a property of van der Waals bonding and of graphite, than of graphene. The radius of carbon in tetrahedral bonds is given as 0.77 A that would give a layer thickness h$' = 0.154$ nm as an alternative. The 2D Young's modulus hY was quite directly measured by Lee *et al.* (2009), who find hY (E^{2D} in their notation) to be 342 N/m ($\pm$50 N/m). Their statement is that this result gives $Y = 1.02$ TPa if the thickness is taken as 0.34 nm. On the other hand, the physics suggests that the right elastic thickness h is perhaps $t/3$, giving Y around 3 TPa for graphene. A review of several experiments/simulations that bear on this is given by Gupta and Batra 2010, who conclude that reasonable values for single layer graphene are $Y \sim 3.4$ TPa and $h \sim 0.1$ nm.

A second characteristic length, the "elastic thickness," $h_e = (\kappa/Y)^{1/2} = (\kappa/hY)^{1/2}$ is also used. In order of magnitude, for graphene if $Y = hY = 340$ N/m (measured by Lee *et al.* [2009] and others), and $\kappa = 1$ eV, we find $h_e = 0.22$ A = 0.022 nm. This is quite a bit smaller and brings into question the numerical value for $\kappa = 1$ eV. An alternative route to find the "elastic thickness" is given below, in terms of the Poisson ratio (eqn [2.69]), giving an estimate h$_e = 0.1$ nm.

Formally, when an isotropic membrane is of thickness h, the relations between the bulk and 2D parameters are as follows: (Witten, 2007) (Here 2D parameters appear italic.)

$$\lambda = 2h\mu/(1 + 2\mu/\lambda) = h\,Y\nu/(1 - \nu^2) \tag{2.64}$$

$$\mu = h\mu = hY/2(\nu + 1) \tag{2.65}$$

$$\kappa = h^3\mu \left[1 + (\mu/\lambda)^2\right]/3(1 + 2\mu/\lambda) = h^3 Y/12(1 - \nu^2) \tag{2.66}$$

$$Y = h\mu(3 + 2\mu/\lambda)/(1 + \mu/\lambda) = hY \tag{2.67}$$

$$\nu = 1/2(1 + \mu/\lambda) = \nu, \tag{2.68}$$

where ν is Poisson's ratio, which takes the same value in 2D and 3D forms. In addition, the elastic thickness is

$$h_e = (\kappa/Y)^{1/2} = h/[12(1 - \nu^2)]^{1.2}. \tag{2.69}$$

For conventional materials, where $1/3 < \nu < 1/2$, h_e is in the range 0.3 h to 0.33 h. If eqn (2.69) is taken for graphene, using 0.165 for Poisson's ratio and taking h as 0.34 nm, then $h_e = 0.0995$ nm = 1 Å. Values in the range 0.7 Å to 1 Å are quoted in recent practical work of mechanical properties of graphene by Gupta and Batra (2010) (see also Lee *et al.*, 2009).

2.7.2 Curvature and Gaussian curvature

Further aspects of the curvature are now addressed. In order for the curvature field to represent a surface with a well-defined displacement field *f(x)*, the curvature tensor must be "curl free" (Millman and Parker, 1977):

$$\partial_i C_{jk} = \partial_k C_{ji}. \tag{2.70}$$

Note that κ_G is taken as the coefficient of the *difference* $[(\text{Tr }\mathbf{C})^2 - \text{Tr}(\mathbf{C})^2]$. This difference is known as the Gaussian curvature C_G, it is the product of the two principal curvatures, or eigenvalues of C. The Gaussian curvature of a sphere is positive, of a cylinder is zero and of a hyperboloid of revolution negative. The Gaussian curvature plays a special role. The associated κ_G energy is strongly constrained by the *Gauss-Bonnet theorem* (Millman and Parker, 1977), The theorem is a purely geometric property of any smooth surface. It says that the integral of C_G over any surface is unchanged by smooth deformations of the surface that do not involve the boundary or its neighborhood. For weakly-deformed surfaces whose energy is described by $E(\mathbf{r}) = S(\boldsymbol{\gamma}) + B(\mathbf{C})$, the κ_G energy is unaffected by distortion of the surfaces. Though the Gaussian curvature energy has no direct impact on the membrane configuration, it has a large indirect impact. This is because the Gaussian curvature necessarily produces a strain field $\boldsymbol{\gamma}$: the curvature and strain fields are not independent. To illustrate this dependence, we suppose that there are equal principal curvatures C_1 and C_2 in the 1 and 2 directions. Then with an appropriate choice of spatial co-ordinates $\mathbf{r}$, we may write

$$\mathbf{r} \cdot \mathbf{n} = 1/2 \, C_1 x_1^2 + 1/2 \, C_2 x_2^2. \tag{2.71}$$

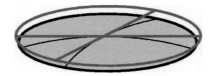

Fig. 2.15 Illustration of how Gaussian curvature leads to strain. A disk of radius ε on a flat sheet is shown (upper lines). The sheet is then curved downward with radial lines held fixed in length. The resulting spherical cap has a reduced perimeter shown as a heavy black line, indicating a compressive strain in the azimuthal direction. (From Witten, 2007. © 2007 by the American Physical Society).

We suppose that the radial lines along x_1 and x_2 do not change their length under this deformation. However, a circle originally of radius ε *contracts* by fraction $(\varepsilon C)^2$ as shown in Fig. 2.15.

Evidently when C_1 and C_2 are both nonzero, the curvature is a source for strain. A uniform Gaussian curvature leads to a strain that grows quadratically with the size of the region. The general statement of this connection is Gauss's fundamental theorem of surfaces (Millman and Parker, 1977):

$$c_1 c_2 = 2\partial_1\partial_2\gamma_{12} - \partial_1\partial_1\gamma_{22} - \partial_2\partial_2\gamma_{11} \qquad (2.72)$$

The scaling of the above energies with thickness h has strong implications about the relative importance of bending and stretching. The ratio of bending to stretching energy for any membrane distortion is proportional to κ/λ and κ/μ. Both of these ratios are proportional to the square of the thickness h. Thus, thin membranes have a relatively high energy cost to stretch relative to their cost of bending. That is, a very thin membrane responds to external forces almost like an unstretchable sheet. This leads to examination of membrane behavior in the unstretchable limit, when surface strains are constrained to remain zero. Such membranes (Witten, 2007) are called inextensible or "isometric."

2.7.3 Isometric distortions of a soft inextensible membrane

Following Witten (2007), we now thus allow only distortions for which strain $\gamma = 0$, so there is no stretching. Therefore the only energy comes from curvature C. However, the curvature must be constrained, to avoid stretching the membrane, hence, the Gaussian curvature must be zero at each point. Thus, at least one of the two principal curvatures must be zero everywhere. If one follows the uncurved direction along the surface, the resulting line is straight in space.

An illustration of this situation from ordinary life may be helpful. A wedge cut from a circular pizza pie is composed of a soft membrane supported by a rigid arc, the crust. If one lifts the wedge by the crust, the membrane droops in the radial direction, because the membrane has small bending rigidity and this has little energy cost. If, before lifting, an azimuthal wrinkle is introduced by breaking the crust, a strong azimuthal curvature is established, a kink running between the crust and the tip. Now radial drooping (appearance of a second, radial component of curvature, to create a Gaussian curvature $C_1 C_2$), would come with a large energy cost, because it would stretch the membrane, which, though soft, is nearly inextensible. Now the wedge can be lifted with no radial drooping, because in the new configuration the drooping would stretch the material. So a new form of rigidity against radial bending appears, without any inherent bending rigidity of the membrane.

The impossibility of bending in all directions without stretching lies at the heart of the anomalous rigidity of *thermally fluctuating membranes.* Isotropic bending necessarily produces energetically costly stretching. Thus the bending is inhibited. The inhibition couples bending and stretching fluctuations. As we have seen, the amount of inhibition depends on the length scale of the fluctuation relative to the thickness. Arriving at a mutual consistency between bending and stretching fluctuations results in a macroscopic bending stiffness and stretching moduli that depend on length scale via subtle power laws (Bowick and, Travesset, 2001).

These membranes may alternatively avoid the stretching cost by another dramatic accommodation: the surface may become macroscopically straight in one direction (Radzihovsky and Toner, 1998).

In either of these scenarios, the system retains its unperturbed size in at least one direction. Thermal fluctuations do not lead to collapsed states like those imposed by a confining sphere. The necessity of a straight line passing through every point imposes severe restraints on confining an isometric membrane. For example, if one wishes to bend the disk-shaped membrane in order to fit inside a sphere, the diameter of the sphere may be no smaller than that of the disk. One may allow somewhat greater confinement by allowing stretching near an isolated point. The Gaussian curvature may occur near this "vertex." This allows a distortion where the uncurved directions converge to the exceptional point, forming a cone-like structure. At the vertex elastic strain is concentrated, without damage to the material because of its strong bonding. Such effects in graphene, appearing as corners, may be seen in Fig. 1.1 and in Fig. 7.5.

2.7.4   Vibrations and waves on elastic sheets and ribbons of graphene

Whether graphene is limp or rigid is a matter of the scale of observation as discussed in Chapter 1 where the effective spring constant K of a square of edge L and thickness t was described as $K = Y t^3 / L^2$. Only on the smallest scales (e.g., below 1 µm) is it rigid as suggested by the large Young's modulus, $Y = 1$ TPa. Measurements show that the conventional theory of waves and vibrations of elastic plates and beams are accurate and useful for micrometer and smaller sized samples of graphene. Since it is found that graphene is usefully described as a classical elastic medium, the details of oscillations and waves expected in such a medium are important.

For a 2D plate of thickness h, Poisson ratio ν, Young's modulus Y and density ρ, lying in the x–y plane, the equation of motion for a vertical displacement $\zeta(\mathbf{r}, t)$ is (Landau and Lifshitz, 1959, p. 111)

$$\rho \partial^2 \zeta / \partial t^2 + Y h^2 / [12(1 - \nu^2)][\partial^2 \zeta / \partial x^2 + \partial^2 \zeta / \partial y^2] = 0. \tag{2.73}$$

Assuming $\zeta(\mathbf{r}, t) = \zeta_0 \exp[i(\mathbf{k} \cdot \mathbf{r} - \omega t)]$,
one finds

$$-\rho\omega^2 + Yh^2k^4/[12(1-\nu^2)] = 0. \tag{2.74}$$

This gives the dispersion

$$\omega = k^2[Yh^2/12\,\rho(1-\nu^2)]^{1/2}. \tag{2.75}$$

Here the frequency is proportional to the square of the wave vector, rather than linear, as would be found in a three-dimensional solid. These waves are referred to as flexural modes, a feature of two-dimensional elastic solids. The components of the velocity $\mathbf{v}$ of the wave are given as $\partial\omega/\partial\mathbf{k}$ valid for small wavevector k. The result is

$$\mathbf{v} = \mathbf{k}\,[Yh^2/3\rho(1-\nu^2)]^{1/2}. \tag{2.76}$$

This velocity of propagation is proportional to the wave vector and not constant, as it is for waves in an infinite medium in three dimensions.

Similar results are obtained for bending waves in thin rods or ribbons. Denoting X the vertical displacement (transverse motion) for a wave moving in the z direction along a rod of width w and thickness t, thus rectangular cross section $S = wt$, the equation of motion becomes

$$\rho S \partial^2 X/\partial t^2 = YI_y \partial^4 X/\partial z^4. \tag{2.77}$$

Assuming

$$X(z, t) = X_0 \exp[i(kz - \omega t)], \tag{2.78}$$

one finds

$$\omega = k^2[YI_y/\rho S]^{1/2}, \text{ where } I_y = wt^3/12. \tag{2.79}$$

In the case of a rod or ribbon clamped at both ends, an appropriate substitution in the equation of motion is

$$X(z, t) = X_0(z)\,\cos(\omega t + \alpha) \tag{2.80}$$

that leads to the equation

$$d^4X_0/dz^4 = \kappa^4 X_0, \tag{2.81}$$

where

$$\kappa^4 = \omega^2 \rho S/YI_y. \tag{2.82}$$

Following Landau and Lifshitz (1959) this has general solution

$$X_0 = A\cos \kappa z + B\sin \kappa z + C\cosh \kappa z + D\sinh \kappa z. \qquad (2.83)$$

If the beam is clamped at both ends, the boundary conditions are $X_0 = dX_0/dz = 0$ at $z = 0$ and $z = L$.

The resulting distortion is

$$\begin{aligned} X_0 = A\,[&(\sin \kappa L - \sinh \kappa L)(\cos \kappa z - \cosh \kappa z) \\ &+(\cos \kappa L - \cosh \kappa L)(\sin \kappa z - \sinh \kappa z)] \end{aligned} \qquad (2.84)$$

This gives minimum frequency

$$\omega_{\min} = 22.4/L^2\,[YI_y/\rho S]^{1/2}\,, \qquad (2.85)$$

where $I_y = wt^3/12$ and with ρS the mass per unit length $= \rho\,wt = m/L$

In the case of a cantilever, one end clamped and one end free, similar analysis gives

$$\begin{aligned} X_0 = A\,[&(\cos \kappa L + \cosh \kappa L)(\cos \kappa z - \cosh \kappa z) \\ &+(\sin \kappa L - \sinh \kappa L)(\sin \kappa z - \sinh \kappa z)] \end{aligned} \qquad (2.86)$$

and a minimum frequency

$$\omega_{\min} = 3.52/L^2[YI_y/\rho S]^{1/2}. \qquad (2.87)$$

The frequency of the doubly clamped beam is thus a factor $22.4/3.52 = 6.36$ higher than the cantilever, as mentioned in Chapter 7 (see eqns 7.16 and 7.41) in connection with measurements of oscillations of graphene membranes.

It is argued on physical grounds that a free beam of length L, as a xylophone key, would oscillate with a frequency of a cantilever of length $L/2$. Thus the minimum frequency of the xylophone key is

$$\omega_{\min} = 14.08/L^2[YI_y/\rho S]^{1/2}. \qquad (2.88)$$

This is lower than the frequency of the doubly clamped beam because the outer ends are free, while they have curvature due to clamping in the former case.

A related question is the elastic stability against buckling of a rod of cross section $S = wt$ and moment of area $I = wt^3/12$, under compression force T. If the ends of the rod are clamped, the critical compression force

$$T_{\mathrm{cr}} = 4\pi^2\,YI_y/L^2\,, \qquad (2.89a)$$

while if the ends are hinged the value is $1/4$ as large. This is referred to as the Euler buckling instability, treated by Golubovic *et al.* (1998).

Fig. 2.16 Sketch indicating buckling of a rod, mounted with fixed angle at each end, under compressive force. An extremely small force produces buckling because of the small bending modulus compared to the compression modulus. (After Landau and Lifshitz, 1959).

As an exercise we consider the size L of a square of graphene mounted vertically by clamps on top and bottom as shown in the sketch, Fig. 2.16.

What is the size L such that the rod (sheet) of dimensions $L \times L \times t$ will buckle under a force $F = mg$ equal to its own weight (assumed applied at the top mounting). The condition comes to

$$L^3 = 4 \ \pi^2 \mathrm{Y} t^2 / 12 \rho g,$$

yielding $L = 0.258$ mm.

We can ask what strain results from this compressive force $F = mg$, using the relation (Landau and Lifshitz, 1959, p. 15)

$$u_{zz} = F'(1 + \nu)(1 - 2\nu)/\mathrm{Y}(1 - \nu) \tag{2.89b}$$

where ν is Poisson's ratio, 0.165 for the basal plane of graphite (Blakslee *et al.*, 1970).

Taking F' as mg/Lt, this strain u_{zz} comes to 5.4×10^{-12}. This indicates the high energy cost of compression of the basal plane, relative to its elastic buckling. Graphene buckles extremely easily. This estimate applies only to free graphene, excluding other forces as commonly come from a substrate.

Graphene as grown on SiC is known to withstand compressive strain about 1%, from the observation that it remains unbuckled when grown on SiC whose lattice constant is 1% smaller than that of graphene. From the comparison with the number above, it seems likely that the adhesive energy of the graphene layer to the SiC substrate, a van der Waals attraction, plays a large role in keeping the graphene on SiC from buckling out of the plane. There is now a larger energy cost associated with buckling, related to moving against the van der Waals attraction.

Following Landau and Lifshitz (1959) p. 113 for a plate of thickness t with supported edges of lengths a and b, the solutions of eqns (2.77) and (2.81) are of the form

$$\zeta_0 = A \sin(m\pi x/a) \sin(n\pi y/b), \tag{2.90}$$

where m and n are integer numbers of half-waves across the rectangle's sides.

The corresponding frequencies are

$$\omega = [\,Yt^2/12\rho(1 - \nu^2)]^{1/2}\pi^2[m^2/a^2 + n^2/b^2]. \tag{2.91}$$

As an exercise let us estimate the lowest frequency $m = n = 1$ for a square of graphene of side L clamped on four sides. Setting $L = 1$ μm we find $\omega = 41.2 \times 10^6$ rad/s. For $L = 1$ mm, the radian frequency is 41.2 s^{-1}, while for $L = 1$ m the radian frequency is 41.2×10^{-6} that corresponds to a period of 42.3 hours. This is less than the 270 hours estimated above for a one meter square clamped on two opposite sides.

2.7.5 An excursion into one dimension

Suppose we return to the case of the "xylophone key" (or ribbon), two free ends, with ρS the mass per unit length,

$$\omega_{\min} = 14.08/L^2[YI_{\rm y}/\rho S]^{1/2} \tag{2.92}$$

and investigate the amplitude of thermal oscillation. To put this in the form

$$\omega_{\min} = 14.08/L^2[Ywt^3/12\,\rho wt]^{1/2} = (K^*/m)^{1/2} \tag{2.93}$$

with $m = \rho wtL$
we find $K^* = 16.5\,Ywt^3/L^3$ (where 16.5 replaces 32 in the formula with two ends clamped). Setting $1/2 K^* x_{\rm rms}^2 = 1/2 k_{\rm B}T$,
we find $x_{\rm rms}/L = (L/L_{\rm cr})^{1/2}$, with "crumpling length"

$$L_{\rm cr} = 16.5\,wt^3\,Y/k_{\rm B}T. \tag{2.94}$$

Setting $w = 10\ t = 3.4$ nm, to crudely resemble a carbon nanotube, and $T = 300$ K, this "crumpling length" comes to 530 μm = 0.53 mm. This nanoribbon is predicted to vibrate chaotically with vibration amplitude comparable to its length, a credible form of "crumpling." Its motion will fill a volume on the order of L^3. The characteristic frequency is 358 rad/s.

A more realistic assessment of carbon nanotubes of diameter 1.0 to 1.5 nm (the diameter of the [10,10] carbon nanotube is 1.35 nm) rather than a nanoribbon, can be deduced from the excellent experimental work of Krishnan *et al.* (1998). For the tube clamped at one end, the rms motion of the tip is found by Krishnan *et al.* to be (their eqn 30)

$$\sigma = 0.849 \, L^3 \, k_{\mathrm{B}} T / [\, Y W G (W^2 + G^2)] \tag{2.95}$$

where W is the diameter of the tube and G is the thickness of its wall that we can take as 0.34 nm. At $L = 0.53$ mm and $Y = 1$ TPa this rms excursion for 300 K comes to 4.56×10^{-6} m. Using the formula of Krishnan *et al.* the critical length, such that the excursion equals the length, for a (10,10) carbon nanotube of diameter 1.35 nm, is 15.4 mm, about a factor 30 larger than the estimate above. This is reasonable and confirms that 1D samples are unstable and will have huge thermal oscillations, at modest lengths. This is not true for 2D samples.

In two dimensions, e.g., a square of side L (with ends free) nothing of the sort occurs, we find

$$x_{\mathrm{rms}}/L = (T/T_0)^{1/2} \tag{2.96}$$

with $T_0 = 16.5 \, Y \, t^3 / k_{\mathrm{B}}$ that evaluates as $T_0 = 4.7 \times 10^6$ Kelvin for graphene. There is no chance that the thermal excursion will approach the length of the sample, as there is in one dimension. For a square of side $L = 1$ m and $T = 300$ K, the predicted thermal motion is 2.53 mm. The thermal motion of unit cells on a modest size sample at modest temperature will easily exceed the lattice spacing, a failure of crystallinity in a strict sense, but the system will not crumple. It does seem more reasonable to describe such a result as an oscillating crystal rather than as a failure of crystalline order.

We now turn to more detailed treatment of graphene on a microscopic level, starting with carbon atoms.

3

Carbon in atomic, molecular and crystalline (3D and 2D) forms

---

The carbon atom is the basic unit of organic chemistry, and biological molecules, notably deoxyribonucleic acid (DNA), the code for all terrestrial life. Carbon crystallizes into diamond and graphite, and also forms nanotubes and fullerene molecules. The variety of forms of carbon reflects the several ways in which the four $n = 2$ electrons can arrange themselves. The fact that the electron energy depends only upon the principal quantum number n (in the Bohr model and in the one-electron Schrödinger treatment) means that 2s and 2p electrons have nearly the same energy within about 4 eV.

The details of the carbon atom are important to an understanding of graphene. A realistic picture starts with the Schrödinger equation.

$$[-(\hbar^2/2m)\nabla^2 + U(r)]\psi = \mathscr{H}\psi = E\psi \tag{3.1}$$

Here the (first) kinetic energy term appears with the assignment of $-i/\hbar \; \partial/\partial x$ to p_x, etc., in Cartesian coordinates x, y, z, and the potential energy term $U(r)$ is $-k_C Ze^2/r$, with r the radius from the electron to the nucleus of charge Ze. For carbon the value of Z is 6. The probability density for the electron is $\psi^*\psi$ and $k_C = 1/4\pi\varepsilon_0$, with $\varepsilon_0 = 8.85 \times 10^{-12}$ F/m, the permittivity of free space.

3.1 Atomic carbon C: $(1s)^2(2s)^2(2p)^2$

The carbon atom is basically spherical, since the potential energy U of the electron in the electric field of the nucleus depends only on the radius r. It is clear that spherical polar coordinates are appropriate. The Schrödinger Equation (3.1) is more complicated when expressed in these coordinates, and it is frequently no longer possible to easily solve the equations for a given situation.

In spherical coordinate notation, where $x = r \sin\theta \cos\varphi$, $y = r \sin\theta \sin\varphi$, $z = r \cos\theta$, with θ and φ, respectively, the polar and azimuthal angles, the Schrödinger equation becomes:

$$\frac{-\hbar^2}{2m}\frac{1}{r^2}\frac{\partial}{\partial r}\left(\frac{r^2\partial\psi}{\partial r}\right) - \frac{\hbar^2}{2mr^2}\left[\frac{1}{\sin\theta}\frac{\partial}{\partial\theta}\left(\sin\theta\frac{\partial\psi}{\partial\theta}\right) + \frac{1}{\sin^2\theta}\frac{\partial^2\psi}{\partial\varphi^2}\right] + U(r)^\psi = E^\psi. \tag{3.2}$$

It is found, because of the symmetry, that the equation separates into three equations in single variables r, θ and φ, by setting

$$\psi = \mathrm{R}(r)\mathrm{f}(\theta)\mathrm{g}(\varphi). \tag{3.3}$$

3.1.1. Wavefunctions for principal quantum numbers $n = 1$ and $n = 2$

The solutions are conventionally described as the quantum states $\Psi_{n,l,m}$, specified by integer quantum numbers n, l, m, and the separation constants are the origin of the peculiar properties that distinguish quantum mechanics from classical mechanics.

The principal quantum number n, that solely governs the energy of a state, is associated with the solutions

$$\mathrm{R}_{n,l}(r) = (r/a_\mathrm{o})^1 \exp(-r/na_\mathrm{o})\mathcal{L}_{n,1}(r/a_\mathrm{o}) \tag{3.4}$$

of the radial equation. Here $\mathcal{L}_{n,1}(r/a_\mathrm{o})$ is a Laguerre polynomial in $\rho = r/a_\mathrm{o}$, and the radial function has n-l-1 nodes, with l the orbital quantum number limited in values from $l = 0$ to $n - 1$. The Bohr radius

$$a_\mathrm{o} = \hbar^2/k_\mathrm{C}me^2 = 0.0529 \text{ nm}, \tag{3.5}$$

is identical to its value in the Bohr planetary model, but it no longer signifies the exact radius of an orbit. The energies of the electron states of the one-electron atom are $E_n = -Z^2 E_\mathrm{o}/n^2$, where

$$E_\mathrm{o} = k_\mathrm{C}\, e^2/2a_0 = 13.6 \text{ eV}. \tag{3.6}$$

The energy can be expressed as $E_\mathrm{n} = -k_\mathrm{C}Ze^2/2r_n$, where $r_n = n^2 a_\mathrm{o}/Z$.

The lowest energy wavefunctions $\Psi_{n,l,m}$ of the one-electron atom are listed in Table 3.1 (Pilar, 2001). As before, to represent the carbon atom, set $Z = 6$. The $n = 1$ groundstate is spherically symmetric, according to

$$\Psi_{100} = (Z^{3/2}/\sqrt{\pi})\exp(-Zr/a_\mathrm{o}). \tag{3.7}$$

The probability of finding the electron at radius r is given by

$$\mathrm{P}(r) = 4\pi r^2\Psi_{100}^2, \tag{3.8}$$

a smooth function easily seen to have a maximum at $r = a_\mathrm{o}/Z$. This is not an orbit of radius a_o, but a spherical probability cloud where the electron's most probable radius is a_o. There is no angular momentum in this wavefunction.

Table 3.1 One-electron wavefunctions in real form.

Wavefunction designation	Wavefunction name, Real form	Equation for real form of wavefunction*, where $\rho=Zr/a_o$ and $C_1 = Z^{3/2}/\sqrt{\pi}$
Ψ_{100}	1s	$C_1\,e^{-\rho}$
Ψ_{200}	2s	$C_2\,(2-\rho)\,e^{-\rho/2}$
$\Psi_{21,\cos\varphi}$	2p$_x$	$C_2\,\rho\,\sin\theta\,\cos\varphi\,e^{-\rho/2}$
$\Psi_{21,\sin\varphi}$	2p$_y$	$C_2\,\rho\,\sin\theta\,\sin\varphi\,e^{-\rho/2}$
Ψ_{210}	2p$_z$	$C_2\,\rho\cos\theta\,e^{-\rho/2}$
Ψ_{300}	3s	$C_3\,(27-18\rho+2\rho^2)\,e^{-\rho/3}$
$\Psi_{31,\cos\varphi}$	3p$_x$	$C_3\,(6\rho-\rho^2)\,\sin\theta\,\cos\varphi\,e^{-\rho/3}$
$\Psi_{31,\sin\varphi}$	3p$_y$	$C_3\,(6\rho-\rho^2)\,\sin\theta\,\sin\varphi\,e^{-\rho/3}$
Ψ_{310}	3p$_z$	$C_3\,(6\rho-\rho^2)\,\cos\theta\,e^{-\rho/3}$
Ψ_{320}	3d$_{z^2}$	$C_4\,\rho^2\,(3\cos^2\theta-1)\,e^{-\rho/3}$
$\Psi_{32,\cos\varphi}$	3d$_{xz}$	$C_5\,\rho^2\,\sin\theta\,\cos\theta\,\cos\varphi\,e^{-\rho/3}$
$\Psi_{32,\sin\varphi}$	3d$_{yz}$	$C_5\,\rho^2\,\sin\theta\,\cos\theta\,\sin\varphi\,e^{-\rho/3}$
$\Psi_{32,\cos2\varphi}$	3d$_{x^2-y^2}$	$C_6\,\rho^2\,\sin^2\theta\,\cos2\varphi\,e^{-\rho/3}$
$\Psi_{32,\sin2\varphi}$	3d$_{xy}$	$C_6\,\rho^2\,\sin^2\theta\,\sin2\varphi\,e^{-\rho/3}$

*$C_2 = C_1/4\sqrt{2}$, $C_3 = 2C_1/81\sqrt{3}$, $C_4 = C_3/2$, $C_5 = \sqrt{6}C_4$, $C_6 = C_5/2$.

The $n = 2$ wavefunctions start with Ψ_{200}(2s) that exhibits one node in r, but is spherically symmetric like Ψ_{100}. The first anisotropic wavefunctions are:

$$\Psi_{210} = R(r)f(\theta)g(\varphi) = C_2\rho\cos\theta e^{-\rho/2} \tag{3.9}$$

$$\Psi_{21,\pm1} = R(r)f(\theta)g(\varphi) = C_2\rho\sin\theta e^{-\rho/2}\exp(\pm i\varphi), \tag{3.10}$$

where again $\rho = Zr/a_o$.

The functions $\Psi_{21,\pm1}$ are the first two wavefunctions to exhibit orbital angular momentum, here $\pm\hbar$ along the z-axis. Generally

$$g(\varphi) = \exp(\pm im\varphi), \tag{3.11}$$

where m, known as the magnetic quantum number, represents the projection of the orbital angular momentum vector of the electron along the z-direction, in units of $\hbar$. The orbital angular momentum **L** of the electron motion is described by the quantum numbers l and m.

The orbital angular momentum quantum number l has a restricted range of allowed integer values:

$$l = 0, 1, 2, \ldots \text{n} - 1. \tag{3.12}$$

The length of the orbital angular momentum vector **L** is $[l(l+1)]^{1/2}$ $\hbar$. Equation (3.12) confirms that the ground state, $n = 1$, has zero angular momentum. In the literature the letters s, p, d, f, g, respectively, are often used to indicate $l = 0, 1, 2, 3$ and 4. So a 2s wavefunction has $n = 2$ and $l = 0$, and the wavefunctions for $n = 2$ and $l = 1$ are called the 2p wavefunctions.

The allowed values of the *magnetic quantum number* m depend upon both n and l according to the scheme

$$m = -l, -l + 1, \ \ldots \ , (l - 1), l. \tag{3.13}$$

There are $2l+1$ possibilities. Again, m represents the projection of the angular momentum vector along the z axis, in units of $\hbar$. For $l = 1$, for example, there 3 values of m: -1, 0 and 1, and this is referred to as a "triplet state." In this situation the angular momentum vector has 3 distinct orientations with respect to the z-axis: $\theta = 45°$, 90°and 135°. In this common notation, the $n = 2$ state (containing 4 distinct sets of quantum numbers) separates into a "singlet" (2s) and a "triplet" (2p).

For each electron there is also a spin angular momentum vector **S** with length $[s(s+1)]^{1/2}$ $\hbar$, where s $= {}^1/_2$, and projection

$$m_{\rm s} = \pm({}^1/_2)\hbar. \tag{3.14}$$

In cases where an electron has both orbital and spin angular momenta (for example, the electron in the $n = 1$ state of the one-electron atom has only **S**, but no **L**), these two forms of angular momentum combine as $\mathbf{J} = \mathbf{L} + \mathbf{S}$, that has a similar rule for its magnitude: $\mathbf{J} = \sqrt{(j(j + 1))}\hbar$.

As mentioned, the $n = 2$, $l = 1$, $m = \pm 1$ wavefunctions (3.10) are the first two states having angular momentum. A polar plot of $\Psi_{21,\pm1}$ has a node along z, and resembles a doughnut flat in the x–y plane.

3.1.2 Linear combinations of $n = 2$ wavefunctions

The sum and difference of these states are equally valid solutions to Schrödinger's equation, for example

$$\Psi_{211} + \Psi_{21-1} = C_2\rho\sin\theta e^{-\rho/2}[\exp(i\varphi) + \exp(-i\varphi)] = C_2\rho\sin\theta e^{-\rho/2}\,2\cos\varphi. \tag{3.15}$$

This is just twice the 2p$_{\rm x}$ wavefunction in Table 3.1. This linear combination, eqn (3.15), is exemplary of real wavefunctions in Table 3.1, where linear combinations have canceled the angular momenta to provide a preferred direction for the wavefunction and resulting probability cloud.

A polar-plot of the 2p$_{\rm x}$ wavefunction (3.15), shows a node in the z-direction from the sinθ and a maximum along the x-direction from the cosφ, so it resembles a dumbbell at the origin oriented along the x-axis. Similarly the 2p$_{\rm y}$ resembles a dumbbell at the origin oriented along the y-axis.

These real wavefunctions, where the $\exp(im\varphi)$ factors have been combined to form $\sin\varphi$ and $\cos\varphi$, are more suitable for constructing bonds between atoms in molecules or in solids, than are the equally valid (complex) angular momentum wavefunctions. The complex wavefunctions with the $\exp(im\varphi)$ factors are essential for describing orbital magnetic moments, as occur in iron and similar atoms. The electrons that carry orbital magnetic moments usually lie in inner shells of their atoms.

To complete the description of the electron state, one must add its spin projection $m_s = \pm^1/_2$. It is useful to separate the space part $\phi(x)$ and the spin part χ of the wavefunction, as

$$\psi = \phi(x)\chi. \tag{3.16}$$

For a single electron $\chi = \uparrow$(for $m_s = {}^1/_2$) or $\chi = \downarrow$(for $m_s = -^1/_2$).

3.1.3 Two-electron states as relevant to covalent bonding

Bonding of the covalent type that occurs in molecular hydrogen and benzene and graphene involves two electrons. To understand this we consider the possible complete states for two electrons, including the spins. For two electrons there are two categories, combined spin S = 1 (parallel spins) or S = 0 (antiparallel spins). While the S = 0 case allows only $m_s = 0$, the S = 1 case has three possibilities, $m_s = 1, 0, -1$, therefore referred to as constituting a "spin triplet."

A good notation for the spin state is $\chi_{S,m}$, so that the spin triplet states are

$$\chi_{1,1} = \uparrow_1\uparrow_2, \chi_{1,-1} = \downarrow_1\downarrow_2, \text{ and } \chi_{1,0} = \uparrow_1\downarrow_2 + \downarrow_1\uparrow_2 \quad \text{(spin triplet).} \tag{3.17}$$

For the singlet spin state, S = 0, one has

$$\chi_{0,0} = \uparrow_1\downarrow_2 - \downarrow_1\uparrow_2 \quad \text{(spin singlet).} \tag{3.18}$$

Inspection of these makes clear that the spin triplet (S = 1) is symmetric on exchange of the electrons, and the spin singlet (S = 0) is antisymmetric on exchange of the two electrons.

Since the *complete* wavefunction (for fermions like an electrons) must be *antisymmetric* for exchange of the two electrons, this can be achieved in two separate ways:

$$\psi_A(1,2) = \phi_{\text{sym}}(1,2)\chi_{\text{anti}}(1,2) = \phi_{\text{sym}}(1,2) \quad \text{S} = 0 \text{ (spin singlet)} \tag{3.19}$$

$$\psi_A(1,2) = \phi_{\text{anti}}(1,2)\chi_{\text{sym}}(1,2) = \phi_{\text{anti}}(1,2) \quad \text{S} = 1\text{(spin triplet)} \tag{3.20}$$

3.1.4 Pauli principle and filled states of the carbon atom

The one-electron-atom quantum states are a start at understanding the states of the carbon atom. We have seen that the energy of the one-electron-atom states scale with nuclear charge Z as Z^2. If we imagine creating a carbon atom by adding electrons one-by-one to the $Z = 6$ nucleus, the first electron will be bound by Z^2 13.6 eV = 489.6 eV. The second electron to be added will be bound by a smaller but still similar energy, diminished because the two electrons now repel each other, but this repulsion is minimized by their tendency to avoid each other. [Actually, the measured ionization energies of the one-, and two-electron ions of carbon are 490 eV and 392.1 eV. This difference, 97.9 eV, can be compared with the electrostatic repulsion of two electrons separated by twice the $Z = 6$ Bohr radius, 2 $a_0/6$ = 17.6 pm (since the Bohr model predicts orbit radius $r_n = (n^2/Z) \, a_o$), namely 81.7 eV).] The filling of the one-electron states in the many-electron atom is governed by the "building-up principle" that describes the shell structure of atoms, and the chemical table of the elements. This comes from the Pauli exclusion principle, applying to Fermi particles, that only one electron is allowed per fully specified state. The two electrons of the $n = 1$ (1s) state differ only by their spin projections, $m_s = \pm^1/_2$.

The levels that are filled in carbon are indicated schematically in Fig. 3.1. The levels shown are important in conceptual understanding of the bonding of carbon. These one-electron levels however are a convenient mental construct and not the most fundamental description of the carbon ground state, that is designated as $1s^2 \, 2s^2 \, 2p^2 \, {}^3P_0$. (Bransden and Joachain, 2003)

Numbers that can be used to gain understanding of these levels are the successive ionization energies for the carbon atom, that are (in eV) 11.26, 24.38, 47.88, 64.49, 392.1 and 490.0 (Lide, 2004, pp. 10–183).

The first guess for the binding energy of 2s state of the C^{3+} ion would be that for an effective $Z = 4$, assuming the $n = 1$ electrons at smaller radius shield the nucleus. The estimate is, then, $E_n = Z^2 E_0/n^2 = 66.4$ eV, close to the measured ionization energy 64.49 eV. The difference between the ionization energies of the 2+ and 3+ ions, 64.49 − 47.88 = 16.6 eV, can again be roughly understood as the electrostatic repulsion of two electrons spaced by twice the $Z = 4$, $n = 2$ Bohr radius. Thus,

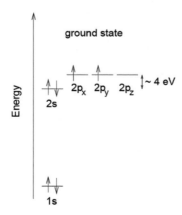

Fig. 3.1 Atomic levels filled in the carbon atom, schematically. (From Goerbig, 2011a).

$\Delta r = 2a_0 n^2 / Z = 2\,a_0 2^2 / 4 = 2a_0$ and $\Delta E = k_c\,e^2 / 2a_0 = 13.6$ eV, again quite close to the measurement. The binding energies of the sixth and fifth electrons are shown as 11.26 eV and 24.38 eV, respectively. The difference of these two, 13.12 eV, is again similar to 13.6 eV, as would come from the previous estimate of electrostatic repulsion. A crude approach to estimate the second ionization energy, 24.38 eV, is to view the fifth electron as in an $n = 2$ state outside a filled spherical shell of effective charge Z^* that might be close to $2 = 6-4$. Thus, we set 24.38 eV $= E_0(Z^*)^2 / n^2 = 13.6$ eV $(Z^*)^2/2^2$, that gives $Z^* = 2.67$. This is 33% larger than $Z = 2$, and suggests that the fifth electron (of the five-electron ion C^+) is not completely shielded from the nucleus by the first two $n = 2$ electrons. It appears that the fifth electron in its motion penetrates significantly into the $1s^2\,2s^2$ shell created by the first four electrons, to see an effective charge $Z^* = 2.67$, greater than 2. The radius for outer two $n = 2$ electrons in carbon then can be estimated from the Bohr formula $r_n = n^2\,a_0 / Z = 4\,(0.0529)/2.67$ nm $= 79.3$ pm.

In considering molecules and graphene formed by carbon, the four outer electrons hybridize into linear combinations of 2s and 2p states. In CH_4 the four electrons go into tetrahedrally-directed sp^3 hybrids, fully localized states. These linear combinations are facilitated by the fact that all $n = 2$ states have roughly the same energy (the s state is about 4 eV lower in energy), so that turning a 2s into a 2p (going from $2s^2 2p^2$ to $2s2p^3$) with an energy cost near 4 eV, will occur if the external potential favors such a modification. In graphene the outer three electrons go into trigonal sp^2 bonding states leaving one $2p_z$ electron free to join the delocalized extended states that form the conduction and valence bands of graphene, the superior electronic conductor.

3.2 Molecular carbon: CH₄, C₆H₆, C₆₀

The formation of molecular (and crystalline) carbon has two aspects. One is the choice of linear combinations of the one-electron wavefunctions to point in particular directions, principally the trigonal sp^2 and the tetrahedral sp^3 combinations. These are called hybrid because they combine s states ($l = 0$) with p states ($l = 1$) with similar but not identical energies. The second aspect is the two-electron combinations of wavefunctions based on the separate nuclei that form the bond. We start by examining the second aspect, the basic notion of the covalent bond. This in turn has two aspects: that of binding, by electron hopping or tunneling; and that of the strong influence on the allowed energies by the Pauli principle in the case of two electrons.

3.2.1   Covalent bonding in simple molecules

A diatomic molecule, starting with H_2, but including atmospheric gases oxygen and nitrogen, is bound by "electron exchange," an effect that is purely quantum in its nature, although final result is an electrostatic attraction. The covalent bond can involve one, two or more electrons. The underlying effects are exemplified in the simplest cases, the one-electron bond in H_2^+, and the two-electron bond in H_2. Electron exchange or hopping (tunneling) between sites is basic in these effects.

3.2.1.1 One-electron bond in $H_2{}^+$ ion: bonding by electron exchange. The physics of the hydrogen molecular ion, the simplest one-electron covalent bond, is inherently quantum-mechanical. For the system of one electron and two protons at large spacing there are two obvious wavefunctions: $\psi_a(x_1)$ and $\psi_b(x_2)$, that represent, respectively, the electron on the left proton and on the right proton. For large spacing these states will be long-lived, but for smaller spacing they will be unstable. An electron starting in $\psi_a(x_1)$, say, will tunnel to $\psi_b(x_2)$, at a frequency f.

To find the ground state we make use of the idea, central to quantum mechanics, that a linear combination of allowed solutions is also a solution. A general solution is

$$\Psi = A\,\psi_a(x_1) + B\,\psi_b(x_2) \tag{3.21}$$

where $A^2 + B^2 = 1$.

The linear combinations that are stable in time are the symmetric and anti-symmetric forms

$$\Psi_S = 2^{-1/2}[\psi_a(x_1) + \psi_b(x_2)], \quad \Psi_A = 2^{-1/2}[\psi_a(x_1) - \psi_b(x_2)]. \tag{3.22}$$

These states are stable, because the electron is equally present on right and left, and the tunneling instability no longer occurs. This is a two-level system, analogous to spin $1/2$, that in principle can be used as a "qubit" in a quantum computer.

It is easy to understand that the symmetric combination Ψ_S has a lower energy than Ψ_A, because the probability of finding the electron at the midpoint is nonzero, while that probability is zero in Ψ_A. The midpoint is an energetically favorable location for the electron, because it sees attraction from both protons. The energy difference between the symmetric and anti-symmetric cases is

$$\Delta E = hf, \tag{3.23}$$

where f is the frequency of an electron started on one side for tunneling to the other side. The value ΔE is about twice the binding energy, that is 2.65 eV for $H_2{}^+$ (Pauling and Wilson, 1935). So the tunneling rate is 1.28×10^{15}/s, corresponding to a residence time 0.778 fs. Exchange of an electron between the two sites is also referred to as "hopping" or "resonance" and a more detailed treatment will give us the "hopping integral" that determines the rate.

Consider two protons (at sites a and b, assume they are massive and fixed) a distance R apart, with one electron. If R is large and we can neglect interaction of the electron with the second proton, then, following the discussion of Pauling and Wilson (1935), we have

$$[-(\hbar^2/2m)\nabla^2 + U(r)]\psi = \mathscr{H}\psi = E\psi \tag{3.24}$$

with r the electron position. This gives solutions, with no interactions,

$$\psi = \psi_a(x_1) \text{ and } \psi = \psi_b(x_2) \tag{3.25}$$

at energy $E = E_o$.

The interactions of the electron and the first proton with the 'second' proton,

$$\Delta E = -k_c e^2 (1/r_{a,2} + 1/R) \tag{3.26}$$

are now considered. The attractive interaction, primarily occurring when the electron is between the two protons, and is attracted to both nuclear sites, stabilizes H_2^+.

We can write the interaction as

$$\mathscr{H}_{\text{int}} = k_c e^2 [1/R - 1/r_{a,2}], \tag{3.27}$$

where the first term is the repulsion between the two protons spaced by R. Following Pauling and Wilson (1935), one finds

$$E - E_o = (k_c e^2 / D a_o) + (J + K)/(1 + \Delta) \text{ for } \Psi_S \tag{3.28}$$

$$E - E_o = (k_c e^2 / D a_o) + (J - K)/(1 - \Delta), \text{ for } \Psi_A, \quad \text{where} \tag{3.29}$$

$$K = \iint \psi_b^*(x_2)[-k_c e^2(1/r_{a,2})]\psi_a(x_1) d^3 x_1 d^3 x_2 = -(k_c e^2/a_0)e^{-D}(1 + D) \tag{3.30}$$

$$J = \iint \psi_a^*(x_1)[-k_c e^2(1/r_{a,2})]\psi_a(x_1) d^3 x_1 d^3 x_2 = (k_c e^2/a_o)[-D^{-1} + e^{-2D}(1 + D^{-1})] \tag{3.31}$$

$$\Delta = \iint \psi_b^*(x_2)\psi_a(x_1) d^3 x_1 d^3 x_2 = e^{-D}(1 + D + D^2/3), \quad \text{where } D = R/a_o \tag{3.32}$$

K is known as the resonance or exchange or hopping integral, and measures the rate at which an electron on one site moves to the nearest neighbor site. Frequently this quantity is denoted t or β, in carbon systems it typically is of order 2 eV to 3 eV. One sees that its dependence on spacing is essentially $e^{-D} = e^{-R/a}$, as one would expect for a tunneling process, and that the basic energy (the prefactor to the exponential term) is $-(k_c e^2/a_o) = -2E_o = -27.2$ eV.

The binding energy of the symmetric case is shown in eqn (3.28). The major negative term is K, and this term changes sign in eqn (3.29), the anti-symmetric case. So the difference in energy between the symmetric and anti-symmetric cases is about 2K, that amounts to about 2×2.65 eV $= 5.3$ eV for H_2^+.

3.2.1.2 Covalent single bond created by two electrons. The prototype single covalent bond occurs in molecular hydrogen, where it joins two 1s states ($n = 1, l = 0$). The essential features are retained in more general cases, for example bonds between 2s and 2p states. Consider two protons (a and b, assume they are massive and fixed) a distance R apart, with two electrons. If R is large and we can neglect interaction between the two atoms, then, following the discussion of Tanner (1995), where the r's represent the space coordinates of electrons 1 and 2, we have

$$[-(\hbar^2/2m)(\nabla_1^2 + \nabla_1^2) + U(r_1) + U(r_2)]\psi = (\mathscr{H}_1 + \mathscr{H}_2)\psi = E\psi. \tag{3.33}$$

Solutions to this problem, with no interactions, can be

$$\psi = \psi_a(x_1)\psi_b(x_2) \text{ or } \psi = \psi_a(x_2)\psi_b(x_1) \tag{3.34}$$

(with $\psi_{a,b}$ the wavefunction centered at proton a, b) and the energy in either case is $E = E_a + E_b$.

The interaction between the two atoms is the main focus. The interactions are basically of two types. First is the repulsive interaction $k_c e^2 / r_{1,2}$, with $r_{1,2}$ the spacing between the two electrons. Second is the attractive interaction of each electron with the 'second' proton, $-k_c e^2(1/r_{a,2} + 1/r_{b,1})$. The latter attractive interaction, primarily occurring when the electron is in the region between the two protons, can derive binding from both nuclear sites at once, to stabilize the hydrogen molecule (but destabilize a ferromagnet).

Altogether the interatom interaction is

$$\mathcal{H}_{int} = k_c e^2 [1/R + 1/r_{1,2} - 1/r_{a,2} - 1/r_{b,1}]. \tag{3.35}$$

To get the expectation value of the interaction energy the integration extends over all six relevant position variables q. (Spin variables are not acted on by the interaction.)

$$<E_{int}> = \int \psi^* \mathcal{H}_{int} \psi dq. \tag{3.36}$$

The appropriate wavefunctions have to be overall antisymmetric. Thus, following eqns (3.17) and (3.18), the symmetric $\phi_{sym}(1,2)$ orbital for the antisymmetric S = 0 (singlet) spin state, and the antisymmetric $\phi_{anti}(1,2)$ orbital for the symmetric S = 1 (triplet) spin state.

The interaction energies are (Tanner, 1995)

$$<E_{int}> = A^2(K_{1,2} + J_{1,2}) \quad S = 0 \text{ (spin singlet)} \tag{3.37}$$

$$<E_{int}> = B^2(K_{1,2} - J_{1,2}) \quad S = 1 \text{ (spin triplet), where} \tag{3.38}$$

$$K_{1,2}(R) = \iint \phi_a^*(x_1)\phi_b^*(x_2)\mathcal{H}_{int}\phi_b(x_2)\phi_a(x_1)d^3x_1 d^3x_2 \tag{3.39}$$

$$J_{1,2}(R) = \iint \phi_a^*(x_1)\phi_b^*(x_2)\mathcal{H}_{int}\phi_a(x_2)\phi_b(x_1)d^3x_1 d^3x_2 \tag{3.40}$$

The physical system will choose, for each spacing R, the state providing the most negative of the two interaction energies. For the hydrogen molecule, the exchange integral $J_{1,2}$ is negative, so that the covalent bonding occurs when the spins are anti-parallel, in the spin singlet case.

To summarize, the result is that in the spin-singlet case the orbital wavefunction is symmetric, allowing more electron charge to locate halfway between the protons, where their electrostatic energy is most favorable. A large change in electrostatic energy

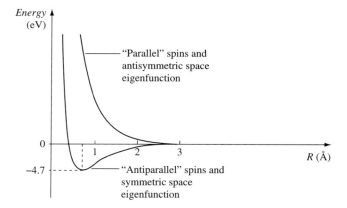

Fig. 3.2 Energy curves for bonding and antibonding states of the hydrogen molecule. The bonding state requires antiparallel spins. The equilibrium separation is 0.074 nm. (From Wolf, 2009).

(about $2J_{1,2}$)is linked (through the exchange symmetry requirement) to the relative orientation of the magnetic moments of the two electrons. This effect can be formally stated as the exchange interaction

$$\mathscr{H}_{e} = -2J_{e}\mathbf{S_1} \cdot \mathbf{S_2}. \tag{3.41}$$

In the case of the hydrogen molecule, J is negative, giving a negative *bonding* interaction for *antiparallel* spins. The parallel spin configuration is repulsive or *antibonding*. The difference in energy between the bonding and anti-bonding states is about 9 eV for the hydrogen molecule at its equilibrium spacing, $R = 0.074$ nm, as indicated in Fig. 3.2. The bonding energy is 4.7 eV in what is a prototype for a single covalent bond.

This appears as a magnetic effect, but actually arises as a combination of fundamental symmetry and electrostatics, to make it much stronger. The covalent bond as described here is a short-range effect, because it is controlled by overlap of the exponentially decaying wave functions from each nucleus. This is true even though the underlying Coulomb force is a long range effect, proportional to $1/r^2$.

3.2.2 Methane CH₄ : tetrahedral bonding

Methane is a covalent tetrahedral molecule involving the $2s2p^3$ electrons of carbon. The 2s and 2p wavefunctions are listed in Table 3.1. In forming the hybrid bonds pointing to the corners of the tetrahedron, it is seen that the needed coefficients of $2p_x$, $2p_y$ and $2p_z$ are $(1,-1,1)$, $(-1,1,1)$, $(1,-1,1)$, $(1,1,-1)$, and $(-1,-1,-1)$.

Covalent bonds exist not only between two s-states ($l = 0$), as in hydrogen, but also between directed combinations of p-states as we discussed in conjunction with the carbon atom. Such bonds are termed hybrid. The linear combination of s and, e.g., p_x, points in the positive x-direction, because $p_x \sim (\rho \cos\varphi)\sin\theta\, e^{-\rho/2} = x\sin\theta\, e^{-\rho/2}$, reverses sign for negative x, where it subtracts wavefunction amplitude from the spherical s-wavefunction amplitude. The bonds in methane CH_4 (and also in diamond and in silicon crystals) are of the tetrahedrally oriented sp^3 type. In all of these cases there is rotational symmetry about the axis joining the two atoms, and this case is referred to as a sigma bond. Sigma bonds based on sp^3 hybrids occur also in ethane, C_2H_6, that can be described as two methanes sharing one sigma bond between the carbons. The next two in the series of "linear alkanes" are propane, C_3H_8, and butane, C_4H_{10}, all based on single sigma bonds of the sp^3 type.

Methane is a gas commonly used in the chemical vapor deposition CVD of graphene on a metal substrates such as copper.

3.2.3   Benzene C_6H_6 : sp^2 and π bonding

Planar carbon bonding of trigonal symmetry occurs in benzene rings, graphite, and graphene. The sigma- portion of these bonds is based on sp^2 hybrids, that are planar with bond angles near $120°$. These combinations are sketched in Fig. 3.3. Note the offset of each lobe from the center, that results from the signed nature of the p orbitals mentioned above, as these wavefunctions are added to the symmetric 2s wavefunction.

The hybrid combinations of $n = 2$ orbitals that form the in-plane sigma single bonding for benzene and graphene are sketched. In this sketch, the $n = 2$, p_z orbital would be out of the page. These two additional p_z orbitals, form a π bond, adding to the sigma bond, to result in a "double bond." The linear combinations for the in-plane sigma single bonds, marked 2, 3, 4 in Fig. 3.3, are as follows:

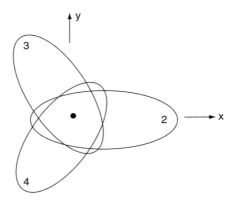

Fig. 3.3 sp^2 bonding sketched in the x–y plane. The p_z orbital is out of the page. (After Baym, 1969).

$$|1> = |p_z>$$ (3.42a)

$$|2> = \sqrt{1/3}\,|s> + \sqrt{2/3}\,|p_x>$$ (3.42b)

$$|3> = \sqrt{1/3}\,|s> - \sqrt{1/6}\,|p_x> + \sqrt{1/2}\,|p_y>$$ (3.42c)

$$|4> = \sqrt{1/3}\,|s> - \sqrt{1/6}\,|p_x> - \sqrt{1/2}\,|p_y>$$ (3.42d)

In hexagonally-configured planar benzene C_6H_6, mentioned in Chapter 1, each carbon atom contributes three of its four valence electrons ($n = 2$) to sigma (rotationally symmetric) bonds to its two carbon neighbors, plus one hydrogen neighbor.

The remaining $2p_z$ valence electrons of each carbon, oriented perpendicular to the carbon–carbon axis, forms a π bond localized between the two carbons. Together the sigma and π bonds constitute a carbon–carbon double bond. The π bonds, that form only when the perpendicular p_z-orbitals are themselves parallel, make the structure planar. Thus, π bonds characteristically do not exhibit rotational symmetry about the bond axis. Carbon–carbon double bonds of this type (combined sigma and π bonds) are characteristic of alkenes (also called olefins), molecules of the form C_nH_{2n}.

Benzene is of special importance because it can be regarded as the origin of graphene. The π bonding in benzene C_6H_6 is anomalous because there are sufficient electrons to permit a π (2-electron) bond only on one side of each carbon. Benzene forms a hexagonal ring of six carbon atoms, with equal spacings 139.9 pm (Lide, 2004, referred to as HCP), so that in some way the one π bond per carbon (making the full double bond) "resonates" from one side to the other.

The positions in the hexagonal ring where the double-carbon bonds appear can be exchanged without changing the structure. So there is a "resonance" between two possible electronic states of the ring (known as the Kekulé structures), interchanging the locations of the single and double bonds in the ring. Experimentally all of the ring bonds are exactly equivalent, with a length 139.9 pm, intermediate between the lengths of single and double bonds of carbon. How this happens has long been of interest. Similar rings, including some with five and some with seven members, occur widely in organic compounds, so the importance is greater than that of benzene specifically. Aspects of this bonding extend to graphite, graphene and carbon nanotubes.

A carbon–carbon triple covalent bond occurs in acetylene, C_2H_2. This bond is based on sp hybrids that are linear with 180° bond angles. Each carbon devotes one 2s electron to bond its hydrogen, and the remaining three $2s2p^2$ electrons bond to its neighbor carbon. Thus six electrons lie between the two carbons, four participating in π bonds and two in sigma bonds. The carbon–carbon distance in acetylene is reported (Lide, 2004) as about 121 pm, compared to 134 pm in a typical double bond, and 153 pm in a single covalent bond. (For comparison the Bohr radius is 53 pm and the spacing in the hydrogen molecule is 74 pm.)

3.2.3.1 Delocalized electrons in symmetric rings. It is suggested that some of the six carbon $2p_z$ electrons involved in carbon–carbon π bonding in benzene are delocalized and can move around the ring. Support for the idea of delocalization comes from enhanced diamagnetic susceptibility of benzene, from details of the nuclear magnetic resonance of protons and ^{13}C in benzene, as well as from the fact that all of the carbon–carbon distances in benzene are the same, 139.9 pm. This length is intermediate between that expected for single bonds (typically 153 pm) and double bonds (typically 134 pm) of carbon.

Two distinct bond lengths are measured in the substituted benzene ring, $C_6H_4O_2$ (*p*-benzo-quinone), where two oxygens replace hydrogens on opposite carbons. These C–C lengths are 134.4 pm and 148.1 pm, representing distinct double and single bonds. So the delocalization occurs only in a symmetric arrangement. In the chemical literature delocalization of electrons in closed rings is described as "aromaticity" (Schleyer and Jiao, 1996). These substituted atoms break the symmetry of the ring and pin down a single "Kekulé" structure.

In a symmetric ring, a simple argument for delocalization can be made. Namely, confining the pi electrons in a "box" of $L = 134$ pm (the typical carbon–carbon double-bond spacing) will cost more energy than allowing their motion to extend six times that distance, thinking of the ring as a cyclic "box." The confinement energy in one dimension is $h^2/8\,mL^2$, about 20.9 eV for $L = 134$ pm. By comparison, a "confinement" energy about 1.05 eV will be estimated later [eqn (3.54c)], if the electron is delocalized into a six-fold periodic potential similar to the dimension of the benzene ring. Delocalization is important to understand the behavior in six-fold carbon rings embedded in larger bonding networks in graphite, graphene, carbon nanotubes and C_{60}. In the last case there are also five-fold pentagonal carbon rings.

Two benzene molecules sharing one side form naphthalene, and three rings sharing sides in a linear array form anthracene. Coronene has seven benzene rings, six being hexagonally located around a central ring. The carbon atoms that are shared between two rings will have bonds to two carbons in one ring plus a bond to one carbon in the adjoining ring. Hydrogen atoms on the interior of the 2-d networks are released, the C–H bonds being replaced by C–C bonds. This replacement can destroy the six-fold symmetry of electron motion around the individual rings. The largest flat molecule composed of benzene rings to date is C_{222} (Simpson *et al.*, 2002). Limiting planar arrays arising from benzene molecules include the graphene sheet, devoid of hydrogens, with one 2p electron per atom delocalized; graphane, where the hexagonal infinite array retains one hydrogen at each carbon site; and fluorographene, with one F atom at each carbon site. The latter is an insulating crystal with band gap 3.0 eV, see Section 3.2.5.

3.2.3.2 Diamagnetism of delocalized orbitals. The benzene ring is important because its diamagnetism demonstrates an electric current carried through covalent bonds. Diamagnetism is the property of generating, in response to an applied magnetic field B, a small opposing response field B'. This is $B' = \chi_m \mu_o B$ (in SI units) where χ_m and μ_o, respectively, are the magnetic susceptibility and the permeability of free space. This occurs through an induced electric current in the diamagnetic system. Using covalent bonds as "wires" is a central idea of molecular electronics. So it is worth discussing how we know the benzene ring can carry current around its closed loop. The measured diamagnetism, of benzene, in excess of the diamagnetism of its atomic cores, also offers an approach to determining how many of the six available p_z electrons in the ring are actually free to move around the ring.

To study the current around the ring, we use a simplified model for the ring based on free electrons. This model is an alternative to standard treatments of the benzene ring based on linear combinations of the carbon wave functions on the six atoms. It turns out that the free electron model is helpful in describing diamagnetism and ring currents, and also helps to understand the bonding between carbon atoms in the ring.

Diamagnetism is a basic effect that occurs for an electron in a closed orbit subjected to a perpendicular magnetic field, B. The diamagnetic susceptibility is a measure of the effect, and its value for benzene is relatively large. Organic chemists (Schleyer and Jiao, 1996) tabulate a "diamagnetic susceptibility exaltation," a measure of *excess* diamagnetic susceptibility in "aromatic" compounds, including five- and six-fold rings.

We now consider the mechanism that allows a diamagnetic system to produce an opposing field, B', following an analysis given by Feynman *et al.* (1964).

Suppose a charge q orbits at radius r, and thus enclosing an area πr^2. (This is a current loop, with $I = dq/dt$). Imagine that a perpendicular magnetic field, $B(t)$ slowly increases from value zero. Faraday's Law says that an electric field E will appear, so that the resulting voltage around the loop will equal $-d\Phi_m/dt$, the rate of change of the magnetic flux through the loop:

$$V = 2\pi r E = -\pi r^2 dB/dt, \tag{3.43}$$

$$E = -(r/2)\, dB/dt. \tag{3.43a}$$

The electric field E acts on the orbiting charge to change its angular momentum, because it exerts a force, and thus a torque, $\tau = qrE$. The angular momentum changes at a rate,

$$dL/dt = \tau = qrE = -(qr^2/2)\, dB/dt \tag{3.44}$$

After a time t, when the field reaches value B, the accumulated angular momentum, ΔL, is

$$\Delta L = -(qr^2/2)B. \tag{3.45}$$

It is known that the gyromagnetic ratio relating a magnetic moment μ to an angular momentum L is $\mu = (q/2m)\, L$. Thus, the change in magnetic moment $\Delta \mu$ brought about by applying the field B is

$$\Delta \mu = (q/2m)\, \Delta L = -(q^2 r^2/4m)B, \tag{3.46}$$

with r the orbit radius. The basic magnetic moment formula is $\Delta\mu = \Delta IA$, where A is the fixed area of the loop carrying current ΔI. Explicitly, for Z electrons,

$$\Delta I = \Delta\mu/\pi r^2 = -Z(e^2/4\pi m)B. \tag{3.47}$$

It is clear that this is a persistent current, typically on the scale of nanoamperes per Tesla (nA/T), that exists as long as the magnetic field is present. Note that r^2 in eqn (3.46) is the radius from an axis through the atom parallel to B, so that if B is in the z direction then $r^2 = x^2 + y^2$. If we consider a spherical atom, where $r^2_{\rm av} = x^2 + y^2 + z^2$, this will be 3/2 times the quantity we want, and we should adjust the factor 4 in the denominator of eqn (3.46) to 6, if we use the usual spherical average, $r^2_{\rm av}$. So this formula is usually expressed as

$$\mu = -(Ze^2/6m)r^2_{\rm av}B, \tag{3.48}$$

where Z is the number of orbiting electrons, and the average $r^2_{\rm av}$ includes orbits randomly oriented with respect to B. Finally, the diamagnetic susceptibility $\chi_{\rm m} = \mu_{\rm o}M/B$ (SI units) for a medium that contains N induced magnetic moments per m^3 (spherical atoms or molecules) is

$$\chi_{\rm m} = -\mu_{\rm o}N(Ze^2/6m)r^2_{\rm av}. \tag{3.49}$$

Now we turn to the case of a planar molecule, such as a benzene ring, in a liquid or gas, so that the current loop quickly assumes random orientations. The magnetic field B is in the z direction, the angle θ between the z-axis and the normal $\mathbf{n}$ to the ring rapidly assumes all values between zero and 180°. The area A is fixed, but the flux intercepted is $BA\cos\theta$. The induced moment is always along $\mathbf{n}$, but the projection of the induced magnetic moment along z (again, proportional to $\cos\theta$) is what is important in calculating the susceptibility. So $\Delta\mu_{\rm z} = -(Ze^2r^2/4m)B\cos^2\theta$, and the average value of $\cos^2\theta$ is $^1/_2$. So the average z-axis contribution of the ring of radius r, as it samples all directions in the liquid is

$$\Delta\mu_{\rm z} = -(Ze^2r^2/8m)B. \qquad \text{(planar molecule in liquid)} \tag{3.50}$$

If there are N benzene rings per unit volume, the magnetization (the magnetic moment per unit volume) M is $-N(Ze^2r^2/8m)B$, and the susceptibility,

$$\chi_{\rm m} = \mu_{\rm o}M/B = -N\mu_{\rm o}(Ze^2r^2/8m). \qquad \text{(planar molecule in liquid, SI units)} \tag{3.51}$$

These formulas apply only to the extended orbit contribution, so that to compare with a handbook value one would need to add the contributions of individual atoms in the molecule, using eqn (3.49).

To compare to the compilations of magnetic properties in the *Handbook of Chemistry and Physics* (HCP) in cgs (centimeter, gram, second) units, we need to display the basic formula eqn (3.49) (spherical case) in cgs units:

$$\chi_m = M/B = -N(Ze^2/6mc^2)r_{av}^2. \text{ (cgs units)} \tag{3.51a}$$

In cgs units, the electron charge is $e = 4.8\ 10^{-10}$ esu, $m_e = 9.1\ 10^{-28}$ g, and c, the speed of light, is 3×10^{10} cm/sec.

3.2.3.3 Diamagnetic susceptibility of benzene. If we apply eqn (3.49) to carbon in the form of diamond, taking $Z = 4$ independent electrons (neglecting the $1s^2$ core electrons) and take the measured value, $\chi_m = -5.9 \times 10^{-6}$ cc per mole, we infer $r_{av.}$ as 72.2 pm. This is a reasonable radius for the outer four electrons in carbon, a bit smaller than the value, 79.3 pm, estimated in Section 3.1.4 from the second ionization energy of carbon using the Bohr model.

We now apply this approach [eqn (3.51)] to the benzene ring, in order to estimate the effective number, Z_π, of current carrying electrons. The handbook (HCP) values of diamagnetic susceptibility, on a molar basis, for benzene (Lide, 2004), *p*-benzo-quinone $C_6H_4O_2$, 1,4 cyclohexadiene C_6H_8 and diamond, are, respectively, -54.9×10^{-6}, -36×10^{-6}, -48.7×10^{-6}, and -5.9×10^{-6} (all in cc/mole). Benzoquinone and cyclohexadiene, are both substituted benzene rings, where we would expect the delocalization to be interrupted, as is confirmed by measurement of distinct C–C single and C=C double-bond distances in these molecules. In diamond we assume the measured diamagnetic susceptibility comes from individual carbon atoms. We can identify an excess of diamagnetic susceptibility for benzene over $C_6H_4O_2$ of -18.9×10^{-6} per mole, and, for cyclohexadiene, -6.1×10^{-6} per mole. These excess susceptibility values of benzene, over its substituent atoms, differ, but average to -12.45×10^{-6} cc per mole.

A thorough study of many such excess diamagnetic susceptibility values for benzene by an expert leads to an accepted experimental value of the *excess diamagnetism* (exaltation, or ring contribution) for benzene of -13.4×10^{-6} cc per mole (Schleyer and Jiao, 1996, see their Table I). This describes only the ring contribution, so we can use it in eqn (3.51) for N planar molecules in liquid or gas, now expressed in cgs units. Assume that the number Z_π of circulating $2p_z$ (π) electrons in C_6H_6 is unknown (Z_π can be at most 6, one electron from each carbon atom in the ring).

$$\chi_m = M/B = -N(Z_\pi e^2 r^2/8m_e^* c^2) \text{ (planar molecule, in liquid, cgs units).} \tag{3.52}$$

We allow an effective mass m_e^*, for the motion of the electrons around the ring, and use the experimental χ_m and $r_{av.}$ values to determine the unknown Z_π/m_e^*. For one mole, taking $r_{av.}$ as 190 pm, and solving eqn (3.52) for $Z_\pi(m_e/m_e^*)$:

$$Z_\pi(m_e/m_e^*) = \chi_m 8m_e c^2/(N_{Avog.}e^2 r^2)$$
$$= (13.4 \times 10^{-6} \times 8 \times 9.1 \times 10^{-28} \times 9 \times 10^{20})/ \tag{3.52a}$$
$$[6.02 \times 10^{23}(4.8 \times 10^{-10})^2 3.61 \times 10^{-16}] = 1.753 = 2/(1.14).$$

We interpret this to mean that C_6H_6 has two circulating π electrons, with effective mass $m_e^*/m_e = 1.14$.

To obtain this we have chosen a rather large radius, $r_{av.}=190$ pm, for the circulating electrons. (We can see in Fig. 3.4(b), evidence for such a large radius.) The radius of the ring of carbon atoms (hexagon) is 139.9 pm, and the C–H bond length is 110 pm. Thus, we take the current path to lie halfway out along the C–H bond, thus 55 pm beyond the carbon. This is not unreasonable, since we estimated from the diamagnetic susceptibility of carbon that the radius of the $n = 2$ electrons in carbon is 72.2 pm. Actually, the $2p_z$ orbitals lie above and below the plane of the ring, that may allow more freedom in the radius value. We may also expect the orbiting electrons to find some attraction from the protons (H), pulling the orbiting electrons out beyond the geometrical ring of carbons, whose radius is 139.9 pm.

We can now ask how large is the screening current, having found that the number of electrons is two. Applying eqn (3.47) in SI (or MKS) units

$$\Delta I = \Delta\mu/\pi r^2 = -Z(e^2/4\pi m)B. \tag{3.47}$$

to a single isolated benzene ring, with a perpendicular magnetic field, we find that the induced current for $B = 1$T, is 3.9 nA! This current is the only plausible origin of the observed diamagnetism, and also of another effect, the chemical shift of the nuclear

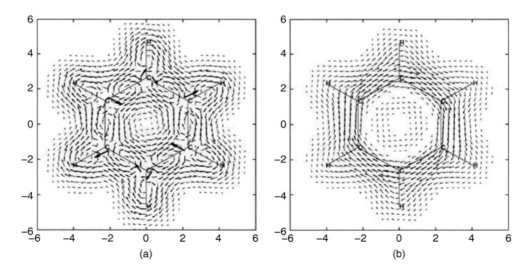

Fig. 3.4 The induced current density in benzene. (a) Current density in the plane of the atoms. (b) Current density calculated one Bohr radius above the plane. Axes are marked in units of Bohr radius, $a_o = 0.0529$ nm. Magnetic field is out of the page, clockwise current corresponds to diamagnetic response. (Reprinted with permission from Juselius *et al.*, 2004. © 2004 AIP).

magnetic resonance (NMR), produced by the resulting magnetic field, B'. This current is in the category of a supercurrent, a term that applies to any orbiting electron in an atom. The orbital motion does not die away, because this would deny the obvious stability of atoms, recognized originally by Bohr.

3.2.3.4 Chemical shifts in the NMR of benzene. A check on the screening current involving two of the six p_z electrons in the benzene ring can be obtained from the nuclear magnetic resonance NMR of carbon (^{13}C) and hydrogen nuclei (protons) in the ring. To use these data, we need to find the response field B' created by the induced screening current in the loop. Ampere's Law applied to a circular current loop gives B' along the axis of the loop, and at a location displaced by z, the value

$$B' = \mu_{\rm o}(\Delta I/2)\,[r^2/(z^2 + r^2)^{3/2}]\; = \; -\mu_{\rm o}[r^2/(z^2 + r^2)^{3/2}]\,Z(e^2/8\pi m)B, \qquad (3.53)$$

using eqn (3.47). B' is a small field, and is usually expressed as a fraction B'/B, conveniently expressed in parts per million (ppm):

$$B'/B = -\mu_{\rm o}Z_\pi(e^2/8\pi m)[r^2/(z^2 + r^2)^{3/2}] \quad \text{(On axis of ring of radius } r, z \text{ is height)} \tag{3.53a}$$

This effect gives a *chemical shift* (of the nuclear magnetic resonance at $\omega = g\mu_{\rm N}B/\hbar$, with $\mu_{\rm N}$ the nuclear magneton), because it shifts the magnetic field for resonant absorption of nearby nuclei. Along the axis, a value 11.2 ppm has been reported for location $z = 100$ pm. (Because the induced field *opposes* the applied field, a bigger applied field is needed, hence the nominal positive sign, 11.2 ppm.) The C₆H₆ protons, outside the ring, experience a *positive* ("deshielding," in chemists' terminology) field B' from the current loop (so the applied field at resonance shifts negatively). There are reliable experimental numbers for both effects. For hydrogen nuclei on the ring, the shifts are in the vicinity of -7 to -8 ppm.

We first estimate the shift at 100 pm above the ring, using parameter values inferred from the susceptibility. (In SI units.)

$$\begin{aligned} B'/B &= -4\pi 10^{-7}\,Z_\pi(e^2/8\pi m_{\rm e}^*)\,[r^2/(z^2 + r^2)^{3/2}] \\ &= -4\pi 10^{-7}\,2\,[e^2/(8\pi\,1.14\,m_{\rm e})]\,[r^2/(z^2 + r^2)^{3/2}] = -8.9 \text{ ppm.} \end{aligned} \tag{3.53b}$$

For $z = 0$ this would become -13.0 ppm. (The measured shifts in the NMR would have opposite signs, and a measured value is 11 ppm.)

To estimate the induced field B' at the hydrogen nuclei in C₆H₆, only a rough approximation is possible. We have modeled the current loop at radius 190 nm. This locates the protons, assuming a C–H bond length of 110 pm, to be only about $a = 55$ pm beyond from the current loop. Viewed from the protons (hydrogens in benzene) the ring current can roughly be approximated as long and straight at radius

a, giving for the proton a response field $B' = \mu_\text{o} I / 2\pi a$. Substituting for I, and setting $a = 55$ pm, the relative shift becomes

$$B'/B = \mu_\text{o} \, Z(e^2/4\pi m_\text{e})/2\pi a$$
$$= (\mu_\text{o}/4\pi)Z(e^2/2\pi m_\text{e}a) = 10^{-7} \quad 2(e^2/2\pi m_\text{e})/a = 14.3 \text{ ppm.} \tag{3.53c}$$

This value should be multiplied by the average of $\cos^2\theta$, leading to a value -7.15 ppm that is close to experimental values, taken from liquid benzene, in the vicinity of -7 to -8 ppm.

So this simplified model, a circular benzene ring, assuming $Z_\pi = \textit{two}$ circulating 2p_z electrons, fits data. The excess diamagnetic susceptibility is closely obtained, and the two measurable chemical shifts are predicted reasonably.

3.2.3.5   *Theoretical chemistry applied to benzene ring current.*   As a check on our simplified model, sophisticated theoretical chemistry results are available. Mathematical methods originating in solid-state physics allow systematic consideration of "self-consistent fields" and electron correlations. One class of computational methods now used in theoretical chemistry is known as "Hartree–Fock self-consistent-field methods." Such calculations of the induced ring currents in benzene have been reported by Juselius *et al.* (2004), for the plane of the ring, and also for planes one Bohr radius above and below the plane, more appropriate to the location of the p_z orbitals.

Figure 3.4 shows "current-vector fields" for benzene, with the x- and y-axes measured in Bohr radii. The magnetic field is applied perpendicular to the ring. Clockwise circulation in the figure is the diamagnetic response. The Fig. 3.4(b) (for the in-plane currents) must reflect largely the trigonal bonding electrons, since the p_z electrons have a node in the plane of the molecule. A substantial diamagnetic screening is provided by in-plane electrons outside the ring, but is canceled, to some extent, by opposite circulation inside the ring. In detail, local circulations are seen around the expected locations of C–C and C–H bonds. Figure 3.4(b) shows the calculated diamagnetic current distribution for the 2p_z electrons, which we have considered. This calculation explicitly considers the current distribution above (and below) the ring at a height of one Bohr radius (the 2p_z functions extend perpendicularly from the ring plane).

It is seen that the calculated current distribution extends in radius past the midpoint of the C–H bonds, consistent with the radius 190 pm that we inferred (it is clear that a single orbit radius, as we have assumed, is not realistic). The authors conclude that the "ring current is induced in the 'π-electron cloud' and that significantly less current is flowing along the σ-bond framework."

Figure 3.4(a) shows the current-vector fields calculated in the plane of the ring. In this plane, $z = 0$, diamagnetic flows occur outside the ring and paramagnetic currents flow inside the ring. It appears that these two contributions largely cancel, consistent with the statement of Juselius *et al.* (2004) that the main induced current is in the p_z electron system (right panel).

In the plane of the ring, in addition to the oppositely directed (paramagnetic) screening seen on the inside of the ring, Fig. 3.4(a) shows whirls of current around the midpoints of both C–C and C–H bonds. [We will see below a model that gives some credence to the shift of origin for "resonating valence bond wavefunctions" eqn (3.55) by $a'/2$ or $30°$, to a point midway between carbons, as mentioned in the text prior to eqn (3.55).]

In Table I of their paper, Juselius *et al.* (2004) give a range of induced ring current values, 8 nA/T to about 13 nA/T, depending on the calculation method. These values are two to three times what we found from fitting our simple model to susceptibility data. It is not clear that the theory approach used in their work would allow discussion of how many electrons are involved in the screening current, although it appears that contributions from a wide set of states have been summed. The authors state "Although the current density is a proper quantum mechanical observable, it has not been directly observed experimentally . . . current density plots do not provide any quantifiable measures of the current strengths. . . ." It is not clear, either, if comparison with experimental values of chemical shift or susceptibility has been carried out.

The treatment of diamagnetism of benzene (and related molecules) has been extended by Fliegl *et al.* (2009), who now find that the screening current density actually peaks in the plane of the benzene molecule, at a radius approximately 170 pm along a line perpendicular to a C–C bond (a larger radius would be found in a direction through a carbon atom). Fliegl *et al.* (2009) predict that a larger portion of screening current flows in the in-plane sigma bonds that increases the total diamagnetic screening current to 16.7 nA/T. This current is countered by an opposite current inside the ring, to bring a total of 11.8 nA/T for benzene, and shifts the main screening current location from above and below the plane, to z = 0. These authors suggest that the idea that the screening current is primarily carried by π electrons is incorrect. This suggestion that the trigonal sigma-bonding network contributes strongly to the magnetic susceptibility is brought into question by the fact that in graphene it is clear that only the p_z electrons are needed to accurately describe both electrical conduction and the diamagnetic susceptibility.

We extend the simple model to consider the behavior and participation in bonding of the remaining four electrons in the ring. The question is how the bonding is strengthened from the sigma single bond value, to give the equal 139.9 pm spacings between the ring carbons.

3.2.3.6 Modeling delocalized and bonding electrons in the benzene ring. Assume, initially, that the six p_z electrons in the ring are free (have constant potential energy) along a circular track of length $L = 2\pi r$. (Our simplified approach suggests that two are free and four are bound.)

For a quantum treatment of this circular trap, consider the angular part of Schrödinger's equation, eqn (3.1), and specify θ as $90°$ to confine motion to the x–y plane. This leaves a single variable, the azimuthal angle, φ, conventionally measured from the positive x-axis. In general we expect a dependence of potential energy U on the angle φ, but we will start with $U = $ constant. So the equation for the delocalized electron becomes

$$\frac{\hbar^2}{2mr^2} \partial^2 \psi / \partial \varphi^2 + U(\varphi)\psi = E\psi. \tag{3.54}$$

If we try a solution of the form

$$\psi = A \exp(im'\varphi), \tag{3.54a}$$

the physical requirement is that ψ be single-valued, so adding $m'2\pi$ to the value of φ leaves the value ψ unchanged. This implies $m' = 0, \pm 1, \pm 2, \ldots$ (This corresponds to integer multiples of electron wavelength around the loop.)

Inserting this function (for $E > U$) into the equation, we find

$$\frac{\hbar^2}{2mr^2} m'^2 = E - U. \tag{3.54b}$$

If $U = 0$, then the allowed energies are

$$E = m'^2 E_1, \quad \text{where } E_1 = \frac{\hbar}{2mr^2} \approx 1.05 \text{ eV and } m' = 0, \pm 1, \pm 2, \ldots. \tag{3.54c}$$

The angular momentum involved, as projected on the z-axis, is $m'\hbar$.

Here, the state $m' = 0$ has no angular motion, and can be occupied by two electrons, one of spin up and one of spin down. These two states are reasonably identified with the two delocalized electrons in the benzene ring that was inferred above from the diamagnetic susceptibility and chemical shift data. If a magnetic field is turned on, one can imagine that these delocalized states would develop the rotation that was described above.

The states for integer m, $\exp(im\varphi)$ and $\exp(-im\varphi)$, are two distinct states, and each could be occupied by two electrons, one for spin up and one for spin down. The state $A \exp[im(\varphi - \varphi_0)]$ is also a solution with the same energy for any φ_0. Linear combinations of $\exp[im(\varphi - \varphi_0)]$ and $\exp[-im(\varphi - \varphi_0)]$, are $A \cos[m(\varphi - \varphi_0)]$ and $B \sin[m(\varphi - \varphi_0)]$ that are analogous to the combinations shown in Table 3.1 for the $2p_x$ and $2p_y$ states. Setting $\varphi_0 = 0$ but recognizing that a shift of the origin in the angle makes no difference for the energies [eqn (3.54c)], the corresponding probability distributions are $\cos^2(m\varphi)$ and $\sin^2(m\varphi)$, with energies $m^2 E_1$.

The next step is to turn on a potential $U(\varphi)$ representing interaction with the ions. The six equally spaced ions, by angles $\Delta\varphi = 60°$, make this potential six-fold symmetric, with a leading Fourier component of the form $U_6 = -U_6 \cos^2(3\varphi)$, if we assume that one ion is located at $\varphi = 0$. We assume that the electron has an attraction to its source ion, so an electron probability distribution $\cos^2(3\varphi)$ would be expected to have a strong binding energy, while the combination $\sin^2(3\varphi)$ would correspond to anti-bonding. The $\cos^2(3\varphi)$ distribution, part of the $m = 3$ state, while strongly benefiting from its commensurate nature, will suffer from a large energy prefactor $9E_1$ and will accommodate only two electrons. The system will choose among the available states based on the lowest energy, and the energy will be affected by a commensurate relation between the probability peaks in the electron probability distribution and the peaks in the potential U as well as by the $m^2 E_1$ value.

The situation of having enough electrons for only three double bonds for six bond locations prompted Kekulé to posit two equivalent dimerized states, each consisting of three dimers being formed (under the three double bonds), leaving a larger spacing between carbons on the other three sites. There is a body of work on one dimensional systems like polyacetylene where it was proved quantum mechanically that a linear array will dimerize. There is a connection between a ring and a linear system that can be imagined to connect into a loop with a cyclic boundary condition. If we assume there is a tendency for the carbon atoms in the ring to dimerize, then the potential $U_3 = -U_{03}\cos^2(3\varphi/2)$ would appear. A solution in this case could involve two wavefunctions:

$$\psi = A \cos(3\varphi/2) \pm iB \sin(3\varphi/2) \tag{3.55}$$

where we imagine the origin of φ to be shifted by 30°, to an angle midway between two carbon atoms. So now $\varphi = 0$ corresponds to a dimer location. Each of these wavefunctions [eqn (3.55)] is occupied by 2 electrons. The probability distribution for each of these two wave functions is P = $\psi^*\psi$ is

$$P(\varphi) = A^2\cos^2(3\varphi/2) + B^2\sin^2(3\varphi/2) \tag{3.55a}$$

This distribution totally contains four electrons, two (spin up and spin down) in each of the forms of eqn (3.55). The first term in the distribution peaks at angles 0°, 120°, 240°, midway between atoms 1–2, 3–4, and 5–6, while the second term in the distribution peaks at 60°, 180°, and 300°, midway between atoms 2–3, 4–5, and 6–1. In going from the $\cos^2(3\varphi)$ distribution to the $\sin^2(3\varphi)$ the dimer locations shift by one bond length, thus the double bond moves from one side of an atom to the other side. The interpretation of eqn (3.55a) is that with probability A^2 the four electrons are equally distributed between atoms 1–2, 3–4, and 5–6, and with probability B^2 the four electrons are equally distributed between atoms 2–3, 4–5, and 6–1. This adds ~1.3 electrons to each of the three dimer locations promoting the single sigma bonds to double bonds. The choice $A = B$ is natural and means that it is equally likely to have a double bond on either side of an atom which means the atomic spacing is not distorted. The electrons in these states do not have angular momentum and do not contribute to the diamagnetism that was discussed above. These wavefunctions do not fit into the prescription eqn (3.54) unless one were to allow $m' = 3/2$. These wavefunctions suggest the two Kekulé states, account for the four available electrons, and explain the single observed spacing. In the end the carbons remain fixed and the electronic structure around the carbons fluctuates between the two different Kekulé states represented by the two terms in eqn (3.55a).

This accounts for four electrons out of the six, the other two are in the angular momentum state $m' = 0$ discussed earlier. The four resonating bond electrons no longer carry angular momentum ± 1, since the linear combinations forming $\sin(kx)$ and $\cos(kx)$ have canceled out the momentum. For this reason these four electrons do not contribute to the diamagnetic susceptibility.

The circular model here described is simplified, but does allow a good understanding of the diamagnetic susceptibility and chemical shifts arising from two mobile electrons, and equalized bond lengths in the benzene ring affected by four pi electrons in localized states. A modern chemist's view of the benzene ring [see Carey and Guiliano (2011), pp. 434–5] describes the six pi electrons as falling into three "highest occupied molecular orbital" (HOMO) bonding states holding two electrons apiece, each characterized by one nodal plane. The lowest state has the nodal plane in the molecular plane, so that π electron clouds above and below the plane are free to screen a perpendicular magnetic field. The next two degenerate HOMO states at higher energy have nodal planes at right angles that intersect the molecular plane, prohibiting circulation around the ring. This picture coincides with the idea of two circulating electrons for the screening current that would reside in the lowest energy HOMO state. The next two degenerate single-node HOMO states, distributing charge at different angles φ, having orthogonal nodal planes, and do not correspond to the two Kekulé states, since those are displaced in angle by 120°. Carey and Guiliano (2011) cite no references, and in fact there seems to be dispute in the chemical literature on the electronic structure of the benzene ring [see for example, Pilar (2001), pp. 434–7, and Cooper *et al.* (1986)]. It appears that the molecular orbital choice for a given molecule is not unique, and among allowable orbitals for benzene are Kekulé-like choices (See Table 13.11 on p. 436 of Pilar) plus many that do not conform to conventional ideas of bonds. It appears that diamagnetism predictions have not been considered in selecting the appropriate molecular orbitals. Neither book cited has an index entry for diamagnetism or susceptibility. The latter two cited sources question superiority of the molecular orbital method and suggest that a valence bond method, based on the "linear combination of atomic orbitals" (LCAO) is equally useful or superior. The latter two sources state that a linear combination of Kekulé states, such as suggested in eqn (3.55a) is reasonable for benzene.

Going back to a fundamental view of the six atoms, taking the $2p_z$ atomic orbital functions $u_{2pz}(\mathbf{r} - \mathbf{R_j})$ as a basis, a set of linear combinations that are consistent with the six-fold rotational symmetry of the ring are (Bransden and Joachain, 2003, pp. 533–7) :

$$\Phi_\mu(\mathbf{r}) = 6^{-1/2}\Sigma_{(j=1 \text{ to } j = 6)} \exp(i\pi\mu j/3) \, u_{2pz}(\mathbf{r} - \mathbf{R_j}) \qquad (3.56)$$

where

$$\mu = 0, \pm 1, \pm 2, 3. \qquad (3.56a)$$

The phase factors in eqn (3.56) are such that the probability density $\Phi_\mu^*(\mathbf{r})\Phi_\mu(\mathbf{r})$ is unchanged by rotation by any multiple of 60°. Such a symmetry requirement qualifies the state eqn (3.56) as a molecular orbital. The energies of these six linear combination states are obtained, using approximations explained by Bransden and Joachain (2003), as

$$\boldsymbol{E}_\mu = <\Phi_\mu|\mathrm{H_e}|\Phi_\mu> \; = \alpha + 2\beta\cos(\pi\mu/3) \tag{3.57}$$

The values for the energy coefficients α and β are both negative, so that the lowest state is the $\mu = 0$ state where the wavefunctions add in phase. This state corresponds to the $m' = 0$ state in eqn (3.54a) above. The values for the energies are approximately $\alpha = -7$ eV and $\beta = -24$ eV. On this basis the $\mu = 0$ state is at $\alpha + 2\beta = -55$ eV and the degenerate $\mu = \pm 1$ states are at $E = \alpha + 2\beta\cos(\pi/3) = -31$ eV, 24 eV higher than the ground state. Each of the linear combination states indexed by μ accommodates two electrons. The occupied states are then $\mu = 0$, $\mu = 1$ and $\mu = -1$. The $\mu = 0$ state would allow circulation of charge with application of a magnetic field, to give the diamagnetic screening current as discussed above. The $\mu = \pm 1$ states probably would be combined in forms

$$\Psi_\pm = 2^{-1/2}[\Phi_1(\mathbf{r}) \pm \mathbf{i}\Phi_{-1}(\mathbf{r})] \tag{3.58}$$

without angular momentum to accommodate four electrons and would correspond to Kekulé-like states suggested in eqn (3.55) above. These states would not contribute to diamagnetic screening.

An example of the LCAO method to predict energies for benzene and molecules formed by aggregations of benzene rings is given by Fukui *et al.* (1952). This method is also the basis for successful treatments of graphene. In graphene one p_z electron per carbon is assumed delocalized, the trigonal bonding network is ignored for purposes of electrical conduction and diamagnetism, and good agreement is found with experiments. Graphene seems simpler and better understood from a theoretical point of view than is benzene.

3.2.4   Fullerene C₆₀

The C_{60} "Fullerene" molecules are empty spherical shells, containing exactly 60 carbon atoms in five- and six-fold rings. There are 32 rings, comprising 20 hexagons and 12 pentagons. The molecule might be viewed as a graphene sheet distorted to fit onto a spherical surface. The pentagonal and hexagonal carbon rings are located in a similar fashion as pentagonal and hexagonal panels in a soccer football, and of course similar to icosahedral geodesic domes made by architect R. Buckminster Fuller. A view of the C_{60} molecule, emphasizing five-membered rings, is shown in Figure 1 of Elser and Haddon (1994). Each hexagonal benzene ring shares sides with three hexagonal and three pentagonal rings. The bonds have resonance or hopping integrals near 2.44 eV, according to the authors.

These molecules, present in tiny amounts in nature, are very stable and covalently bond the four valence electrons of each carbon atom. They spontaneously form in high temperature oxygen-free atmospheres containing free carbon atoms. Free carbon atoms are present, for example in an arc drawn between two carbon electrodes, an extremely high temperature vapor, in early times used as a light for a motion picture projector. There are reports of several other "magic numbered" molecules, notably C_{70}. It is possible to condense C_{60} molecules onto plane surfaces, and even to make such deposits superconducting, by adding stoichiometric amounts of alkali metals.

3.2.5   Graphane and Fluorographene

It was found by Elias *et al.* (2009), that an exfoliated graphene crystal can be transformed into crystalline graphane $[(CH)_N$ or $(C_6H_6)_N]$ by two-hour exposure to a cold discharge of atomic hydrogen, after annealing in Ar for four hours at 300°C. This profoundly changes the bonding, from sp^2 to sp^3, as suggested in Fig. 3.5. The material no longer has metallic π bands but is a semiconductor with a bandgap. It is found that the hydrogen atoms alternate on the sides of the layer, *A* sublattice on one side and *B* sublattice on the other. (One-sided graphane may also be possible.) The carbon atoms are now pulled out of the plane, the local bonding is more diamond-like, but of course the single sheet is still two dimensional. Remarkably, annealing in Ar atmosphere at 450°C for 24 hours restores nearly completely the graphene properties. The lattice constant of graphane is measured as 0.242 nm, with some distribution of values, that is close to, but slightly smaller than, that of graphene, $3^{1/2}$ $a = 3^{1/2}$ 142 pm = 0.246 nm. A modeling of graphane as the channel of a field-effect transistor has recently been given by Fiori *et al.* (2010). It appears that graphane is not fully stable at common conditions, tending to lose hydrogen.

Fluorographene, fully fluorinated graphene, has been reported by Nair *et al.* (2010). This is a hexagonal lattice with one fluorine at each carbon, having a lattice constant 0.248 nm. It is an excellent insulator with an optical bandgap of 3 eV, with mechanical properties similar to graphene. It can be obtained by exposing graphene on both sides to XeF_2, and also, in lower quality, by cleaving fully-fluorinated graphite, that typically has many lattice defects. Since the fluorine attachment to carbon is stronger than that

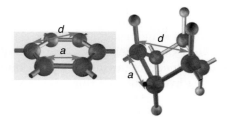

Fig. 3.5 Transformation from benzene (graphene) to graphane. (From Elias *et al.*, 2009, with permission from AAAS).

of hydrogen, fluorographene is stable to 200°C and inert, and is suggested as a two-dimensional version of Teflon®. Teflon®, CF, is one-dimensional, and comprises chains of carbon with one fluorine per carbon. It seems possible that fluorination might be used locally to induce a tunneling barrier in a ribbon of graphene, but the adverse effects of fluorine on the substrate, in addition to its effect on the graphene, may be a problem. Because of the impermeability of graphene to gases, the fluorination and also the hydrogenation, to be complete, must be applied to both sides of the graphene sheet.

3.3 Crystals: diamond and graphite

Diamond, with a cubic structure and tetrahedral bonding common to silicon, is actually metastable with respect to graphite at atmospheric pressure. Diamond was formed deep in the earth at high pressure and temperature, where it is more stable than graphite. Diamond is now available as micrometer-size crystals and films grown by chemical vapor deposition (CVD). An excellent review is given by DeVries (1987).

The word Graphite comes from the Greek *grapho*, meaning to write. Graphite is pure carbon, and, like diamond, is principally obtained by mining from the earth.

3.3.1 Mined graphite

In the earth graphite can be regarded as a limit in the progression from bituminous (soft coal), toward hard coal, anthracite. The density of graphite, 2.27 g/cc, however, is well beyond the range given for anthracite, 1.3–1.4 g/cc. Graphite is also the final product of pyrolysis (destructive distillation, dry distillation, or carbonization) of hydrocarbons. Heating to extreme temperature in the absence of oxygen releases hydrogen, and leads, at temperatures above 2000°C, to the planar graphite structure. Graphite is a refractory material, with a sublimation temperature 3642°C. A temperature 3000°C is used, for example to *graphitize* petroleum coke (a step in the pyrolysis of oil) to make graphite electrodes for carbon-arc furnaces. According to Dresselhaus *et al.* (1996), natural flakes of graphite with extent of several millimeters, but with thickness generally less than 0.1 mm, are found in diverse locations, including Russia, Madagascar and the Ticonderoga region in New York State. Mined graphite of varying quality is the primary source for industrial applications, including carbon to make steel from iron, and carbon electrodes for electric-arc furnaces to produce steel.

3.3.2 Synthetic "Kish" graphite

Synthetic single-crystal graphite known as "Kish" graphite has become a preferred source for scientific purposes, providing flakes that are similar but typically larger in diameter than natural graphite flakes. (Another bulk form of synthetic graphite is produced by heating SiC at 4150°C.) "Kish graphite" is a byproduct of steel production, as excess carbon (the eutectic composition of iron and carbon at 1147°C is 4.3% carbon by weight) precipitates out, producing waste including large single crystal flakes of graphite. According to Dresselhaus *et al.* (1996), the highest quality graphite single crystals available for research are smaller size Kish flakes with diameter around 1 mm.

Suppliers of Kish graphite include Toshiba Ceramic Co. and natural graphite flakes are available, e.g., from Crestmore quarry, California, USA.

3.3.3   Synthetic HOPG: Highly Oriented Pyrolytic Graphite

Pyrolysis, as noted above, is heating in the absence of oxygen and will lead to graphite from carbonaceous feed materials. "Pyrolytic graphite" is also described as partially ordered polycrystalline material produced by thermal decomposition of hydrocarbon gas on a hot substrate. The degree of order of such materials can be improved by heating under pressure, the products earlier being called "compression-annealed pyrolytic graphite" and now known as "highly oriented pyrolytic graphite" (HOPG). Ubbelohde *et al.* (1963) describe annealing a disk of pyrolytic graphite at pressures of 4000 to 5000 psi directed along the c-axis at temperature from 2800°C to 3000°C. Physical properties of such resulting "compression-annealed pyrolytic graphite" were investigated by Blakslee *et al.* (1970).

Such products were characterized by angular spread of crystallites along the c-axis of less than 1 degree. In more recent commercial production the alignment of the highest grades of HOPG graphite is improved by stress annealing the pyrolyzed graphite above 3300°C. The resulting material remains polycrystalline, with lateral in-plane crystal size about 1 µm and along the c-direction about 0.1 µm. In this material there is limited rotational order of the crystallites perpendicular to the planar directions. It is used for monochromators for neutrons and x-rays and also by researchers of graphene.

Niimi *et al.* (2006) report using HOPG graphite (Super Graphite, grade MB) from Matsushita Electric Industrial Co., Ltd, that was synthesized by chemical vapor deposition (CVD) followed by heat treatment under high pressures.

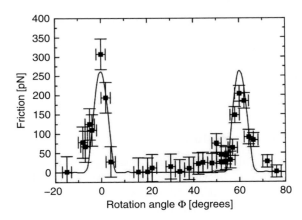

Fig. 3.6 Measured angle dependence of critical rotational torque on HOPG material gives evidence of locking of adjacent layers in graphite at commensurate angles. (From Dienwiebel *et al.*, 2004. © 2004 by the American Physical Society).

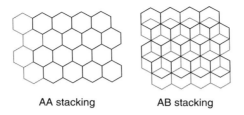

AA stacking   AB stacking

Fig. 3.7 Bernal stacking (right panel) in graphite and bilayer graphene is labeled AB stacking.

A recent study of Kish and HOPG graphites using the scanning tunneling microscope (STM) (Matsui *et al.*, 2005) concluded that "Kish graphite is bulk graphite and HOPG is graphite with a finite thickness of 40 layers." These researchers confirmed that the HOPG graphite contains a higher stacking fault density then either natural or Kish graphite. They measured the ratio α of the in-plane to out-of-plane electrical conductivity, finding 190 for HOPG graphite and 3100 for Kish graphite.

Single crystal graphite, that is approached by the HOPG material, has AB packing of adjacent layers. It appears that rotations between layers are a common defect, and these rotations are necessary for the commonly noted lubricating behavior of graphite to appear. The phenomenon of "superlubricity" as indicated in Fig. 3.6 is based on unlocking the angles between adjacent planes, introducing incommensuration between layers as essential to the lubrication. The authors Dienwiebel *et al.* (2004) used Highly Oriented Pyrolytic Graphite grade ZYA, rated with lateral grain size up to 10 mm. The superlubricity is a result of the lack of stacking order in the HOPG product, and is not expected to occur in more highly ordered graphite.

The sketch of the stacking patterns for graphite is shown in Fig. 3.7.

Graphite is the basis for the commercial Grafoil®, a sheet that is 97–99.8% pure carbon used as a sealant with applications similar to Teflon® tape, but stable at high temperature. Grafoil® is also used in mounting of heat sinks, for example, in personal computers (PCs) because of its high thermal conductivity. A graphite cloth product similar to Grafoil® is used as a moderator in nuclear reactors. The low mass of carbon atoms makes graphite effective in reducing the kinetic energy of neutrons.

4
Electron bands of graphene

The band structure of graphite (treated as layers of graphene) was worked out by Wallace (1947) in a demonstration of what could be calculated in the field of theoretical solid state physics. This field, based on Schrödinger's equation, was also being developed for the bands of semiconductors by Shockley (1950, 1953). In the recent developments, the reduction of Schrödinger theory to Dirac theory, as a consequence of the symmetry of graphene, has revealed unusual properties of electrons in graphene, as mentioned in Chapter 1. In large part, the recent theoretical work reduces to recognizing subtle features that were inherent in the original paper of Wallace (1947). One such subtle feature, the two-component wavefunction, leads to unusually high electron mobility as backscattering is canceled on symmetry grounds.

4.1   Semimetal vs. conductor of relativistic electrons

The basic electron bands in graphene were correctly predicted by Wallace in 1947 based on the honeycomb lattice (Fig. 4.1). He used the tight-binding theory (see for example, Kittel, 1986) to treat the band of energies formed by the $2p_z$ electrons of carbon in graphene and found the unusual linear $E(k)$ near E_F that give a constant carrier speed, an effective speed of light, and zero carrier mass. Wallace pointed out that the spacing of the planes of graphite, 0.337 nm, was so large compared with the hexagonal spacing in the layer, 142 pm that useful calculations of graphite can neglect interaction between planes. [Recent experimental and theoretical work on 3D graphite is offered by Schneider *et al.* (2012).] Thus, the Wallace treatment was of graphene, described as "zone structure of a single hexagonal layer", as a first approximation to graphite.[1] Wallace also pointed out that of the four valence electrons of principal quantum number 2, three were involved in the sp^2 trigonal bonding and played no role in the conduction bands of interest. One needs to consider only bands arising from one $2p_z$ electron per

---

[1] We will see (Section 4.6), in discussion of bilayer graphene, how the weak interaction between electrons in adjacent layers removes the "relativistic" linear $E(k)$ and restores the conventional "semimetal" behavior of parabolic electron bands that touch at E_F. Wallace did not recognize the other "relativistic" feature of monolayer-graphene electrons: the pseudo-spin aspect, arising from the dual sublattices, that makes them analogous to neutrinos with helicity and leads to the useful reduction in backscattering. We return to that subject in Section 8.5. and to the related subject of Klein tunneling in Section 8.7.

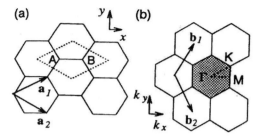

Fig. 4.1 (a) Conventional representation of graphene crystal structure. The unit cell is a rhombus, shown dotted. The whole crystal can be generated by repeated translation by labeled basis vectors $\mathbf{a_1}$, $\mathbf{a_2}$. Note that the repeat distance (lattice constant a) is $\sqrt{3}$ times the carbon bond length $a_{\text{c-c}}$. (b) Conventional description of first Brillouin zone for graphene, shown shaded. (From Saito *et al.*, 1998).

atom, described by orbital $\varphi(\mathbf{r} - \mathbf{r_j})$. Wallace assumed the Bloch-form wavefunction, where N is the number of atoms and the sum is over all lattice points:

$$\psi_{\mathbf{k}}(r) = N^{-1} \sum_j \exp(i\mathbf{k} \cdot \mathbf{r})\, \varphi(\mathbf{r} - \mathbf{r_j}). \qquad (4.1)$$

The Hamiltonian includes nearest-neighbor interactions $t \sim 2.8$ eV, that are of the form $a_i^* b_j$ (since nearest neighbors are on opposite sub-lattices), and next-nearest neighbor (same sub-lattice) interactions $a_i^* a_j$, $b_i^* b_j$ at $t' \sim 0.1$ eV (this is negligible in many cases). So the tight-binding Hamiltonian is

$$H = -t \sum (a_i^* b_j + b_i^* a_j) \; - t' \sum [(a_i^* a_j + b_i^* b_j) + \text{H.C.}], \qquad (4.2)$$

where a_i^* (a_i) is an operator to create (annihilate) an electron at site i on the A sublattice, similarly for b^* (b) on the B sublattice. The term $a_i^* b_j$, for example, annihilates an electron at site j on the B lattice and creates one at site i on the A lattice, a process referred to as hopping. Here "H.C." denotes Hermitian conjugate. The sums are over lattice locations and spin-indices.

The first order energy is obtained by calculating the diagonal matrix elements of the Hamiltonian H of the crystal, where $\boldsymbol{\rho_m} = \mathbf{r_m} - \mathbf{r_j}$:

$$<\mathbf{k}|H|\mathbf{k}> = \sum_{\mathbf{m}} \exp(-i\mathbf{k} \cdot \boldsymbol{\rho_m}) \int \mathrm{d}V \varphi^*(\mathbf{r} - \boldsymbol{\rho_m}) H \varphi(\mathbf{r}). \qquad (4.3)$$

If one considers only overlap with the nearest-neighbor sites separated by $\boldsymbol{\rho}$, such that $\int \mathrm{d}V \varphi^*(\mathbf{r}) H \varphi(\mathbf{r}) = -\alpha$, and $\int \mathrm{d}V \varphi^*(\mathbf{r} - \boldsymbol{\rho}) H \varphi(\mathbf{r}) = -t$, the first order band energy is

$$<\mathbf{k}|H|\mathbf{k}> = -\alpha - t \sum_{\mathbf{m}} \exp(-i\mathbf{k} \cdot \boldsymbol{\rho_m}) \; = E_{\mathbf{k}}. \qquad (4.3a)$$

The basic overlap or hopping energy t (also called the resonance integral), if evaluated between two hydrogen atoms, and given in Rydberg units (Ry $= me^4/2\hbar^2 \approx$ 13.6 eV) exponentially decays with separation ρ as

$$t(\mathrm{Ry}) = 2(1 + \rho/a_0) \exp(-\rho/a_0), \qquad (4.4)$$

where $a_0 = \hbar^2/me^2$(the Bohr radius in Gaussian units) is 0.0529 nm. In the work of Wallace $t = 2.8$ eV is from A sub-lattice to B sub-lattice, and he included a second nearest-neighbor hopping energy $t' \sim 0.1$ eV (on the same sub-lattice). On the honeycomb lattice (two interpenetrating triangular lattices, see Fig. 4.1) Wallace found

$$E_{\mathbf{k}.} = E_{\pm}(\mathbf{k}) = \pm t[3 + \mathrm{f}(\mathbf{k})]^{1/2} - t' \, \mathrm{f}(\mathbf{k}) \qquad (4.5)$$

where + denotes the upper (π^*) and – denotes the lower (π) one-electron band derived from the 2p$_z$ state of carbon, and

$$\mathrm{f}(\mathbf{k}) = 2 \cos(\sqrt{3} \, \mathrm{k}_y a) + 4 \cos(\sqrt{3} \, \mathrm{k}_y a/2) \cos(3 \, \mathrm{k}_x a/2). \qquad (4.6)$$

Here a is the carbon–carbon distance 1.42 A and $\sqrt{3}a$ is the lattice constant.

Referring to Fig. 4.1(b), the shaded portion in the k$_x$, k$_y$ plane is the first Brillouin zone for graphene. Wavevectors k, conventionally measured from the center of this figure, labeled as Γ, reaching points inside the shaded area include all momentum vectors $\mathbf{p} = \hbar\mathbf{k}$ enclosed by a boundary where the first *Bragg reflection* phenomenon occurs. The shaded hexagon in Fig. 4.1(b), the first Brillouin zone, is the generalization of the interval, for lattice constant a,

$$-\pi/a < k < \pi/a \qquad (4.7)$$

that is important for motion in one dimension. The points labeled **K** and **M** represent important directions of high symmetry for motion in graphene. In particular, the point **K** is the point where the conduction and valence bands meet. The point $\mathbf{K'} = -\mathbf{K}$ becomes important when the assumed inversion symmetry is broken.

For some purposes it is useful to represent the honeycomb lattice as the sum of sublattices A and B displaced by one translation $\mathbf{a_1}$ [see Fig. 4.1(a)]. If the graphene layer is disturbed in a manner that removes the inherent inversion symmetry, the point K becomes distinct from its inversion point –K, now denoted K'. This effect occurs if the graphene is grown epitaxially on a substrate and also in the case of two stacked, offset, graphene layers, called *bilayer graphene,* to be discussed in Section 4.6 below. The energy surfaces essentially as predicted by Wallace are shown in Fig. 4.2. In this Figure the vertical scale is about ± 2.5 times basic hopping energy $t \approx 3$eV. The linear energy dispersions near the neutral point are clear in this Figure. It can be seen that the cross sections of the cones are distorted in a trigonal fashion at larger displacements from the six vertices.

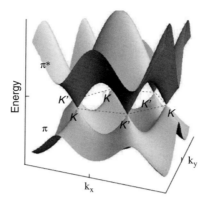

Fig. 4.2 Energy bands essentially as predicted by Wallace (1947). The vertical range of energy in this figure is about ± 2.5 t, where t is the hopping energy. The Dirac-like features are the linear energy dispersions, present near the neutral points K, K′. Wallace did not deal with the dual-sublattice/pseudo-spin aspect of the excitations that lead to absence of backscattering. (From Goerbig, 2011a).

There is a small shift of the neutral point if the overlap energy t' is retained. The full energy dispersion of Wallace, if expanded for $\mathbf{k} = \mathbf{K} + \mathbf{q}$ near the crossing points $\mathbf{K}$, $\mathbf{K'}$, reveals the linear terms, so that, setting $\hbar = 1$,

$$E_{\pm}(\mathbf{q}) \approx 3t' \pm v_{\mathrm{F}} {}_{|}\mathbf{q}| - [9t'a^2/4 \pm 3ta^2/8 \sin(3\theta_{\mathrm{q}})_{|}\mathbf{q}|^2, \tag{4.8}$$

where $v_{\mathrm{F}} = 3ta/2$ and $\theta_{\mathrm{q}} = \tan^{-1}(q_x/q_y)$. Usually the offset t' is ignored, so (with $\hbar = 1$)

$$E_{\pm}(\mathbf{q}) \approx \pm v_{\mathrm{F}} {}_{|}\mathbf{q}| + \mathbf{O}(q/\mathrm{K})^2. \tag{4.8a}$$

The conical features at zero-energy evident in Fig. 4.2 were shown clearly by McClure (1957), in Figure 4 of his paper. Confirmation of this work with emphasis on optical properties is provided by Painter and Ellis (1970).

A further consequence of the conical band structure is that the cyclotron mass of electrons near the Dirac point is anomalous in graphene. It is obtained (Castro Neto *et al.*, 2009) from a definition

$$m^* = (2\pi)^{-1}[\partial A(E)/\partial E], \quad \text{evaluated at } E = E_{\mathrm{F}}. \tag{4.9}$$

The area in q-space is

$$A(E) = \pi q(E)^2 = \pi E^2/E_{\mathrm{F}}^2, \tag{4.10}$$

so that (in units where $\hbar = 1$)

$$m^* = E_{\mathrm{F}}/v_{\mathrm{F}}^2 = k_{\mathrm{F}}/v_{\mathrm{F}}. \tag{4.11}$$

This is related to the electron density n by the relation $n = k_{\mathrm{F}}^2/\pi$, so that

$$m^* = (\sqrt{\pi}/v_{\mathrm{F}}) \sqrt{n}. \tag{4.12}$$

This is in agreement with data as shown by Novoselov *et al.* (2005). The effective mass goes to zero at the Dirac points. Near the Dirac points, and neglecting the smaller hopping term t', the density of electronic states is given by

$$g(E) = 2 A_{\mathrm{c}} \, |E|/\pi \, v_{\mathrm{F}}^2 \tag{4.13}$$

where the area of the cell $A_c = 3 \sqrt{3} \, a^2/2$ with a the interatomic spacing.

Small corrections to the effective mass and slope of the cone leading to the Fermi velocity have been given by Elias *et al.* (2011), in the context of many-electron interactions. Conditions that may lead to merging of the two Dirac points have been discussed by Montambaux *et al.* (2009).

4.2 Linear bands of Wallace and the anomalous neutral point

The work of Wallace (1947) predicted the bandstructure and revealed the essential linear portions of the energy dispersion near the neutral point. The wave equation for the excitations or quasiparticles (and the diamagnetic susceptibility) was given by McClure (1956, 1957), and the similarity to the Dirac equation that resulted from applying the Schrödinger equation to the problem at hand was noted and discussed by Semenoff (1984), and by DiVincenzo and Mele (1984).

4.2.1 Pseudo-spin wavefunction

The Hamiltonian eqn (4.2) can be re-expressed, because of the dual-lattice aspect, into the Dirac form eqn (1.1) leading to the spinor wavefunction. To this end, DiVincenzo and Mele (1984) applied the effective mass theory to the bands near $\mathbf{K}$ (a similar result occurs at $\mathbf{K}'$) in graphene, using the $\mathbf{k} \cdot \mathbf{p}$ wave function

$$\Psi(\mathbf{k}, \mathbf{r}) = \mathrm{f}_1(\boldsymbol{\kappa}) \, \exp(i\, \boldsymbol{\kappa} \cdot \mathbf{r}) \, \Psi_1^{\mathrm{S}}(\mathbf{K}, \mathbf{r}) + \mathrm{f}_2(\boldsymbol{\kappa}) \, \exp(i\boldsymbol{\kappa} \cdot \mathbf{r}) \, \Psi_2^{\mathrm{S}}(\mathbf{K}, \mathbf{r}), \tag{4.14}$$

where $\mathbf{k} = \mathbf{K} + \boldsymbol{\kappa}$. Inserting this $\Psi(\mathbf{k},\mathbf{r})$ into the Schrödinger equation, keeping terms of order $\boldsymbol{\kappa}$ and taking $E_{\mathrm{F}} = 0$ gives the Dirac Hamiltonian eqn (1.1). Thus, DiVincenzo and Mele (1984) found that "the essence of the $\mathbf{k} \cdot \mathbf{p}$ theory (and also of the effective mass theory) is to replace the graphite bands by conical dispersions at E_F"

These authors then also constructed the spinor wave function eqn (1.3).

These steps from eqn (4.2) to the Dirac Hamiltonian and spinor wave function can also be accomplished (Castro Neto *et al.*, 2009) by taking the Fourier transforms of the creation- and annihilation-operators $a^* a$ [see eqn (4.2)] and expanding these around the points **K**, **K'** in the Brillouin zone. Thus, summing on n, counting lattice locations $\mathbf{R_n}$, with N_c the number of cells, the Fourier transform is

$$a_n = (N_c)^{-1/2} \sum \exp(-i\mathbf{k} \cdot \mathbf{R_n}) a(\mathbf{k}). \tag{4.15}$$

New electron operators $a_{1,n}$, $a_{2,n}$ and $b_{1,n}$, $b_{2,n}$, with subscripts 1, 2 referring to **K**, **K'**, are then expressed as

$$a_n = \exp(-i\mathbf{K} \cdot \mathbf{R_n}) a_{1,n} + \exp(-i\mathbf{K'} \cdot \mathbf{R_n}) a_{2,n} \tag{4.16a}$$

$$b_n = \exp(-i\mathbf{K} \cdot \mathbf{R_n}) \, b_{1,n} + \exp(-i\mathbf{K'} \cdot \mathbf{R_n}) b_{2,n}. \tag{4.16b}$$

Following Castro Neto *et al.* (2009), these operators are substituted into the Hamiltonian eqn (4.2), keeping lowest-order terms in $\mathbf{K-k}$, $\mathbf{K'-k}$. This leads to a complicated Hamiltonian expression that can be identified as two copies of the massless Dirac-like Hamiltonian, eqn (1.1), one holding for $\mathbf{p}$ near **K** and one holding for $\mathbf{p}$ near **K'**:

$$H = -i\hbar v_F \int \int \mathrm{d}x \mathrm{d}y \, [\boldsymbol{\Psi}_1^*(\mathbf{r}) \boldsymbol{\sigma} \cdot \mathbf{grad} \boldsymbol{\Psi}_1(\mathbf{r}) + \boldsymbol{\Psi}_2^*(\mathbf{r}) \, \boldsymbol{\sigma}^* \cdot \mathbf{grad} \, \boldsymbol{\Psi}_2(\mathbf{r})]. \tag{4.2a}$$

Here $v_F = 3ta/2\hbar$, $\boldsymbol{\sigma} = (\sigma_x, \sigma_y)$, $\boldsymbol{\sigma}^* = (\sigma_x, -\sigma_y)$ are Pauli matrices and $\boldsymbol{\Psi}_i^*(\mathbf{r}) = (a_i^*, b_i^*)$, $i = (1, 2)$. Comparing to eqn (4.2), this form has dropped terms in t', and is only valid near **K**, **K'** or, in energy terms, within one eV or so of E_F.

The natural and commonly-adopted alternative description, as here justified, is thus the single Dirac-like Hamiltonian acting on a two-component wavefunction, eqn (1.2).

It appears that the same Dirac-Hamiltonian physics had earlier appeared in the literature of narrow band semiconductors (Keldysh, 1964; Aronov and Pikus, 1967). In this literature the effects were known as "interband" or "Landau–Zener" tunneling.

The pseudo-spin aspect of the wavefunction, in the range of energies near the conical crossings, was described in Section 1.2.2, see eqns (1.1) to (1.4). The important theoretical work of Semenoff (1984) and DiVincenzo and Mele (1984) behind this understanding was not widely noticed until the recent upsurge in sophisticated experiments on graphene made possible by the method of Novoselov *et al.* (2004) to get excellent samples. The theory of Semenoff (1984) has been reviewed by Castro Neto *et al.* (2009). It is worth emphasizing that all of the results come from tight-binding theory based on Schrödinger's equation. The appearance of the massless Dirac description of carriers in graphene near the neutral point comes from coupling of orbital motion to the pseudo-spin degree of freedom due to the presence of A and B sublattices.

4.3 L. Pauling: graphene lattice with "1/3 double-bond character"

From an historical point of view, the successful treatment using a Schrödinger tight-binding nearest-neighbor Hamiltonian approach reached beyond what was possible from a chemical-bond-oriented point of view. Pauling intuitively described the graphene lattice in his 1939 book (3rd edn, 1960) as hexagonal carbon where "1/3 double-bond character" was required to account for the four valence electrons per atom. Such an extension to graphene of bond-based methods, previously applied to the benzene molecule (see Section 3.2.3) did not have the predictive power of the tight-binding Hamiltonian approach. A bond-based approach did not reveal features such as the dual-lattice symmetry that forbids carrier backscattering.

In the sketch of Fig. 4.3 the four valence electrons of each carbon atom are used to form bonds with its three neighbors; the system resonates among many valence-bond structures such as shown in Fig. 4.3. In this way, according to Pauling, each carbon–carbon bond achieves "one-third double bond character" This characterization is advanced by Pauling on the basis of the progression of carbon–carbon bond lengths. These are 1.504 Å for the pure single carbon–carbon bond, 1.334 Å for the pure double bond distance, with intermediate values 1.397 Å for benzene (50% double bond character) and here 1.420 Å for graphite.

4.4 McClure: diamagnetism and zero-energy Landau level

An example of a property unavailable to a bond-oriented approach is the zero-energy Landau level in graphene predicted by McClure (1956). We have seen earlier that the Landau-level spectrum is anomalous, including a prominent level at zero energy that supports the pseudo-spin wavefunction. The anomalous observed levels can be written (Novoselov, 2011) (see Fig. 1.8) as

$$E_n = \pm v_\mathrm{F}[2e\hbar B(n + 1/2 \pm 1/2)]^{1/2} \qquad \text{where } n = 0,\ 1,\ 2,\ldots \qquad (1.5)$$

In this expression, describing a half-integer quantum Hall effect, with v_F the cone-related Fermi velocity, the $\pm 1/2$ term is related to the chirality of the quasiparticles and

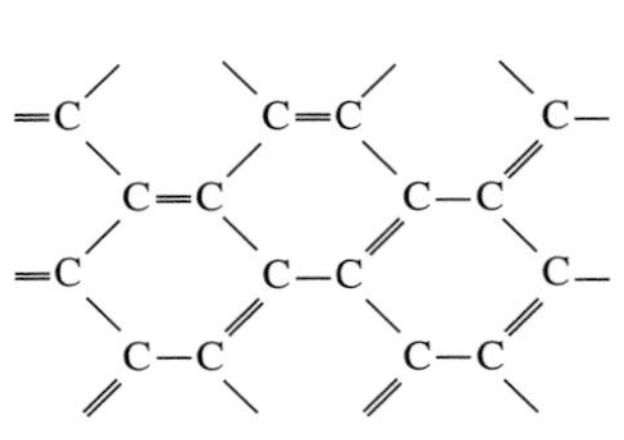

Fig. 4.3 Linus Pauling's suggestion of a snapshot of one of many equivalent bonding configurations, which exhibit "one-third double bond character". The carbon–carbon distance is 1.420 Å, slightly longer than in benzene, at 1.397 Å, representing one half double bond character. In such a situation the electron bonds fluctuate quantum-mechanically, leaving the interatomic distances precisely equal. (From Pauling, 1960).

ensures the existence of two energy levels (one electron-like and one hole-like) at exactly zero energy, each with degeneracy half that of all other Landau levels. (McClure, 1956, 1960). The diamagnetic susceptibility obtained for pure graphite (considering only the 2p free-electron motions) by McClure (1956) (his eqn 3.15) is

$$\chi = -0.0014 \, t^2 \, T^{-1} \operatorname{sech}^2(E_F/2k_B T) \text{ emu/gram} \tag{4.17}$$

(that evaluates as about –0.001/T in emu/gram) where t is the hopping integral expressed in eV, T is the temperature in Kelvin and E_F is the Fermi energy.

This susceptibility is close to the experimental value choosing $t = 2.6$ eV. This agreement suggests negligible contribution from the trigonal sigma-bonding network, counter to the recent suggestion of Fliegl *et al.* (2009), mentioned in Section 3.2.3.5.

Finally, we mention details of the energy surfaces near the Dirac points, that are conical near the crossing. The so-called "trigonal warping" of the energy surfaces, from the $\sin(3\theta_q)$ term in eqn (4.8) becomes more pronounced at energies increasing above and below the crossing. The tripling of the angle θ_q comes from the underlying three-fold lattice symmetry. Accurate cross-sectional cuts showing this effect (for displacements in energy up to about 2 eV) are given in Figure 7a of Goerbig (2011a).

4.5 Fermi level manipulation by chemical doping

Molecular and atomic doping of graphene has been described by Wehling *et al.* (2008), McChesney *et al.* (2010) and Zhao *et al.* (2011). Intercalation compounds of graphite are mentioned following Fig. 5.2. Many metals can be introduced between the layers of graphite, introducing carriers into the graphite planes. A long and highly cited review of graphite intercalation compounds is given by Dresselhaus and Dresselhaus (2002). Examples are KC_8 and CaC_6 that suggest that K and Ca, if deposited on graphene, will make it n-type, as was confirmed by McChesney *et al.* (2010). Silicon bonding to graphene has been studied by Zhou *et al.* (2012). A class of essentially insulating boron–carbon–nitrogen BCN compounds (that can be viewed as substitutional impurities, rather than surface adsorbates), and may be suitable for introducing barriers in graphene films, is discussed in Section 5.2.5. Mechanisms of doping graphene have been surveyed theoretically by Pinto *et al.* (2010). A similar careful theoretical review of doping by Giovanetti *et al.* (2008) is discussed in Chapter 9, following Fig. 9.5. In making CVD graphene on copper, it is reported that adding NH_3 to the gas mixture will make the film n-type, with details of nitrogen's electron donation described by Zhao *et al.* (2011). Further discussion of doping by molecules is given in Section 7.1.5.

4.6 Bilayer graphene

Bilayer graphene is two layers stacked in the A-B Bernal stacking indicated in Fig. 3.7. Early reports of high-quality bilayer graphene were given by Berger *et al.* (2006) and Ohta *et al.* (2006).

Two graphene layers (the bilayer) can be isolated approximately as easily as a single layer, both by extraction from a single graphite crystal and by epitaxial/catalytic growths, that would be the basis for device applications.

Ohta *et al.* (2006) and Berger *et al.* (2006) have grown high quality graphene and bilayer graphene in ultra-high vacuum using a particular type of commercially available SiC single crystal [6H polytype with (0001) orientation] as the substrate. This process (see section 5.2.1), referred to as vacuum graphitization, is basically a brief high temperature (1400°C) vacuum anneal, so that Si atoms rapidly leave the SiC surface, leaving a C layer that is based on (epitaxial to) the same crystal lattice. This method can be used to produce single- and double-layer graphene.

Low energy electron diffraction (LEED) by Ohta *et al.* (2006) revealed large epitaxial domains of bilayer graphene, lattice constant 246 pm (corresponding to C–C bond length 142 pm) rotated by 30° with respect to the SiC lattice, of parameter 307 pm. With this rotation, the atoms of C and of SiC line up sufficiently to foster long-range order in the graphene. Long-range order was definitely observed using electron diffraction.

These epitaxial layers exhibited very high electron mobility (Berger *et al.* (2006) quote 2.75 m^2/ Vs at 300 K), indicating long mean free path for electrons in the epitaxial graphene. Ohta *et al.* (2006) found that high currents could be passed through these two atom thick layers, corresponding approximately to 1 nA per graphene C atom. (Under the conditions of this measurement, current up to 400 mA through a graphene sheet of dimensions 5 mm × 15mm, the underlying SiC was highly insulating.)

Ohta *et al.* (2006) found, under conditions effectively applying a voltage between the two graphene layers that the conduction and valence energy bands shifted to create an energy gap of about 200 meV. This forms a semiconductor from what was initially a semimetal. The Ohta *et al.* (2006) experiment was carried out in ultra-high vacuum, and the technique of angle-resolved photoemission spectroscopy (ARPES, see Chapter 6) directly shows the energy bands and their shift as an effective voltage across the bilayer was applied. The effective voltage comes from deposition of potassium atoms in monolayer amounts onto the exposed graphene bilayer. The single valence electron of the potassium atom is taken up by the graphene layer, providing a substantial carrier concentration and a dipole electric field. See Fig. 4.4(b) for schematic depiction of the energy gap of about 200 meV. Note that the energy band structure for the bilayer has two conical energy surfaces in its conduction band and valence band, and the effect of the applied voltage is to create an energy gap between. (The definitive measurements in support of this picture are not shown here.)

At least in principle, bilayer graphene can be manipulated with gate voltage to provide a variable energy gap. The conceptual device is shown in Fig. 4.4(a).

Appearance of an energy gap in the bilayer graphene as carrier concentration is increased (e.g., by a gate electrode) is well understood from a theoretical point of view. The calculation of McCann (2006) (see Figure 1b of his paper) shows values up to 120 meV as carrier density approaches 10^{13}/cm^2. The carrier concentration is controlled by gate voltage in a typical situation. For comparison, the energy gap in Si is about 1000 meV. McCann (2006) neglects additional weak couplings such as A1–B2 (termed γ_3) that results in trigonal warping and is relevant at low density n $\sim 1 \times 10^{11}$ cm^{-2}.

While graphene is approximately a semimetal with linear dispersion near the neutral point, bilayer graphene has parabolic bands (as a semimetal) but a gap can be

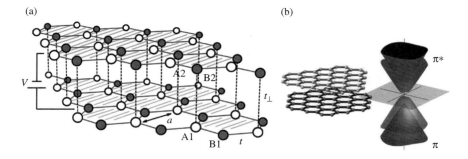

Fig. 4.4 (a) Crystal structure of **bilayer** graphene (Castro *et al.,* 2007), showing equivalent voltage bias V. Honeycomb lattice comprises sublattices A and B in each layer. Hopping integral is *t* ($\sim$3.1 eV) within each layer, and t_p ($\sim$0.22 eV) between layers. (b) A second view of the crystal structure of asymmetric bilayer graphene, and depiction of the energy bands near K showing energy gap resulting from an effective voltage (Ohta *et al.,* 2006) between layers.

introduced whenever the symmetry between the layers is broken, as by an electric field. (McCann, 2006; Novoselov *et al.,* 2006). More recent experimental data of the dependence of the gap induced in bilayer graphene by an electric field are shown in Fig. 4.5, after Zhang *et al.* (2009). This effect has been of interest toward making field-effect transistors with larger On/Off ratio, that can be accomplished by increasing the band gap in the material forming the channel. We will return to this in Chapter 9.

Strain can also induce an energy gap in bilayer graphene, according to an analysis by Verberck *et al.* (2010).

More complicated behavior for bilayer graphene has been predicted by McCann and Fal'ko (2006), suggesting that, at lower carrier density, a parabolic minimum in the density of states is replaced by four minima. Mayorov *et al.* (2011), using superior suspended and annealed bilayer samples, measure extremely high mobility. They also find evidence for a different band structure, replacing the usually observed parabolic minima, with two minima having linear dispersion. These authors suggest that electron–electron interactions drive a change away from the situation described by McCann and Fal'ko (2006) to a "nematic phase," with reduced rotational symmetry, as predicted by Lemonik *et al.* (2010) and by Vafek and Yang (2010).

As we have seen, a gap can be introduced in bilayer graphene by breaking the symmetry between the two layers (McCann, 2006; Ohta *et al.,* 2006; Castro *et al.,* 2007; Oostinga *et al.,* 2007; Zhang *et al.,* 2009a; Zhang and MacDonald, 2012).). Excellent mobility values for graphene and its bilayer are reported by Morozov *et al.* (2008). A detailed study of the infrared absorption of bilayers is given by Kuzmenko *et al.* (2009).

Mayorov *et al.* (2011) report that suspended and annealed bilayer graphene samples have shown carrier mobility up to 10^6 cm^2/Vs. These samples were mechanically cleaved and placed on oxidized silicon wafers. Wet-etching in buffered hydrofluoric

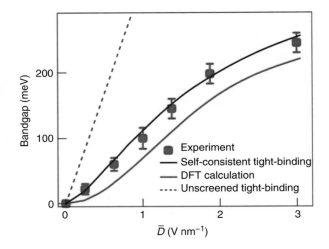

Fig. 4.5 Electric-field dependence of tunable energy bandgap in graphene bilayer. Experimental data (solid squares) are compared to theoretical predictions based on self-consistent tight binding (upper solid line), *ab initio* density functional (lower solid line) and unscreened tight binding calculations (dashed line). The error bars are from interpreting the absorption peaks in the measured spectra. (From Zhang *et al.*, 2009a, by permission from Macmillan Publishers Ltd., © 2009).

acid was used to remove some of the oxide under the bilayer graphene, to suspend the sample (Du *et al.*, 2008). Passage of current of the order of 1 mA/μm heated the film. The heating step much improved the electrical characteristic, leaving a very sharp resistance peak almost exactly at zero voltage, corresponding to residual doping $\sim 10^8/\mathrm{cm}^2$.

The Landau level structure of bilayer graphene was discussed by McCann and Fal'ko (2006), whose prediction is shown in Fig. 4.6. Here the prediction of the quantum Hall conductivity $\sigma_{xy}(\mathrm{N})$, with N the carrier density, for bilayer graphene is the solid line, and the dashed line is monolayer graphene.

Both curves were observed by Novoselov *et al.* (2005). The bilayer spectrum is dominated by the zero field level arising from doubly degenerate orbital states. This gives a large $8e^2/h$ step in $\sigma_{xy}(\mathrm{N})$ at zero energy. According to McCann and Fal'ko (2006), the zero energy Landau level in the bilayer is eight-fold degenerate, including spin and valley degeneracies. This compares with four-fold degeneracy for all other bilayer Landau levels, and with four-fold degeneracy of all Landau levels in monolayer graphene. For bilayer graphene, the authors find a quadratic energy spectrum at low energy that changes to a linear spectrum, at high energy. Thus at low energy

$$\varepsilon = p^2/2m, \tag{4.18}$$

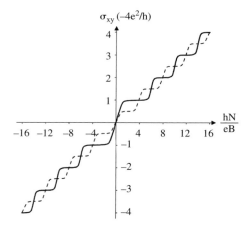

Fig. 4.6 Solid line is $\sigma_{xy}(N)$ predicted for bilayer, dashed is predicted monolayer conductivity, with N the carrier density. These curves are similar to experimental curves given by Novoselov *et al.* (2005). (From McCann and Fal'ko, 2006. © 2006 by the American Physical Society).

with

$$m = \gamma_1^2/2^{2v}, \tag{4.19}$$

and, at high energy,

$$\varepsilon = vp. \tag{4.20}$$

Here the inter-layer coupling γ_1 is much smaller than the intralayer couplings t, t' mentioned earlier. If γ_1 is estimated from properties of crystalline graphite, then m is small, $m \approx 0.054 m_e$. A crossover between the two behaviors occurs at $p = \gamma_1/2v$, leading to carrier concentration

$$N^* = \gamma_1^2/(4\pi\hbar^2v^2) \approx 4.4 \times 10^{12}\text{cm}^{-2}. \tag{4.21}$$

Further discussion of bilayer graphene is given in Section 8.6.

5

Sources and forms of graphene

Graphene is obtained chiefly from exfoliating or cleaving crystalline graphite, chemically or mechanically, or by direct synthesis in a chemical vapor deposition apparatus from carbon containing gases such as acetylene or methane. We include as a category of graphene several commercial products that are partially exfoliated, reduced-density, carbon obtained in various ways by "expanding" graphite. Chemical processes are used to produce a large-surface-area product we might call a "cloth." Perhaps the best known of these is Grafoil®. The major advance initiated by Geim and Novoselov was to find a practical method for getting good single- or few-layer samples of graphene from graphite crystals, as was described in Section 3.3.

5.1 Graphene single-crystals, flakes and cloths

The highest quality graphene crystals are obtained from graphite, using the micro-mechanical cleavage method clearly described in the Nobel Lecture of Novoselov (2011). The source graphite in these cases is typically either graphite mined from the ground or Kish graphite that appears at the surface of molten iron in steel-making, as the precipitation of excess carbon. Graphene single crystals are also now being grown by vapor deposition on a suitable catalytic surface, usually copper or nickel, although other metals have been used.

5.1.1 Micro-mechanically cleaved graphite

Following Novoselov (2011), graphene creation in the simplest fashion is to start with graphite and micro-mechanically cleave it into individual planes using Scotch tape. (A long history of chemical methods for exfoliating graphite, capable of large volume but leading generally to low quality graphene, will be considered in Section 5.1.2. The word exfoliation implies in the literature a chemical process.) Graphite is a layered material and can be considered as a stack of individual graphene layers. While high quality crystalline graphite requires high temperatures [above 3000 K in treating highly oriented pyrolytic graphite (HOPG), Fe/C eutectic near 1100°C for Kish graphite], micro-mechanical cleavage can be done at 300 K. An earlier bibliography of attempts at micro-mechanical cleavage is given by Novoselov (2011), but these evidently produced a minimum thickness of 20 layers. The important advance was made by Novoselov *et al.* (2004) to find an optical method to obtain and clearly identify single layers. This method is also known as the Scotch tape method. The ease of cleavage of graphite had been earlier mentioned, e.g., by Blakslee *et al.* (1970), Lu *et al.* (1999) and the same approach was used on the layered superconductor BSCCO ($Bi_2Sr_2CaCu_2O_{5+\delta}$)

by Terashima *et al.* (1991), but no effective means of identifying the single layers was available until the work of Novoselov *et al.* (2004).

The top layer of the high-quality graphite crystal is removed by a piece of adhesive tape, which—with its graphite crystallites—is then pressed against the substrate of choice. If the adhesion of the bottom graphene layer to the substrate is stronger than that between the layers of graphene, a layer of graphene can be transferred onto the surface of the substrate. This can leave extremely high quality graphene monolayer crystals via a simple procedure. A wide variety of substrates are in principle suitable. It was importantly found by Novoselov *et al.* (2004) that optical microscopy was the most suitable technique to locate suitable micrometer size graphene crystals once transferred to the substrate. Novoselov and colleagues found that some substrates, and particularly Si/SiO_2 with a 300 nm SiO_2 layer, can produce an optical contrast of up to 15% for some wavelengths of light, to locate a single monolayer. The physics of this was explained by Abergel *et al.* (2007) and Blake *et al.* (2007), and made Si/SiO_2, with either a 100 nm or a 300 nm SiO_2 layer, the basis for many experiments.

The important use of hexagonal boron nitride (h-BN) as a substrate came later (Dean *et al.*, 2010) with the disadvantage that thicknesses on h-BN could not easily be determined optically. Once a single graphene monolayer was verified on the SiO_2 surface, it was learned by Reina *et al.* (2008) how to release it to be transferred elsewhere. (Their transfer method was importantly extended to large area graphene grown on Ni and Cu by chemical vapor deposition.) For some important measurements the micro-mechanically cleaved graphene layers were laid across trenches etched into the Si surface to permit study of unsupported sections of single-layer graphene (Bunch *et al.*, 2007). Such samples have better electrical properties in general because they do not see random electric fields as from charged impurities in the SiO_2, but they are harder to use because of the weakness of large areas of unsupported graphene against bending. They can also more readily be cleaned of weakly adsorbed surface impurities by current induced heating, to improve their electrical characteristics.

Novoselov *et al.* (2005) extended their micro-cleavage technique to release single crystal monolayers of additional layered compounds BN, $NbSe_2$, MoS_2 and the high temperature superconductor BSCCO ($Bi_2Sr_2CaCu_2O_{5+\delta}$). A similar method was used on MoS_2 by Mak *et al.* (2010) and on Bi_2Te_3 by Teweldebrhan *et al.* (2010).

5.1.1.1 Role of the substrate.   The original substrate for exfoliated graphene sheets was oxidized Si. It has been subsequently discovered that graphene devices on standard SiO_2 substrates are somewhat disordered (see Fig. 5.1), have local fluctuations in Fermi energy (known as puddles) and exhibit relatively low carrier mobilities. These aspects are inferior to the expected intrinsic properties of graphene. Although suspending the graphene above the substrate leads to substantial improvement, this is basically impractical for devices. It has been found that hexagonal boron nitride h-BN, available in single crystal layers with lattice constant similar to graphene, is a superior substrate for graphene, exhibiting a lower density of charged impurities than SiO_2.

The works of Dean *et al.* (2010) and of Xue *et al.* (2011) show the dramatic effect of the substrate on the topography of a deposited single mechanically exfoliated layer of graphene. It is evident that the graphene adheres closely to the substrate and that

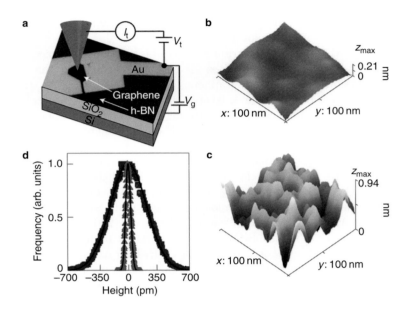

Fig. 5.1 Effect of substrate on roughness of micro-mechanically cleaved graphene. (a) Measurement setup. Optical microscope image of graphene monolayer with h-BN beneath and gold electrodes contacting it above. The measurement circuit of the STM tip and back-gate voltage is indicated. (b) STM topographic 100 nm × 100 nm image of monolayer graphene in h-BN showing the underlying surface corrugations. Tip voltage −0.3 V tunneling current 100 pA. (c) STM topographic image of monolayer graphene on SiO$_2$ showing markedly increased corrugations. Tip voltage −0.5V, tunneling current 50 pA. (d) Histogram of the height distribution for graphene on SiO$_2$ (squares) and graphene on h-BN (triangles) with Gaussian fits. (From Xue *et al.*, 2011, by permission from Macmillan Publishers Ltd., © 2011).

roughness perceived on the graphene layer chiefly arises from the roughness of the substrate. The comparison here of graphene on h-BN vs. oxidized silicon shows that h-BN is a better substrate. It appears that the roughness imparted from the silica originates in the amorphous nature of the grown oxide and from trapped charge that may occur in the silica. The importance of the substrate to the STM topography of graphene micro-mechanically cleaved flakes was earlier studied by Geringer *et al.* (2009), who subjected the samples mounted on Si to extended anneal at 150°C before ultra high vacuum (UHV) STM measurement. They found, similarly to that shown in (c) in Fig. 5.1, that the graphene surface largely conforms to the underlying SiO$_2$, but also found weak evidence for spontaneous ripples. It is now believed that the latter are an artifact arising from the interaction of nearly inevitable mounting-induced strain with ubiquitous atomic adsorbates not completely removed at 150°C. More discussion of the adsorbate induced corrugation is given in Section 7.1.4.

5.1.2 Chemically and liquid-exfoliated graphite flakes

There is a long history of chemical methods to exfoliate graphite into flakes, platelets or sheets of various thicknesses. An example of such a process is that given by Shioyama (2001). In a paper entitled "Cleavage of graphite to graphene" is detailed a chemical method of releasing the individual graphene layers from highly oriented pyrolytic graphite (HOPG). The method is suggested in Fig. 5.2a. The HOPG was transformed into the first stage Graphite Intercalation Compound (GIC) with potassium: K-GIC with chemical formula KC_8. An intercalation compound is "Stage 1" if the number of intercalation layers equals the number of host layers. If intercalate fills every third layer it would be a Stage 3 intercalation compound. (Another well-known Stage 1 GIC is CaC_6 that is a superconducting metal below 11.5 K.) Shioyama found that heating K-GIC (KC_8) at room temperature and at a pressure of 67 kPa in the vapor 1,3-butadiene (with similar results for the vapor of styrene) led to the reaction, evident by the expansion of the K-GIC along its c-axis. The reaction proceeded until all the butadiene vapor was used up. Shioyama suggested that linear polymers, as indicated in Fig. 5.2a, were growing, forcing the graphite planes apart. Heating the resulting black elastic polymer above 400°C resulted in complete release of the potassium polymers, leaving a residue of pure graphitic carbon, completely exfoliated.

This method may be traced back to original works of Brodie (1859) and Staudenmaier (1898), who treated graphite with acids, leading to graphite oxide (originally called "graphon"), see also Wen *et al.* (1992).

Following the recent summary of Novoselov (2011), graphite oxide can be regarded as graphite intercalated with oxygen and hydroxyl groups, thus hydrophilic and easily dispersed in water. This leads to extremely thin, even monolayer, oxide flakes that can be subsequently reduced, e.g., using hydrazine, to pure carbon, according to Ruess and Vogt (1948), Hummers *et al.* (1958) and Boehm *et al.* (1961). Such samples are now regarded as low-quality graphene, according to Novoselov (2011). As we will see, the oxidation followed by reduction leads to a large density of in-plane defects. For

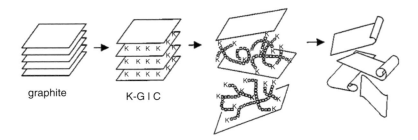

graphite   K-GIC

Fig. 5.2a Schematic depiction of steps in one approach to chemically exfoliating graphite. This approach starts by intercalating the graphite with potassium, by heating in the presence of K vapor. The third panel suggests growth of linear polymers when the intercalated graphite is exposed to a vapor such as styrene or butadiene. (From Shioyama, 2001, with kind permission from Springer Science and Business Media).

example, Gomez-Navarro *et al.* (2007) find that the electrical conductivity and mobility in chemically-reduced graphite oxide flakes are two to three orders of magnitude lower than in mechanically exfoliated graphene flakes. Since mechanically exfoliated flakes are impractical in any production context, the chemical vapor deposition (CVD) and chemical processes will be dominant in applications. The high throughput of the chemical exfoliation processes, despite the resulting lower conductivity and small flake size compared to CVD films, may drive their application in such areas as solar cells, light emitting diodes, as well as in mechanically reinforced and highly conductive composites. Recent methods of chemical exfoliation are described by Wu *et al.* (2010), who suggest application of spin-coated and vacuum reduced graphene films in solar cells and transparent conductors for various types of organic light-emitting diodes.

A notable and well-cited article describing *single graphene sheets* obtained by thermal exfoliation of graphite oxide (releasing oxygen) was published by Schniepp *et al.* (2006). These single graphene sheets were described as "functionalized," i.e., chemically reacted, suitable for specific applications, and were found to be electrically conducting. In substantial part, the functionalizing sites were structural defects, such as kinks and 5-8-5 defects remnant of the successive oxidizing and reduction reactions. Functionalizing sites also include remnant chemical impurities like C–O–C (epoxy) or C–OH groups (within the graphene planes) or C–OH and –COOH groups at the edges.[1] According to Schniepp *et al.* (2006), of the huge literature of reduced graphite oxide platelets before 2006, none had achieved single-layer graphene. Their work starts with graphite flakes that are reacted for 96 hours in an oxidizing solution of sulfuric acid, nitric acid and potassium chlorate (similar to the above-cited process of 1898). The authors find that the 96 hours in the oxidizing solution is needed to completely remove, in x-ray examination, the 0.34 nm interplanar spacing characteristic of graphite, to supplant that spacing with a $0.65-0.75$ nm spacing characteristic of graphite oxide in its solid phase. The following exfoliation of the solid graphite oxide flakes is akin to an explosion releasing CO_2 gas, this is accomplished by placing a sample of completely dried solid-phase graphite oxide, in a quartz tube purged with argon gas, into a furnace preheated to 1050°C. The rapid (>2000°C/min) heating splits the graphite reaction product into single sheets through the evolution of CO_2 gas. Atomic force microscopy (AFM) was among the methods used to establish the single layer nature of the resulting graphene. An increase in AFM-indicated thickness of the single films was attributed to ripples in the graphene, structural defects and these were modeled convincingly as shown in Figure 3 of their paper. This careful modern study of the original (Brodie, 1849) inexpensive, bulk chemical exfoliation process and its several variations, concludes that single-layer graphene can result from such processes, but that any such graphene retains defects. The defects occur in the basal plane of the final graphene, as well as at its edges. The defects may have a useful functionalizing

---

[1]Such defect sites are deemed useful in bonding the platelets into a composite to increase its strength or possibly to improve its electrical conductivity, but are definitely deleterious for an application requiring high electrical conductivity, as a silicon-chip-interconnect or a solar-cell electrode.

role in an additive to a composite material, but are definitely a negative factor in graphene intended as a conductive electronic component.

5.1.2.1 Edge-initiated exfoliation.   Bulk exfoliation of graphite continues to be developed to lead to graphene nano-sheet suspensions of higher electrical conductivity at low cost. It appears that exfoliation can be initiated at the edges of graphite platelets, leaving fewer in-plane defects. A recent approach employs ball-milling of graphite with dry ice, to produce edge-carboxylated graphite (ECG) nanosheets, of size $100-500$ nm, as described by Jeon *et al.* (2012). The resulting graphene platelets are superior to those obtained by chemical exfoliation, because the interior hexagonal planes are undisturbed by the processing. The process breaks up the graphite from the edges, leaving flat hexagonal lattice planes, without the wrinkles and in-plane defects that were noted as a result of chemical exfoliation. The carboxylated edges make the ball-milled product directly dispersible in water and other polar solvents, in the form of individual planes. The carboxylated edges can later be removed by heating, e.g., after a dispersion has been deposited on the target surface. It is suggested that optically transparent films of superior conductivity can be achieved after depositing such nanosheets on surfaces, including silicon, followed by heating. It appears that this process may be inexpensive in bulk production and that it may further the application of graphene as low cost electrodes. Some aspects of the ECG process are illustrated in Fig. 5.2b.

See also, for development of various chemical methods, works of Stankovich *et al.* (2006, 2007); Dikin *et al.* (2007); Gomez-Navarro *et al.* (2007); Ruoff (2008); and Park and Ruoff (2009). A contemporary assessment of production of higher quality exfoliated graphene, on the scale of tons per year, has been given by Segal (2009).

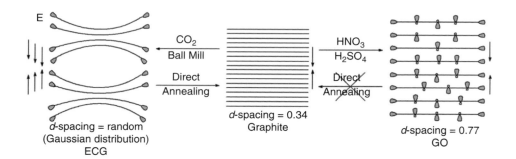

Fig. 5.2b Schematic presentation of solids (l to r) "Edge-Carboxylated Graphite" ECG, Graphite and Graphite Oxide (GO). The figure emphasizes that the ECG leaves hexagonal basal planes undefected in platelets of size $100-500$ nm and can be reversed to pure graphite by direct annealing. In contrast, the basic chemical exfoliation route oxidizes the plane surfaces as well as the edges and cannot be reversed to pure graphite by direct annealing. When explosive exfoliation is applied to GO, reducing it to graphene, defects remain in the planes. (From Jeon *et al.*, 2012).

5.1.2.2   Electrical characterization of chemically exfoliated graphene nanoribbons.
Li *et al.* (2008) find chemical means of producing narrow nanoribbons with smooth
boundaries starting with graphite oxide and go on to closely characterize these nanoribbons by electrical methods. The method of Li *et al.* (2008) is based on a commercial product,[2] "expandable graphite" (Grafguard® 160-50N, Graftech Incorporated,
Cleveland, OH).

"Expandable graphite" is exfoliated (Li *et al.*, 2008) by heating for 60 s at 1000°C in
forming gas (3% hydrogen in argon). This causes violent formation of volatile gaseous
species from the intercalants, and exfoliates the material into a loose stack of few-
layered graphene sheets. The thermal exfoliation step is critical and responsible for
the formation of one- to few-layer graphene and was evidenced to Li *et al.* (2008) by
a visible dramatic volume expansion by ~100 to 200 times after exfoliation.

The resulting exfoliated graphite was dispersed, by Li *et al.* (2008), in a
1,2-dichloroethane (DCE) solution of poly(m-phenylenevinylene-co-2,5-dioctoxy-p-
phenylenevinylene) (PmPV) by sonication for about 30 minutes to form a homogeneous
suspension. (Centrifugation then removed large pieces and the remaining suspension
was surveyed for its content of planes and ribbons of graphene.) The PmPV polymer
non-covalently functionalizes the exfoliated graphene, leading to a homogeneous black
suspension during the sonication process. The authors note that the PmPV was necessary to reach a stable suspension. It appears that the polymer adheres to graphene by
van der Waals forces similar in magnitude to the binding force of graphene in graphite
itself and that otherwise graphene will not disperse even in the organic solvent.

Li *et al.* found that the sonication time to produce ribbons should be optimized,
since ribbons were no longer found after hours of sonication. Extended sonication evidently breaks up ribbons and leads to ever-smaller particles. In general, the graphene
nanoribbon (GNR) content was smaller than that of sheets: the solution after centrifugation contains micrometer-sized graphene sheets and nanoribbons. The survey of the
solution products was carried out using atomic force microscopy (AFM), after removing the PmPV by calcining at 400°C. Particular interest was in finding nanoribbons
in the sonicated solution products.

Li *et al.* (2008) harvested nanoribbons from the sonicated suspension as described
above. The ribbons were built into field-effect transistor-like structures, choosing ribbons with widths W in the range 10 to 55 nm. These ribbons were placed on an
oxidized p^{++} Si wafer that formed the back-gate of the FET. Palladium Pd contacts
were attached to the nanoribbons to act as source and drain. The devices were surveyed as to the On/Off current ratios that were found to rise dramatically at small
nanoribbon widths W, less than 10 nm, as shown in Fig. 5.3. More details on similar
devices regarded as sub-10 nm graphene nanoribbon field-effect transistors were given
by Wang *et al.* (2008a).

---

[2]It appears that this material is graphite oxide, since it is readily expanded by heating as Li *et al.*
(2008) have done. The processing by the manufacturer may involve oxidizing 0.35 mm graphite flakes
by intercalation of oxidizing sulfuric and nitric acid, with oxidation of carbon atoms likely occurring
at the edge, step and defect sites of graphite.

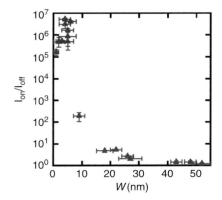

Fig. 5.3 On/Off ratios at 300 K of simple FET devices comprised of nanoribbons of width W directly harvested from sonicated suspension of chemically exfoliated graphite (see text), placed on oxidized p^{++} Si wafer and electroded with Pd. Unknown edge conditions do not disrupt a clear dependence of On/Off ratio on width W. (From Li *et al.*, 2008, with permission from AAAS).

A companion figure (Figure 4b) in the Li *et al.* (2008) paper displays the inferred bandgap E_G in the nanoribbon. The inferred E_G plot [Fig. 5.12(b)] has a similar shape for widths W plotted to 55 nm, and places the band gap E_G value for the cluster of on-off ratios at smallest widths at about 0.38 eV. Those data (not shown) lead to an empirical rule E_G (eV) $\approx 0.8/W$ (nm). The dependence on width W is so strong that variations in the nature of the edges, zigzag vs. armchair or otherwise, are not disruptive in this plot. [Armchair and zigzag boundaries, respectively, can be visualized as horizontal and vertical lines in Fig. 1.2(b).] The data analysis for E_G is based on the idea that the nanoribbons are p-type semiconductors that display diode behavior when contacted with Pd. The diode aspect is a consequence of Schottky barrier formation, due to the high work function of Pd, at its contact to the graphene. On this argument the working formula (Li *et al.*, 2008) is, with E_G the bandgap of the ribbon,

$$I_{On}/I_{Off} \sim \exp(E_G/k_B T). \tag{5.1}$$

The authors find that their inferred gap values lie between theoretical estimates for nanoribbons of zigzag edges and armchair edges (Son *et al.*, 2006). Li *et al.* (2008) find that nanoribbons with homogeneous zigzag or homogeneous armchair edges always have gaps that decrease as the width W increases, on the assumption that carbon atoms at the edges are hydrogen-passivated. These authors state that the energy gaps originate from electron confinement across the width W and from the details of the edge terminations. Experiments on nanoribbons in magnetic field by Oostinga *et al.* (2010) also conclude that nanoribbons always have a band gap, in their view the effects are related to large disorder unavoidably present. The role of disorder in nanoribbon transport was carefully assessed by Kim *et al.* (2010). Theoretical discussion of quantum dot formation in such nanoribbons is given by Sols *et al.* (2007). In practice, more than size quantization is involved, as these authors point out.

(Overlooking the fact that nanoribbons are more likely to have disorder than nanotubes that benefit from the perfection of "large molecules," there is something of a paradox in the experimental fact that GNRs always have a gap, whereas the analogy to a metallic nanotube would suggest that some GNRs, perhaps those that could be rolled up into a metallic nanotube, should be metallic. In fact, Figure 2a in Son *et al.* (2006) shows a metallic armchair GNR as calculated in a tight binding approximation. We will return to this in discussion in Section 5.3 of nanoribbons.) Approaches to creating nanoribbons based on carbon nanotubes have been offered by Kosynkin *et al.* (2009), Jiao *et al.* (2009) and by Kim *et al.* (2010). A facile method of creating nanoribbons from carbon nanotubes (Jiao *et al.*, 2010) leads to nanoribbons of well-defined widths and high quality.

Scanning tunneling microscope (STM) observations of zigzag and armchair edges of graphene have been reported by Kobayashi *et al.* (2005) and, in connection with hydrogen passivation, by Kobayashi *et al.* (2006).

Li *et al.* (2008) estimate that their solution-produced ribbons have hole mobility on the order of 200 cm^2/Vs that they describe as high and consistent with GNR graphene nanoribbons of "high quality, nearly pristine, and free of excessive covalent functionalization." [This mobility is quite low, in fact that may arise in part from their starting with a material similar to Grafoil®, produced from the chemical oxidation/reduction process (see Section 5.1.2.2) established as leaving in-plane defects, in the work of Schniepp *et al.* (2006).] Li *et al.* (2008) consider these GNRs as better suited to electronics than carbon nanotubes because they always have a bandgap.

A sophisticated study of Coulomb blockade and quantum Hall effect behavior in suspended graphene nanoribbons has been recently reported by Ki *et al.* (2012).

An analysis of graphene nanoribbons as interconnects on electronic chips has been offered by Xu *et al.* (2009a), to be discussed further in Chapter 9 below.

A chemical synthesis of brightly fluorescent graphene, based on methanol, has been reported by Parashar *et al.* (2011).

5.1.2.3 Liquid phase vs. chemical exfoliation.   Chemical exfoliation (more generally than via graphite oxide) has been considered at length, see, for example, Dresselhaus and Dresselhaus (1981),Viculis *et al.* (2003) and Chen *et al.* (2004), who use an ultrasonic powdering technique. Liquid phase exfoliation without the oxidation/reduction steps (thus, strictly speaking, not a *chemical* exfoliation but by directly dispersing graphite in other, mostly organic, solvents), has been discussed by Hernandez *et al.* (2008). Their method starts with powdered graphite and leads to dispersions at concentrations up to 0.01 mg/ml. in the solvent. In the case of solvent N-methylpyrrolidone (NMP) a plot of the thickness of the dispersed flakes is given in Fig. 5.4. The authors have compared results with results from three additional organic solvents. They see their scalable method as potentially useful for large area applications from device and sensor fabrication to conductive composites. An earlier review of exfoliation of graphite was given by Chung (1987).

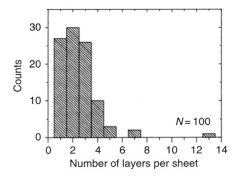

Fig. 5.4 Thickness distribution of graphene sheets from liquid phase exfoliation. Result of electron microscopy of dispersed graphene in direct liquid phase exfoliation using solvent NMP. (From Hernandez *et al.*, 2008, by permission from Macmillan Publishers Ltd., © 2008).

See also Celzard *et al.* (2000), who identify graphite mass-produced products (resulting from reduction of graphite oxide) as Grafoil® (Union Carbide Corp., USA) and Papyex® (Carbone Lorraine Group, France).

A contemporary use of direct liquid phase exfoliation to produce single layer graphene sheets on a metal grid for examination in a TEM is described by Mao *et al.* (2011). In this paper, see references therein, graphene is prepared by liquid phase exfoliation by direct sonication of graphite in the organic solvent, N-methyl-pyrrolidone (NMP).[3]

The resulting graphene dispersion was deposited onto a "holey carbon grid" and vacuum dried before loading into the TEM specimen chamber. The authors confirm the work of Hernandez *et al.* (2008) that such exfoliated graphene fragments are free of oxidation and defects, and approximately 70% contain no more than three graphene layers. About 28% of the flakes were monolayer graphene, see Fig. 5.4.

Complete exfoliation creates a large surface area, potentially useful, e.g., for storage of gases by adsorption. Consider 1 cc of graphite, a mass 2.27 grams. At the c-axis layer spacing, 0.335 nm, there will be $N = 1\,\text{cm}/0.335\,\text{nm}$ graphene planes, thus $N = 2.985 \times 10^7$. If we count the area on both sides of each plane, the area is $2N \times 10^{-4}\text{m}^2 = 5970\,\text{m}^2$. Expanding the crystal along the c-axis will progressively allow access to this limiting area that is 2630 m²/g of graphite.

5.1.2.4 Grafoil® cloth. A commercial chemically exfoliated graphite cloth product known as Grafoil® has density 1.12 g/cc (purity up to 99.8% carbon) and was measured (Schiffer *et al.*, 1993) to have surface area 13.05 m²/g. It is in the form of a flexible

---

[3]In detail, graphite flakes with a size of 1.8–5 mm (from NGS Naturgraphit GmbH, Leinburg, Germany) were incubated in 1 ml N-methyl-pyrrolidone in a 1.5 ml glass vial and sonicated for 3 h in an ultrasonication bath (input 160 W, output 70 W Branson 1510, CT). The sonicated solution was then centrifuged at 500 rpm for 90 min.

cloth used as a gasket for high temperature applications. It appears that Grafoil® is made from "expanded graphite," graphite flakes chemically exfoliated, using acids such as nitric and sulfuric acid in proprietary processes following Staudenmaier (1898). The result is nearly pure carbon but the density is lowered by strongly disrupting the layering. Grafoil® is chemically inert and is rated to be used in an oxidizing atmosphere up to 850°F (454°C). Grafoil® has been used in low temperature physics as a large area substrate for studies of adsorbed gases, including ^{3}He. In the paper of Schiffer *et al.* (1993), the surface area of Grafoil® was measured to be 33.3 m^2 for 2.55 g. This surface area suggests sheet thickness of hundreds of graphene layers. The "Grafoil® GTA Premium Flexible Graphite" sheet has a density of 1.12 g/cc (compared with 2.27 g/cc for single crystal graphite). It is available in a variety of thicknesses.

A report of contemporary production of few-layer graphene, distinct from Grafoil®, on the scale of 15 tons per year by several small companies using various graphite-based chemical methods, is given by Segal (2009). Markets served, or potentially so, include ultracapacitors, lithium batteries as anode materials, transparent conductive electrodes, as replacement for indium tin oxide and electrically conductive composites.

A further review of chemical (exfoliation) procedures has been given by Allen *et al.* (2010).

5.2   Epitaxially and catalytically grown crystal layers

Epitaxial and catalytic growths of graphene on supporting surfaces are important, as potentially manufacturable[4] processes for graphene monolayers or bilayers on device surfaces. The most important epitaxial substrates are SiC (van Bommel *et al.*, 1975; Berger, *et al.*, 2004) and metal crystals/foils such as Ni or Cu. On these metal substrates, however, there is no fixed relation between the crystalline axes of the graphene and the substrate. While the latter chemical vapor deposition (CVD) growth on Cu may sometimes be described as epitaxial growth, a more realistic description is catalytic growth and it is found, for example, that domains of monolayer graphene grow across steps and other defects in the supporting surface.[5] A sophisticated discussion of growth on Cu is given by Chen *et al.* (2012), including a proposed modification of the CVD method to achieve truly epitaxial graphene. We will discuss this in Section 5.2.4. A recent discussion of the roles of Cu and Ni as supporting surfaces is given by Vinogradov *et al.* (2012).

A preliminary report of ultra high vacuum (UHV) growth of graphitic carbon directly onto Si (111) by e-beam deposition was given by Hackley *et al.* (2009). The

---

[4]A "manufacturable" process is one for which large-scale production can be envisioned. Micromechanical cleavage of graphite to give graphene is not such a process. The current manufacturable processes in semiconductor electronics are based on photo-lithography and the depositions include evaporation, chemical vapor deposition, and atomic layer deposition.

[5]The interaction of carbon with copper is much weaker than with carbon (differing from Ru, where epitaxy of C is achieved, with corrugations of the graphene to match the lattice). The Cu does nucleate islands of graphene, because the 111 surface of cubic Cu has six-fold symmetry. The small graphene islands however are found to be most stable if rotated from the Cu directions, and grain boundaries are inevitably the result as explained by Chen *et al.* (2012).

silicon surface temperature was initially 560°C leading to a thin amorphous carbon layer, and then raised to 830°C where graphitic carbon was observed to grow.

SiC is insulating, so that devices can be constructed by patterning graphene deposited on SiC, without concern for electrical currents flowing through the substrate. Generally, it is found that the electrical properties of graphene grown on SiC are somewhat inferior to those of micro-mechanically cleaved planes that are deposited onto oxidized Si. As noted in Fig. 5.1, it is found that hexagonal BN is a substrate for the micro-mechanically cleaved planes that leads to less disorder and also generally to higher carrier mobility. On the other hand, methods for direct deposition or growth on h-BN are not well developed, in contrast to the case of SiC.

5.2.1   Epitaxial growth on SiC: Si face vs. C face

Silicon carbide itself can be viewed as a layer compound, composed of polar SiC layers. The bonding in SiC is principally covalent, as in diamond C and in Si, but in SiC the elemental difference imparts some ionic character. The polar tetrahedrally-bonded layers of importance are arranged perpendicular to the c (0001) direction. There are two types of stacking of the SiC layers, cubic stacking and hexagonal stacking. The important hexagonal "polytypes" 4H and 6H, comprise, respectively, 4 and 6 hexagonally stacked bilayers per unit cell. In 4H-SiC and in 6H-SiC the Silicon polar surface is designated "0001" while the polar Carbon surface is designated [000(−1)]. All of these crystals are insulators, with bandgaps 3.02 eV for 6H-SiC and 3.27 eV for 4H-SiC (Seyller, 2012). Graphene layers can be epitaxially formed by heating SiC in vacuum, as Si atoms leave and the surface reconstructs. A detailed procedure has evolved using the silicon face SiC(0001).

Growth of graphene by heating these surfaces was first identified by Van Bommel *et al.* (1975), confirmed by Forbeaux *et al.* (1998), whose data are shown in Fig. 5.5, and was recently reviewed comprehensively by Seyller (2012). An earlier review of growing graphene and bilayer graphene on SiC was given by Bostwick *et al.* (2009), who

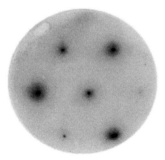

Fig. 5.5 Early LEED pattern of crystalline graphite after annealing 6H-SiC(0001) surface at 1400 C, observed with primary energy 130 eV. Six-fold pattern is characteristic of graphene. (From Forbeaux *et al.*, 1998. © 1998 by the American Physical Society).

also make extensive use of the surface technique ARPES (angle-resolved photoemission spectroscopy, see Chapter 6).

The electronic band structure of graphene grown on SiC has been investigated and found to be nearly ideal by Sprinkle *et al.* (2009). Their methods include angle-resolved photoemission spectroscopy (ARPES) and surface X-ray diffraction (SXRD). Studies of graphene grown on SiC have revealed that even multi-layer growths retain single-layer electronic properties (Hass *et al.*, 2008). It is generally found that graphene on SiC is conductive with electrons, an n-type film with Fermi energy in the order of 0.3 eV. This is discussed in more detail in connection with Figs. 9.5 and 9.6.

Adapting the SiC graphene process toward a production method, Emtsev *et al.* (2009) have shown that it can be carried out at atmospheric pressure, producing wafer-size layers with good results. De Heer *et al.* (2010) (and references therein) have demonstrated an improved form of epitaxial graphene growth: "confinement-controlled-sublimation," including methods to produce step-free surfaces on SiC.

The effect of steps on the SiC face on the resistance of epitaxially grown single layer graphene was investigated by Ji *et al.* (2012).

The Scanning Tunneling Potentiometer device of Ji *et al.* [based on that of Bonnani *et al.* (2008), see also Homoth *et al.* (2009)] deploys three tips: two are current-injecting probes in contact with the surface, typically separated by ∼500 μm, and the third is a scanning tip. Data from this method is given in Fig. 5.6.

The scanning tip measures the topography as well as the local potential, as affected by the imposed current flow. The parameter values relevant to Fig. 5.6 are temperature 72 K, voltage drop between current injecting tips 1.53 V, the estimated local current density in the single monolayer graphene layer in the measured region,

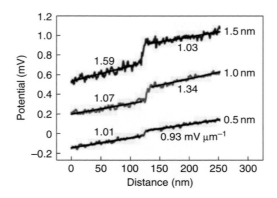

Fig. 5.6 Scanning Tunneling Potentiometry measurements on single layer graphene on SiC in traces (top to bottom) crossing steps of heights 1.5 nm, 1.0 nm and 0.5 nm, respectively. The numbers shown close to the data traces indicate the slopes in mV/μm on the terrace before and after the step is encountered. The authors confirmed that the voltage drops across the steps scale with the injected current. (From Ji *et al.*, 2012, by permission from Macmillan Publishers Ltd., © 2012).

6.4×10^{-6}A/μm. The carrier density in the graphene is estimated as 10^{13} cm^{-2}, and the mobility is estimated as 0.3 m^2/Vs.

The authors find that voltage-drops across steps contribute significantly to the total drop across a sample. A 0.5 nm substrate step contributes extra resistance equivalent to a terrace about ~40 nm wide, while 1.0 and 1.5 nm high steps contribute resistances, respectively, equivalent to terraces ~80 nm wide and ~120 nm wide.

Realistic calculations of the extra electrical resistance that appears when graphene "flows" over a step of height h_s on a SiC substrate, have been carried out by Low *et al.* (2012), as suggested in Fig. 5.7. [This situation occurs in device applications as reported by Lin *et al.* (2011), to be discussed in Chapter 9. More work on related step-geometries has been carried out by Sprinkle *et al.* (2010) and by Hicks *et al.* (2012) to be described in Chapter 9.] Referring to Fig. 5.7, the "step-resistance effect," as interpreted by Low *et al.* (2012), will vary according to the degree to which the substrate in question changes the carrier density of the graphene, as well as the density of steps on the substrate. [Hicks *et al.* (2012) have patterned deeper trenches into SiC, where they believe the graphene—subsequently grown by sublimation heating—on the trench wall, analogous to the "flowing" graphene section in Fig. 5.7, actually develops a bandgap on the order of 0.5 eV. These authors find that sharp curvature in the graphene leads to an observed energy gap.]

The equilibrium graphene spacing from the SiC substrate is taken as 0.34 nm, with an estimate of 0.04 eV per atom binding by van der Waals interaction (Zacharia, *et al.*, 2004).

The lower graphene electron mobility for epitaxial films on SiC compared to mechanically exfoliated graphene placed on oxidized Si, has been a puzzle. The effect is related to the presence of steps on the SiC surface. But the calculations carefully performed by Low *et al.* (2012) suggest that the sharp curvature suggested in Fig. 5.7, as the film flows over the step-edge, is not the source of much scattering. Rather, the extra scattering, to reduce the mobility and increase the resistivity, is due to the

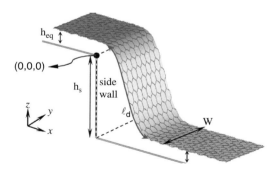

Fig. 5.7 Graphene "flowing" over step edge on SiC leads, in measurements, to a large incremental resistance. The effect comes from the difference in local Fermi level in the bound and free sections of the graphene and, according to these authors, is not influenced strongly by the curvature. (From Low *et al.*, 2012. © 2012 by the American Physical Society).

electrical coupling between the graphene and the substrate that varies sharply in the vicinity of the step. The authors conclude that the morphology affects the resistivity through the tendency of the SiC to dope the graphene, an effect stronger on SiC than on oxidized Si. The physics of electrostatic doping will be discussed in Chapter 8.

5.2.2   Catalytic growth on Ni or Cu, with transfer

It has been known for some time that graphene growth from carbon-containing gases can be catalyzed by various transition metal substrates. [In some cases, for example Ru, the binding to the substrate is strong enough to make the layer epitaxial, see Martoccia *et al.* (2008).][6] On the weakly binding surface of Cu this can be described as catalytic cracking of a hydrocarbon such as methane or acetylene, or precipitation of dissolved carbon onto a metal surface with subsequent graphitization (ordering into the hexagonal lattice with grain boundaries). Reports include Grant and Haas (1970); Gall *et al.* (1987); Nagashima *et al.* (1993); Gall *et al.* (1997); Forbeaux *et al.* (1998); Affoune *et al.* (2001); and Harigaya and Enoki (2002). Limited solubility of C in the Cu substrate can restrict the production to single layers of graphene, a desirable outcome. Large domains of graphene, described as single crystal domains, up to 0.5 mm in size, using low pressure CVD on Cu foil have been reported by X. Li *et al.* (2011), even though the Cu foil was observed as a "highly faceted, rough Cu (100) surface." These authors suggest that the large graphene grains were only weakly coupled to the rough substrate. We will discuss in Section 5.2.4 a proposal (Chen *et al.*, 2012) to modify the CVD process on Cu to make an epitaxial graphene layer without grain boundaries.

An important discovery of Reina *et al.* (2008) was that single crystal micro-mechanically cleaved graphene could be released[7] from the traditional oxidized silicon substrate to be deposited elsewhere.

The flakes in the original work of Reina *et al.* (2008) are on the micrometer scale, obtained from micro-mechanically cleaved graphite.

This basic transfer technique was quickly adapted to much larger area graphene films grown by chemical vapor deposition CVD on transition metals, principally Ni and Cu.

This basic method of growing graphene on copper foil followed by transfer using the Reina method, was extended to the cm scale by Li *et al.* (2009a), whose image of

---

[6]Martoccia *et al.* (2008) find experimentally that 25×25 unit cells of monolayer graphene exactly match (overlay) 23×23 unit cells of Ruthenium on Ru(0001). Their measurements indicate that the binding is stronger than van der Waals and that distortion of the underlying Ru lattice occurs, and that significant electron transfer into the graphene results, making it n-type. This is an example of corrugated graphene arising from incommensurate boundary conditions.

[7]The steps used by Reina *et al.* (2008) include covering (spin-coating) the film with poly(methyl methacrylate) (PMMA, 950 000 molecular weight, $9 - 6$ % wt in anisole), followed by partially etching the surface of the SiO$_2$ in 1 M NaOH aqueous solution. The 300 nm SiO$_2$ does not etch completely, and only a minor etching is enough to release the PMMA/graphene layer. The release is typically accomplished by placing the substrate in water at room temperature, where manual peeling can be used to detach the PMMA/graphene membrane from the substrate. As a result, a PMMA membrane is released with all the graphite/graphene sheets attached to it. This membrane is laid over the target substrate and the PMMA is dissolved carefully with slow acetone flow.

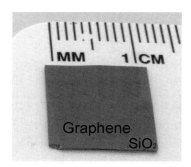

Fig. 5.8 Photograph of monolayer graphene transferred to oxidized silicon wafer. (From Li *et al.*, 2009a, with permission from AAAS).

graphene transferred to a Si substrate is shown in Fig. 5.8 following earlier work by Sutter *et al.* (2008) using Ru.

Reina *et al.* (2009) also extended their method (Reina *et al.*, 2008) to growing graphene on Ni, but it now appears that Cu is more suitable for this purpose. Reina *et al.* (2009) include references to earlier reports of growth of graphene monolayers on single crystalline transition metals, including Co, Pt, Ir, as well as Ru, Ni, and Cu.

Li *et al.* (2009a) use chemical vapor deposition with methane as the carbon source, flowing over copper foil about 25 μm in thickness at temperatures up to 1000°C. It was found that the growth was complete in 30 minutes, and was self-limiting, so that only 5% of the area had more than one layer of graphene. The graphene was found to be continuous, growing across lattice defects in the copper substrate. The mobility in the grown films was measured as high as 0.4 m^2/Vs. The films were transferred to arbitrary substrates by a method[8] similar to that of Reina *et al.* (2008).

The present authors Li *et al.* (2009a) extended their method of graphene transfer to larger area films, as described by Li *et al.* (2009b).

Large 4 × 4 inch graphene films grown on copper and transferred to oxidized silicon were reported by Cao *et al.* (2010). They found carrier mobility about 0.3 m^2/Vs and observed both the quantum Hall effect and weak localization in measurements of the graphene.

A similar method based on Ni substrates has been described by Kim *et al.* (2009). These workers have produced patterned graphene by patterning a Ni deposit on a

---

[8]The transfer method of Li *et al.* (2009a) was to coat the as-grown graphene film on Cu foil with PMMA, diluted in chlorobenzene at concentration of 46 mg/mL that was then cured at 180°C for 1 min. Since graphene grows on both sides of the Cu foil, after PMMA coating, the opposite side of the Cu foil was polished to remove the graphene layer. The 1cm × 1cm × 25μm Cu foil was then etched away in an aqueous solution of iron nitrate (0.05 g/mL) over a period of 12 hours. The resulting PMMA/graphene stack was washed in deionized water, placed on the target substrate, and dried. After transfer to the target substrate, a small amount of liquid PMMA solution was dropped onto the PMMA/graphene to dissolve the pre-coated PMMA. The new PMMA thin film was then slowly cured at room temperature for about 30 minutes and then dissolved in acetone.

Silicon wafer ahead of the CVD growth of graphene. They also described several methods of transferring the Ni grown graphene to other substrates. Growth on Co has recently been presented by Ramon *et al.* (2011).

Plasma-enhanced CVD (PECVD) has been applied to growth of carbon nanostructures by several authors (Wang *et al.*, 2004; Zhao *et al.*, 2006; Chuang *et al.*, 2007; and Kobayashi *et al.*, 2007). It is not clear if the plasma-assisted aspect could be applied in the wide-area growth of graphene on Cu or Ni, to allow a lower temperature deposition that clearly would be desirable. [As we will see later, Z. Li *et al.* (2011) have demonstrated graphene growth on Cu at 300°C by using benzene as the source vapor (Section 5.2.4).]

A substrate-free plasma enhanced growth method for graphene has been described by Dato *et al.* (2008).

A less-developed alternative method for growing graphene reported by Affoune *et al.* (2001) and, more recently, by Amini *et al.* (2010), is to slowly cool a molten metal such as Ni containing dissolved carbon that is found to crystallize on the surface of the melt. This method is somewhat similar to the production of "Kish graphite" as a byproduct of steel manufacture from carbon-rich iron melt, mentioned in Section 3.3 and Section 7.1.

5.2.3   Large area roll-to-roll production of graphene

Production of monolayer graphene by a roll-to-roll method, starting with CVD growth on Cu foil, has been described by Bae *et al.* (2010). The method follows that described earlier by Li *et al.* (2009). The foil is a roll on a cylindrical quartz substrate in the CVD reactor, so that a larger graphene area can be obtained within a tube reactor. Bae *et al.* have achieved deposition on a roll of Cu foil with dimension up to 30 inches in the diagonal direction. An 8-inch-diameter quartz reactor tube of length 39 inches was used. The copper foil was wrapped around a 7.5 inch diameter quartz tube to be inserted into the furnace. Adopting this geometry minimized temperature gradients over the copper area, to promote homogeneity of the deposited carbon. The authors describe a preliminary step of annealing the copper foil in flowing H_2 at 1000°C to increase the grain size of the Cu from a few μm to around 100 μm. After this annealing, methane is added to the gas flow for 30 minutes at 460 mTorr at flow rates 24 and 8 standard cc per minute, respectively, for methane and hydrogen. After this the furnace is quickly cooled to room temperature at 10°C/s in flowing hydrogen at 90 mTorr.

The graphene-coated large area copper foil is attached to a "thermal release tape," a large-area thin polymer supporting sheet, by running the pair of sheets between soft rollers at a pressure ~2MPa. The result, a large-area, "release tape" polymer/graphene/copper sheet is then rolled through a bath to remove the copper. The next step is to place the graphene side of the assembled sheet onto the substrate of final choice, often a roll of 188 μm thick polyethylene terephthalate (PET). Running this sandwich between rollers at mild heat ~100°C transfers the graphene from the "thermal release tape" to the desired support. In typical cases the procedure is iterated to give a stack of four graphene layers (whose azimuthal orientations are random) on the wide area PET supporting sheet and with diagonal dimension as large as 30 inches. This stack is a high quality transparent flexible conductor at best with ~90% optical transmission and low resistance of ~30 $\Omega/\square$. This low value was aided by a step of "chemical doping" with nitric acid HNO$_3$ that makes the film strongly p-type. According to the authors, Bae *et al.* (2010), this transparent conductor is superior to common transparent conductors such as indium tin oxide (ITO) and carbon-nanotube films, reported by J.-Y. Lee *et al.* (2008). Bae *et al.* (2010) describe steps in producing touch screens for electronic devices using their graphene/PET transparent electrodes. The resulting transparent conductors display extraordinary flexibility and it was stated that the electrode functions at up to 6% strain. A review of "chemical doping" has been given by Liu *et al.* (2011).

An intercalation method for enhancing the electrical conductivity of large area, transparent graphene sheets has been described by Khrapach, *et al.* (2012). These authors point to a limitation of graphene as a transparent conductor, namely, its high sheet resistance, typically >30 $\Omega/\square$ with 90% optical transmission for multilayers, compared to transparent electrodes such as ITO (indium tin oxide) that can be obtained at 10 $\Omega/\square$ with 85% optical transmission. (The graphene resistance was lowered in the work of Bae *et al.* by stacking four monolayers and by chemical doping.) Khrapach *et al.* (2012) describe an intercalation with FeCl$_3$ to allow films of 8.8 $\Omega/\square$ with 84% optical transmission, including a carrier density as high as $8.9 \times 10^{14} \text{cm}^{-2}$ with mean free path as large as ~0.6 µm. The authors have extensively characterized the electrical behavior of the samples and find that with up to five intercalated layers the resistance drops noticeably, maintaining large optical transmission near 85%. Their method however thus far is based on mechanically exfoliated graphene placed in solution, certainly not a production method.

5.2.4   Grain structure of CVD graphene films

Direct imaging of grain boundaries in polycrystalline graphene was reported by Kim *et al.* (2011). With serious prospects for CVD graphene films in a variety of electronic devices, recent attention has been placed on analysis of the electrical effects of grain boundaries, clarifying the nature of the grain boundaries, and the optimization of films by control of grain size. Such work is reported by Tsen *et al.* (2012), see Fig. 5.9(a).

The CVD film of graphene is always polycrystalline, as has been determined by Tsen *et al.* (2012), using TEM with a diffraction option. The dimensions of the grains vary from around 1 µm for fast growth rates (Growth A) to more than 50 µm for slow growth (Growth C). The basic $(\rho/\square)$ for the films did not change much with deposition conditions and was on the order of 1000 Ω at a carrier density near $3 \times 10^{11} \text{cm}^{-2}$. The effect of a single grain boundary cutting across a measured section, as shown in the left panel of Fig. 5.9(b), can be modeled as an increase in the length from L to $L + \lambda$. The value of λ is larger if the grain boundary actually contains a physical gap defect, reducing the connectivity, between the two films. The model was established empirically, by measuring R' vs. R for films of dimension $L \times W$, with and without a single grain boundary. The formula $R = (\rho/\square)(L/W)$ for a perfect continuous film, changes, with a single grain boundary, to

$$R' = (\rho/\square)(L/W) + \rho_{\text{GB}}(L/W) = (\rho/\square)(L + \lambda)/W \qquad (5.2)$$

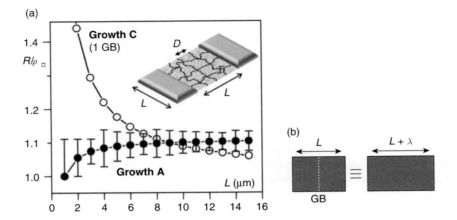

Fig. 5.9 (a) Model of device resistance as a function of device size for CVD graphene growth methods A and C, using empirically determined parameters λ and D. [In this Figure, "(1 GB)" signifies one grain boundary.] Growth method A is rapid, resulting in smaller grain size $D \approx$ 1 µm but better "stitching", i.e., contact between grains, modeled as small value for parameter λ. Growth C is a slow growth with large grain size $D \approx 50$ µm but poor contact between grains, resulting in large λ. When many grains are present the overall device resistance approaches 1.1 times the basic Ohm/square value for the graphene film. (From Tsen *et al.*, 2012, with permission from AAAS). (b) Sketch of modeling of graphene with a grain boundary GB as equivalent to adding a length λ. (After Tsen *et al.*, 2012, with permission from AAAS).

where

$$\lambda = \rho_{\mathrm{GB}}/\left(\rho/\square\right). \tag{5.3}$$

The parameter λ is related to the type (connectivity) of the grain boundary. The values of ρ_{GB} range from the order of 4 kΩ/□ µm near the Dirac point, to about 0.5 kΩ/□ µm at large p-type doping, scaling much as the basic film resistivity with doping. It was found that the resistivity and the mobility in the region with a single grain boundary were reduced by the same factor, $R'/R = 1 + \lambda/L$, and that this factor is not much changed as the carrier density in the graphene is changed by the gate voltage on the underlying Si substrate. Values of λ were found between 200 nm and 1.8 µm, the latter value actually larger than the device length. To apply this model to a square L much larger than the grain size, D, as suggested in the inset to Fig. 5.9, the device resistance is

$$R = \left(\rho/\square\right)\left(L + n\lambda\right)/W, \tag{5.4}$$

where n, the expected number of grain boundaries, is expected to be $n = L/D$. Thus, for large n, the result reduces to

$$R = \left(\rho/\square\right)\left(1 + \lambda/D\right). \tag{5.5}$$

This behavior is shown in Fig. 5.9(a). The authors offer a formula for the conductivity σ_{GB} of a grain boundary, used to estimate that the linear density of defects along a typical grain boundary is $2 \times 10^7/\text{cm}$ and the radius of the perturbing potential at such defects is about 2 nm. The conclusion is, basically that grain boundaries in optimally grown (rapidly grown) films need not be a major problem for devices of size $\geq 5\mu\text{m}$ on a side. Further, they suggest that a growth leading to overlapping (not orientationally related) films, so that one tends to cover the grain boundaries in the other, leads to superior properties, and that such films may be feasible to produce.

Duong *et al.* (2012) have developed optical microscope methods for determining grain structure in graphene grown on copper foil.

Briefly, their method is to selectively oxidize the supporting copper residing directly under graphene grain boundaries by exposure to O and OH radicals produced UV illumination in moist gas. Continued exposure widens the oxidized copper regions that appear only directly under the graphene grain boundaries, and thus mark the boundaries. The resulting pattern of oxidized copper, reflecting the pattern of graphene grain boundaries, is thus amplified until it is visible using an optical microscope. The authors studied the resistivity of copper-grown graphene, after transfer to an insulating substrate. It was found that annealing of the graphene, after transfer from an oxidized copper substrate, restored its initial low resistivity. (They found that the moist atmosphere with UV illumination oxidized the graphene sheet as well, forming OH hydroxyl, C–O–C epoxide and COOH (O=C–OH) carboxyl groups, mostly near defect sites. Neither dry oxygen nor dry ozone O_3 oxidized the graphene at room temperature: moisture and UV illumination were necessary for oxidation to occur.) These authors found that the size of grains in the copper substrate foil [mentioned to fluctuate between $10\,000 - 15\,000\,(\mu\text{m})^2$] did not predict the graphene resistivity, but that the graphene grain size, influenced by substrate temperature of CVD growth, was primary in this regard. The sheet resistance of the graphene decreased to about 300 $\Omega/\square$) as the growth temperature increased to 1060°C, where the graphene grain size was characterized as 72 $(\mu\text{m})^2$. These authors suggest a resistivity formula

$$R = (\rho/\square)_0 \left[1 + (A/A_c)^{-n}\right] \tag{5.5a}$$

where A is the average grain size, A_c is a fitting parameter, and the exponent n was mentioned as 4. The found value $(\rho/\square)_0$ is 230 $\Omega/\square$ from their experimental work. They comment that this is larger than the ideal value 30 $\Omega/\square$ owing to defects in the film such as "point defects, wrinkles and ripples" and to scattering from the substrate. These results would suggest a value 57.5 $\Omega/\square$ for a four-layer stack, reasonably consistent with 30 $\Omega/\square$ as mentioned by Bae *et al.* (2010) for a four-layer stack heavily doped n-type by treatment with HNO_3.

The growth of graphene on Cu foil, typically of (100) surface orientation, is successful but leads to lower mobility than micro-mechanically cleaved graphene. Intuitively it seems that grain boundaries are likely the cause of lower mobility. The characterization of grain boundaries with different growth rates has just been described. A theoretical discussion of growth on Cu (111) has been given by Chen *et al.* (2012) that seems useful even though the copper foils in use have (100) surface orientations. Li *et al.* (2011) report growing "single crystals" of graphene of 0.5 mm size, and comment that their foil displayed "a highly faceted, rough Cu (100) surface with sharp diffraction spots." They also comment that the diffraction pattern of the graphene looked like that of free standing graphene, "perhaps indicating a weak coupling to the rough surface". In fact the mobility in their "single crystals" was 0.4 m^2/Vs, not a single crystal value. A method of growing single crystal graphene on Cu without grain boundaries would be an important advance. Chen *et al.* (2012), presuming Cu (111), note that this surface has hexagonal symmetry matching the graphene benzene ring symmetry. [Cu 111 may be easily available, as it results from evaporation of Cu, as shown by Tao *et al.* (2012), allowing in CVD growth also excellent monolayer graphene.] On Cu (111), however, Chen *et al.* find in simulation that a typical nucleated seven-ring cluster (a ring surrounded by six rings, this is essentially the coronene molecule), initially benefitting from the common hexagonal shape, in fact reduces its energy by rotating 11° from the hexagon of the underlying Cu. For this reason Chen *et al.* al argue that orientational disorders of carbon islands will be abundant in the early stages of nucleation and growth on Cu (111). Such disorders cannot heal with the enlargement of the islands, leading to the prevalence of graphene grain boundaries upon island coalescence. Chen *et al.* go on to suggest that by adding Mn to the Cu, a Mn-Cu (111) surface of hexagonal symmetry could easily be obtained that would exactly stabilize the common "coronene" seven-ring precursor. They in fact suggest that molecular coronene vapor be used in a two-stage CVD process [it has been shown by Z. Li *et al.* (2011) that benzene, when used in the CVD process on Cu, allows graphene growth at 300°C, much lower than the typical 1000°C]. The coronene deposition, followed by de-hydrogenation, would leave a network of like-oriented graphene nuclei that would then be filled in, in a second stage, by depositing from methane or ethylene in the usual way. In this two stage growth, on a modified Cu (111) surface, it appears that grain boundaries would not arise and a graphene single crystal could be envisioned of very large lateral size, as suggested by Chen *et al.* (2012).

5.2.5   Hybrid boron-carbon-nitrogen BCN films

Graphene and (isoelectronic) hexagonal boron nitride have nearly identical lattice constants, but different band structures. BN is a wide-band-gap insulator at about 4.7 eV. It has recently been discovered (Ci *et al.*, 2010; Zhu *et al.*, 2011) that a family of hexagonal layered compounds of the form $(BN)_n(C_2)_m$, with n, m = 1,2,3, ... form, and exhibit different band gaps with different compositions.

An example of such a compound, BCN, is indicated in Fig. 5.10(a). In this figure unmarked vertices are filled with carbon atoms.

In the shown isomer that has NB_3 and BN_3 groupings, the calculated bandgap is 1.8 eV. A difficulty in growing these films is that large domains of pure graphene and pure BN appear, having very different bandgaps. Growth does not result in a uniform composition $(BN)_n(C_2)_m$, on the nanoscale. It is conceivable that deposition schemes might be altered to produce a more homogeneous film with a meaningful average bandgap. A section of this material, if inserted epitaxially in series between

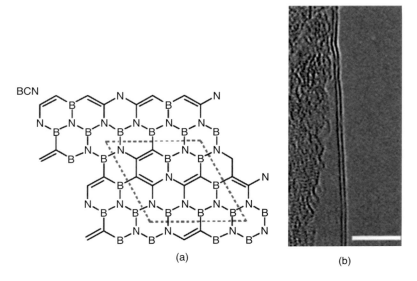

(a)   (b)

Fig. 5.10 (a) An isomer of BCN, in a diagram where unmarked vertices are assumed filled with C atoms. (Reprinted with permission from Zhu *et al.*, 2011. © 2011 American Chemical Society). (b) HRTEM High Resolution Transmission Electron Microscope image (scale bar 5 nm) showing a two layer (folded) h-BCN film deposited by chemical vapor deposition (CVD) onto copper using ammonia borane and methane as source gases. (Other sections of the film were found to be three layers thick.) It was found that the carbon concentration in the films could be varied between 10% and nearly 100%. (From Ci *et al.*, 2010, by permission from Macmillan Publishers Ltd., © 2010).

two sections of pure graphene, could act as a conventional tunnel barrier, a topic that will be considered in Chapter 9. The manufacturability of the BCN family via conventional chemical vapor deposition (CVD) methods has been established (but not yet of a material with homogeneous bandgap) by Ci *et al.* (2010). Figure 5.10(b) shows a two-atomic-layer section of an h-BNC film grown by CVD. The film was grown on copper using a gas mixture with equal parts of B and N, in the form of methane and ammonia borane (NH$_3$-BH$_3$). It was found that the atomic percentage of carbon could be controlled between about 10% to ∼100%. Unlike graphene grown on copper, where the usual thickness is one atomic layer, the h-BCN films were typically two or three atomic layers in thickness. The authors were able to transfer the films grown on copper to other substrates, such as quartz, and it was found that the films could be lithographically patterned and cut into various shapes by exposure to oxygen plasma. This means that the h-BCN atomic films can be easily fabricated into devices.

It had earlier been found that the copper foil substrate was suitable for growing h-BN (Preobrajenski *et al.*, 2005).

Analysis of the electron mobility in chemically modified graphene (and in hexagonal boron nitride) has been given by Bruzzone and Fiori (2011).

Methods for making lateral heterostructures in graphene layers have been given by Levendorf *et al.* (2012). These are potentially important, apart from the need to find a method for making a material with homogeneous bandgap, for introducing tunneling barriers between graphene source and drain electrodes, to be mentioned in Chapter 9, see Fig. 9.22(b).

5.2.6   Atomic layer deposition

The exceptional carrier mobility in graphene that is only one atomic layer thick, is due to the strong covalent crystalline bonding, offering little irregularity to scatter free electrons. In attempts to make an evaporated metal film as thin as a few atomic layers, irregularity appears, even to the point of the film breaking up into islands, a severe limit on miniaturization with conventional metal films. The great advantage of graphene is that not only is it thin, but it is also smooth, continuous and conductive. In making layered devices based on graphene, high electronic quality requires atomic perfection in the interfaces between layers, to avoid scattering. Epitaxial growth by molecular beam epitaxy (MBE) is one well known process. Making insulating layers on graphene is a particular problem in conventional MBE process with thermal evaporation from Knudsen cells. A more general approach to producing an epitaxial growth of more complicated layers of compounds of the form A_nB_m, such as metal oxides, has been developed based on self-limiting surface chemistry. This method is atomic layer deposition (ALD). (George *et al.*, 1996).

This approach has been demonstrated in cases where a binary chemical vapor reaction can be separated into two half-reactions. The first species A must epitaxially grow on the substrate of interest, with capability of stopping the growth at one monolayer. The completed epitaxial layer of A is then flooded with species B that reacts with all available A sites, completing one epitaxial molecular layer. This process is then repeated.

A success of this approach is in the "high kappa" oxide HfO_2 now routinely grown on Si surfaces to make the gate insulator of field-effect transistors. The problem was that the scaling of native SiO_2 was leading to layers so thin that tunnel gate leakage occurred. A requirement of the gate insulator is to allow a sufficiently high capacitance to pull into the conductive channel an adequate density of mobile carriers under the bias available in the device technology. By going from SiO_2, $\kappa = 3.9$, to a heavy-metal oxide such as hafnium dioxide, of larger permittivity $\kappa = 20 - 22$, a thicker layer, sufficient to block tunnel leakage, retains adequate capacitance per unit area. Reactions that are mentioned by George *et al.* (1996), with an A B binary reaction sequence, are

$$HfCl_4 + 2H_2O \rightarrow HfO_2 + 4HCl \tag{5.6}$$

and

$$2Al(CH_3)_3 + 3H_2O \rightarrow 3Al_2O_3 + 6CH_4. \tag{5.7}$$

The gaseous reaction products are released when the surface is held at a suitable temperature. It is possible to supply the first species, in some cases in an inert carrier gas at suitable temperature, using chemical vapor deposition (CVD) methods, rather than in high vacuum from an oven as would be used in molecular beam epitaxy (MBE). To deposit sapphire (Al_2O_3), a workable scheme is to first evaporate a thin layer of Al metal, oxidize it to completion and then proceed with the atomic layer deposition. In some cases on graphene, an organic layer is first deposited to least disturb the bonding/carrier motion in the graphene, and then deposit oxide layers by atomic layer deposition (ALD).

ALD is used in graphene device technology to grow both HfO_2 and Al_2O_3. For example, in Fig. 8.16(a) below the insulating layer between the graphene and the top gate is poly-hydroxysilane (HSQ) and HfO_2. ALD is also used in making a flash memory device as described in the text preceding Fig. 9.17(a).

5.3 Graphene nanoribbons

Ribbons, strips of uniform width, have already been mentioned in Section 5.1.2.2. In Graphene Nanoribbons, the notation GNR, with subdivisions AGRN and ZGNR, is in use for ribbons with armchair and zigzag boundaries, respectively that can be visualized as horizontal and vertical lines in Fig. 1.2(b). The properties of the ribbon, contrasted with an infinite plane of graphene, are modified by the finite width L, with confinement effects in play, and by the nature of the boundary. Confinement in general leads to an energy gap inversely related to the width. (We will mention in Section 5.3.2 something of a paradox, the fact that a large class of carbon nanotubes that in principle can be formed by rolling a GNR along a chosen axis, are in fact metallic.) The propagating wavefunction along a ribbon is

$$\psi(x,y) \sim e^{ikx} \sin(n\pi y/L), \tag{5.6}$$

the latter factor recognizing that the probability of finding the particle at $y = 0$ or $y = L$ must vanish, as a work function barrier confines the particle. Hence $k_y = n\pi/L$ that corresponds, in a conventional metal (or in *bilayer* graphene, with massive electrons in parabolic bands) to a confinement energy

$$(\hbar k_y)^2/2m = h^2/8mL^2. \tag{5.7}$$

(The confinement energy is further discussed in Section 9.3 in connection with quantum dots.) However, in monolayer graphene, as was pointed out by Berger *et al.* (2006), the linear bands lead to a different GNR energy gap. In monolayer graphene, the energy E_n, at the bottom of nth subband state becomes, with ν_F the Fermi velocity,

$$E_n = \nu_F |p| = \hbar \, \nu_F \left(k_x^2 + k_y^2\right)^{1/2} = \left[E_x^2 + n^2 \left(\Delta E\right)^2\right]^{1/2} \tag{5.8}$$

where $n = 1, 2, 3, \ldots$

$$\Delta E = \pi \hbar \nu_{\mathrm{F}}/L \sim 2\,(eV - nm)/L, \qquad (5.9)$$

and $E_x = \hbar\,\nu_{\mathrm{F}} k_x$.

This gives a gap $\sim$100 meV for nanoribbon width $L = 20$nm, quite close to the measurement of Han *et al.* (2007), shown in Fig. 5.12(a). To this basic analysis one then adds consideration of the nature of the edges. In practice it is difficult to achieve a uniform boundary condition (but a notable exception is the synthetic chemical method of Koch *et al.*, 2012) and the size- and boundary-effects are obscured if the material has a short mean free path and if the boundaries are rough.

The fundamental conceptual issues in nanoribbons were addressed by Nakada *et al.* (1996), who discovered an edge state residing at the Fermi energy in zigzag ribbons. Particular half-metallic properties of ribbons with zigzag edges have been addressed by Son *et al.* (2006), while gaps in nanoribbons were addressed by Son *et al.* (2005). Properties of semiconducting graphene nanoribbons have been calculated by Barone *et al.* (2006). These authors have considered effects of hydrogen termination of dangling bonds and have calculated the energy gap vs. ribbon width in several cases, including random boundary terminations.

5.3.1   Zigzag and armchair terminations

Armchair and zigzag boundaries, respectively, can be visualized as horizontal and vertical lines in Fig. 1.2(b). In detail, the properties of a ribbon depend on the atomic configurations at its boundaries. There is a lot to be gained if these configurations can be controlled. One approach to regularizing the edges of a graphene sheet is to heat it. If graphene is raised to a high temperature, the inherently lower energy of the uniform zigzag and armchair edges assert themselves as the system anneals, loses atoms and reconstructs.[9] These effects are seen in the TEM image of Jia *et al.* (2009), Fig. 5.11.

The difficulty remains, however, in achieving regular boundaries on graphene nanoribbons (GNR), because extreme heating is not a practical approach in device fabrication and patterning. Reactive-ion-etching may be a practical approach to make a geometrical edge, but aligning the cut with the crystalline axes of the underlying monolayer graphene, plus uniformly satisfying the dangling bonds, seems in practice impossible. This fundamental difficulty diminishes the prospects of engineering use of nanoribbons at their highest capability in nanoelectronics. Fortunately there are nanoelectronic applications of nanoribbons where the edges need not be exactly specified, for example, in interconnects, some types of FET devices and in the fabrication of graphene flash memory elements. We discuss these nanoelectronic applications, including complementary logic devices, in Chapter 9.

---

[9]These experiments, with graphene heated over 2000 K by current flow on a TEM stage, make clear that graphene is a robust refractory material, contrary to any instability of 2D matter suggested by hasty reading of the HLMPW theorems (Section 2.4). The fact is that graphite sublimates at 3900 K, by release of intact layers of graphene.

Fig. 5.11 Annealed graphene under high electrical current density shows well-defined edges, as marked. 1 nm scale bar. (From Jia *et al.*, 2009, with permission from AAAS).

Somewhat contradictory to this assessment, Li *et al.* (2008) have reported chemically derived, ultrasmooth graphene nanoribbon semiconductors with bandgaps up to 0.4 eV (see Fig. 5.3).

5.3.2 Energy gap at small ribbon width, transistor dynamic range

An early experimental study of the development of a gap in nanoribbons was reported by Han *et al.* (2007). A summary of their measured gaps vs. width W is shown in Fig. 5.12(a). These data were obtained from micro-mechanically cleaved graphene samples. A comparison study based (Li *et al.*, 2008) on chemically-exfoliated small-width nanoribbons, put into transistor-like devices is shown in Fig. 5.12(b).

We described earlier (Section 5.1.2.2) the methods that Li *et al.* (2008) have found for ways of producing ribbons based on chemical exfoliation. The ribbons were selected from a suspension by inspection using the AFM. They found that narrow ribbons of widths ranging from 10 nm to 55 nm, when fabricated into transistors, exhibited On/Off current ratios sharply rising at smaller widths.

These results can be compared with the predictions of Son *et al.* (2006) and Barone *et al.* (2006). The dashed line in Fig. 5.12(a) is a fit of the inferred gaps to the simple form E_G (eV) $\approx 0.8/W$ (nm). The theory of Son *et al.* (2006) predicts that GNR with homogeneous zigzag and homogeneous armchair edges both have band gaps that decrease as the width W of the ribbon increases. The scale parameters however depend on the specific edge terminations. The GNR graphene of Han *et al.* (2007) are estimated to have hole mobility on the order of 200 cm^2/Vs.

These ribbons, unlike nanotubes, were found to always have an energy gap, so none were metallic at low temperature. Carbon nanotubes can be either semiconducting or metallic depending on the diameter and chirality (direction of the bending axis in forming the nanotube from graphene). A simple argument for the existence of metallic nanotubes is based on the structure of graphene, shown in Fig. 1.2(a). In the caption to Fig. 1.2(a) it was shown that the Fermi wavelength in graphene is $\lambda_F = 1/(2\pi k_F) = 3\sqrt{3}a/2 = 369$ pm with $a = 142$ pm the nearest neighbor distance. A criterion for

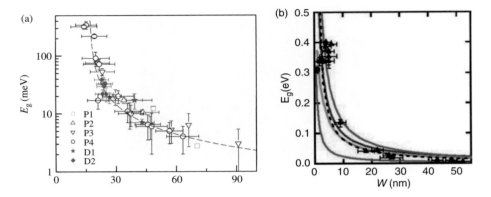

Fig. 5.12 (a) Energy gap vs. width W for six different device sets, all based on mechanically peeled graphene layers placed on oxidized Si used as back gate. The investigation considered the crystallographic orientation and temperature dependent transport, with the graphene near the Dirac point. (After Han *et al.*, 2007). (b) Inferred band gap vs. width of chemically-exfoliated graphene nanoribbons. Data points are inferred from On/Off current ratios of simple FET-like devices, while curves are theoretical estimates based on the paper of Son *et al.* (2006). (From Li *et al.*, 2008, with permission from AAAS).

a nanotube to be metallic is that the circumference be an integer number of Fermi wavelengths, thus $\pi D = n\lambda_F = n3\sqrt{3}a/2$ or $D = n3\sqrt{3}a/2\pi$, with n an integer. In the carbon nanotube literature, the relation between the integers n, m and the diameter D_t is given as

$$D_t = (\sqrt{3}\,a/\pi)\left(n^2 + m^2 + nm\right)^{1/2}, \tag{5.10}$$

with a the nearest neighbor distance, and it is known that the armchair nanotubes, $n = m$, are metallic. In this case, $D_t = m\,(3\,a/\pi)$, that overlaps the criterion above, $D = n3\sqrt{3}a/2\pi$, would require (integer) $m = \sqrt{3}n/2$ that is impossible. However, if the circumference πD is modified to correspond to the length along the bonds, $\pi D^* = \pi D/\cos\,(30°) = (2/\sqrt{3})\,\pi D$, then it appears that armchair metallic nanotubes do correspond to having an integer number of Fermi wavelengths along the circumference, interpreted as the circumferential length along the carbon–carbon bonds. (This corresponds to a vertical zigzag path in Fig. 1.2(b).)

Approximately 1/3 of possible nanotubes are metallic, from experimental and theoretical points of view. Why are there apparently no achievable boundary conditions on GNR nanoribbons that replicate the boundary condition in the metallic nanotubes? This question remains somewhat open, for example, Son *et al.* (2006) state that using Schrödinger's equation in simple tight-binding approximations (that accurately give the graphene band structure) or using the Dirac equation with conical bands and constant Fermi velocity, calculations indicate that armchair AGNR ribbons should be metallic or semiconducting, depending on the width L. In contrast, Son *et al.* find that zigzag ZGNR ribbons are all metallic, with peculiar edge states on both sides of the ribbons, regardless of the width. On the other hand, when these authors do first-principles calculations, where the atom positions are allowed to relax at the edges, the equality of atomic spacings at the edge and in the interior of the ribbon is lost, and all predictions are for gapped semiconducting ribbons. (Some of the gaps are small, however.) The calculations of Son *et al.* (2006) have been confirmed and extended by Yang *et al.* (2007), who find, including electron-electron interaction effects, that quasiparticle band-gaps in GNRs of widths $L = 2$ nm $\rightarrow 1$ nm are significantly larger, 1eV $\rightarrow$ 3eV (assuming all dangling bonds are passivated by hydrogen atoms). While the authors suggest that such ribbons are more suitable for nanoelectronic devices, ribbons of such widths appear not to be manufacturable in large scale.

Analysis of graphene nanoribbon field-effect transistors using boron doped material is given by Marconcini *et al.* (2012).

The experiments show no indication of metallic GNRs, as we have said, and no gaps larger than 0.4 eV. (The data of Fig. 5.12(b) show a gap of 0.4 eV at the smallest attained ribbon width that, however, may still be marginal for a suitably large On/Off ratio in transistors for a logic application.) However, a small gap at room temperature will give quasi-metallic behavior, conductive enough for leads/interconnects in a device context. There is a literature of GNR designed as interconnects for silicon devices to be discussed in Chapter 9. We will also return in Chapter 9 to a proposal for a GNR device logic, analogous to complementary metal-oxide semiconductor (CMOS). In this ambitious proposal the active devices are simply graphene nanoribbons, wide and narrow, acting, respectively, as conductors and as semiconductors based on their width L, the wide ones acting essentially in a metallic fashion, with gate electrodes turning the semiconducting (narrow graphene) channel on and off. It is argued there that the inherent symmetry of the conduction and valence bands in graphene, with essentially identical electron- and hole-masses, is an advantage over silicon in the context of complementary logic, and that such logic could scale to small sizes and compete as a replacement for Silicon at the end of Moore's Law.

A method for tailoring nanoribbons using the scanning tunneling microscope (STM) has been described by Topaszto *et al.* (2008).

5.3.3 Chemical synthesis of perfect armchair nanoribbons

Synthetic chemical routes to making nanoribbons have been pioneered by Cai *et al.* (2010) and by Koch *et al.* (2012) using molecules containing two anthracene units. Anthracene is three benzene rings side-by-side, so the ribbons have that width, about 0.74 nm. These methods lead to completely accurate armchair ribbons, three benzene rings wide, with lengths limited only by the polymerization process. Koch *et al.* report ribbons up to 20 nm, corresponding to around 80 molecular lengths. Schwab *et al.* (2012) start with a different molecule and achieve slightly wider ribbons. The bandgaps estimated for the ribbons are estimated in the range 1.12 eV to 1.6 eV and larger. It seems likely that the ribbons synthesized by Koch *et al.* act as one-dimensional conductors [eqn (5.6)] with a conduction-band edge around 1.3 eV, analogous to a simple picture that is well known in the field of semiconductor heterojunctions: 2D conduction bands appearing above specific band edge energies.

Following the methods of Cai *et al.* (2010), Koch *et al.* (2012) have fabricated long narrow armchair graphene nanoribbons on a gold surface, Au (111). The processing is done in ultra high vacuum (UHV), evaporating the 10,10'-dibromo-9,9'-bianthryl (DBDA) precursor molecules from a Knudsen cell at 470 K.[10] The chemically-synthesized nanoribbon has perfect armchair edges: the ribbon alternates between two and three benzene ring widths. Koch *et al.* used STM to study the "voltage-dependent conductance" as single ribbons are pulled free of the gold surface. While the authors do not describe the nanoribbons as metallic, they did observe currents up to 100 nA through these structures. The large current density suggests that the nanoribbons support metallic conduction in a 1D band above a band-edge energy. The ribbons were mechanically robust and were moved around the gold surface using the STM tip. The methods are attractive in achieving long ribbons of precise width and edge structure, and also in that the highest temperature in the fabrication is only 400°C. (The conventional CVD process for making graphene calls for 1000°C.)

The chemical synthetic route (following Cai *et al.*, 2010), starts with the 10,10'-dibromo-9,9'-bianthryl (DBDA) molecules and polymerizes these molecules at 200°C on the Au surface, freeing the bromine atoms. This polymerization yields non-planar acetylene oligomers potentially of indefinitely large length. A second heating step at 400°C removes the hydrogens and makes the pure graphene nanoribbon that is noted to be somewhat rigid as a long wire, and can be detached, by the STM tip in UHV, from the Au quite easily. The measurements at small bias voltage indicate exponential decay of free carriers in the wire, moving away from the end contacts, so the wire is not metallic. Such exponential decay, rather than metallic conductance, of molecular wires is indeed indicative of a bandgap, as would be expected viewing the object as a nanoribbon of small width. The authors have not deduced an effective energy gap from their measurements. They display in their Figure 3A a conductance spectrum measured by STM as the ribbon lies flat on the Au surface that shows a rise in conductance of three orders of magnitude in the range 1V to ~ 1.3V, to a peak near 1.3 Volts. This could be interpreted as a one-dimensional conductor with the bottom of the conduction band at 1.3 eV.

Again, as in the case of a carbon nanotube, the question would be of relocating such a ribbon to a useful position in a device structure, and whether such a process could be scaled.

It does seem that these nanoribbons are superior to carbon nanotubes, in the sense that they are well defined molecules that do not need to be sorted for metallic vs. non-metallic types, and that they do have the energy gap that the transistor designers find necessary.

It might even be possible, because of the relatively low temperatures needed, to do such a growth process on a silicon wafer, or other practical surface. It is not obvious that the Au (111) surface plays a unique role, and the reactions might well be found

---

[10]The word "bianthryl" denotes a molecule composed of two anthracene molecules.

to occur on other surfaces. It is conceivable that the orientation of the polymer and the nanoribbon might be controlled by putting ridges on the supporting surface.

Early assessments of electronics based on nanoribbons, also including original experimental work, have been provided by Chen *et al.* (2007) and by Avouris *et al.* (2007), who treat carbon electronics more broadly. From a device point of view these workers confirm a role of 1/f noise in GNR devices. They also find energy gaps increasing with smaller widths W, in agreement with the contemporary work of Han *et al.* (2007) and later work of Li *et al.* (2008).

6

Experimental probes of graphene

The basic discovery of the electric field effect (Fig. 1.4), allowed electrical transport measurements including resistivity, and Hall effect that served well to establish the basic electronic properties of graphene. Extended to high magnetic field, the Schubnikov–de Haas effect and quantum Hall effect were central to establish the anomalous properties of the material. The basic experimental geometry is shown in the upper right inset to Fig. 1.4. In this chapter, following a brief review, we describe a few of the experimental methods that are less widely known.

6.1 Transport, angle-resolved photoemission spectroscopy

Electrical four- and six-terminal measurements with variable temperature and magnetic field were essential in establishing the properties of graphene. The six-terminal sample arrangement, with two current electrodes at the ends of the Hall bar, with four potential electrodes along the side of the bar to allow longitudinal and transverse (Hall) voltage measurements, is shown in the inset to Fig. 1.4 and also in Figs. 2.5, 2.6 and 2.7. The unique aspect of such measurements on graphene is that the conductivity type and carrier density is always controlled by a voltage applied to the underlying silicon substrate. A discussion of the early results on graphene has already been given in Chapter 2.

The bandstructure of a metal is directly revealed in the technique of angle-resolved photoelectron spectroscopy (ARPES). This technique benefits from synchrotron radiation as is available, for example, at the Advanced Light Source, at the Berkeley Laboratory of the U.S. Dept. of Energy, and other centers. We have mentioned in Section 4.8 the work of Ohta et $al.$ (2006) that originally determined the bands in bilayer graphene. A review of the bandstructure determination by this Berkeley group was given by Bostwick et $al.$ (2009).

Angle-resolved photoemission spectroscopy (ARPES, see Fig. 6.1, performed in ultra high vacuum) allows measurement of energy and momentum of an excitation, according to rules conserving energy and momentum (Zhou et $al.$, 2007)

$$E_{\mathrm{B}} = h\nu - E_{\mathrm{kin}} - \varphi, \quad \hbar K_{\mathrm{ll}} = (2mE_{\mathrm{kin}})^{1/2}\sin\theta. \tag{6.1}$$

Here $h\nu$ is the photon energy, E_{kin} is the measured kinetic energy of the photo-emitted electron, E_{B} is the binding energy of electrons in the material, k_{ll} is the momentum of electrons in the material parallel to the surface and φ is the work function. $K_{\mathrm{ll}} = k_{\mathrm{ll}} + \mathrm{G}$ is the component of momentum of electrons in the

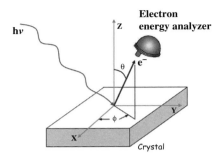

Fig. 6.1 Angle-resolved photoemission spectroscopy (ARPES) measurement. (From Zhou *et al.*, 2007).

sample surface, where G is a reciprocal lattice vector. $K_{\parallel}$ can be calculated from the measured kinetic energy and collector angle θ, see Fig. 6.1. Therefore, by measuring the intensity of the photoemitted electrons as a function of their kinetic energy, at different emission angles, the energy and momentum of the electrons in the sample surface can be measured directly. The density of states, but not the angular information, is available in scanning tunneling spectroscopy (STS) (see Section 6.3). In comparison, STS is a local probe, while ARPES necessarily integrates information from a large area of sample surface.

6.2　Optical, Raman effect, thermal conductivity

Interband transitions make the optical absorption of graphene strong on a per-atom basis, but the overall absorption of a monolayer nonetheless is only about 2%. We will see in Chapter 9 that optical devices based solely on graphene are broadband and fast, but do not afford gain. In conjunction with quantum dots, high gain photoconductive devices are possible.

Figure 6.2 shows optical absorption spectra in the infrared, arising from holes in graphene, at three different values of magnetic field B. These normalized data were obtained by dividing spectra taken at filling factors $\nu = -2$ and $\nu = -10$. Two Landau level resonances are denoted by T_1 and T_2 and are interpreted as arising, respectively, from transitions indicated by the outer-two, and the middle-two, arrows in the inset. The dashed lines are Lorenztian fits to the data, while experimental artifacts are noted. The inset shows a schematic set of Landau levels, with allowed transitions indicated by the arrows. The data feature labeled T_2 arises from the middle two transitions in the inset; the optical absorption of graphene is indicated in Fig. 6.3.

The Raman spectra of graphene and graphite are similar. Careful comparison reveals small characteristic shifts in the Raman lines, going from single-layer graphene to multiple-layers, as illustrated in Fig. 6.4 (Ferrari *et al.*, 2006). The peak near $2700 \text{ cm}^{-1} (\sim 8065 \text{ cm}^{-1}/\text{eV})$, known as the "2D peak," arises from phonons near the K-point in the Brillouin zone [see Fig. 1.2(a)]. In the Raman process, a photon from

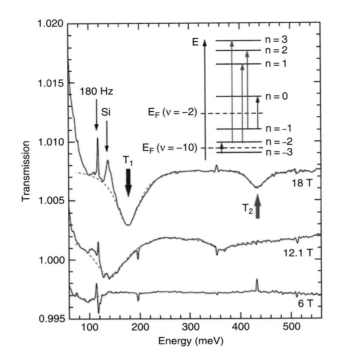

Fig. 6.2 Infrared absorption confirming Landau levels at magnetic fields B of 6, 12.1 and 19 T in graphene. The experimental setup is sketched in the inset to Fig. 8.11. (From Jiang *et al.*, 2007. © 2007 by the American Physical Society).

the 514 nm laser is absorbed and re-emitted with a shift of about two phonon energies, for single-layer graphene (Ferrari *et al.*, 2006). These energies represent characteristic "molecular" vibrations of six-fold carbon, for example a breathing mode for the benzene ring. The underlying phonon energy is then about $1350 \, \text{cm}^{-1} = 2700/[2 \times 8065] = 0.167 \, \text{eV}$. This value is close to the observed phonon energy near the K point in the neutron spectroscopy of Nicklow *et al.* (1972) (see Fig. 7.8.) The details of the structural shifts in the spectra with layer thickness are explained in some detail by Ferrari *et al.* (2006). These authors conclude that graphene's electronic structure is revealed by the Raman spectrum that clearly evolves with the number of layers. Raman fingerprints for single-, double- and few-layer graphene reflect changes in the electronic structure and electron–phonon interactions and allow unambiguous, nondestructive identification of graphene layer thicknesses. The authors point out that in AFM measurements it is, in practice, only possible to distinguish between one- and two-layer graphene if the films contain folds or wrinkles. [The AFM situation is similar to the TEM situation, revealed in Fig. 5.10(b) where the thickness is only evident where the field of view includes a fold.] A review of Raman work on graphene is given by Malard *et al.* (2008).

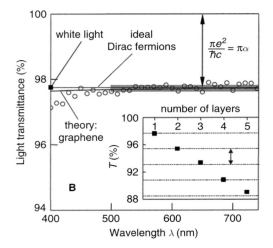

Fig. 6.3 Optical absorption of graphene is about $\pi\alpha = 2.3\%$ per layer ($\alpha = 1/137$ is fine structure constant). (From Nair *et al.*, 2008, with permission from AAAS).

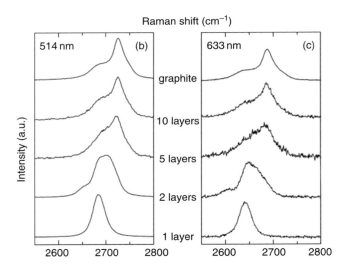

Fig. 6.4 Demonstration of small changes with layer thickness on the Raman lines observed at 514 nm and 623 nm, left to right. (From Ferrari *et al.*, 2006. © 2006 by the American Physical Society).

The thermal conductivity κ of supported graphene has been studied by Seol *et al.* (2010). Measurements on supported graphene reveal values κ near 600 W/(m−K). This value, from graphene on the traditional SiO_2 on Si substrate, exceeds that of common electronic materials, such as copper. This value κ = 600 W/m−K, however, is much lower than values in the range 4840−5300 W/m−K inferred (Balandin *et al.*, 2008) from Raman measurements on suspended graphene, and 3500 W/m−K on a single wall carbon nanotube (Pop *et al.*, 2006).

These large thermal conductivity values, exceeding those of diamond and graphite, are consequences of the strong bonding of the light carbon atoms. The reduction of thermal conductivity, for samples mounted on amorphous quartz, is attributed to phonon–substrate scattering. In more detail, phonons leak from the graphene into the substrate, and the flexural modes that are predicted to contribute strongly to the phonon conductivity, are strongly scattered by the substrate.

6.3   Scanning tunneling spectroscopy and potentiometry

The scanning tunneling microscope STM allows atomic resolution topography because the electron density outside a metal tip falls off very quickly, on an Ångström scale. In favorable cases the acting part of an STM tip may be a single atomic orbital, perhaps a 3d ($n = 3, l = 2$) atomic state that allows tunnel transitions to the metallic surface under inspection. This process is exponentially sensitive to the spacing between the two atomic orbitals, and this allows topography to be measured at the scale of Bohr radii. A good example of atomic resolution STM topography was shown in Fig. 2.12, see also Fig. 5.1. Such images are obtained by a servo mechanism that keeps the tunneling current constant as the tip is scanned across the sample, thus lifting the tip as it encounters the atom on the surface. The motion of the STM tip is controlled by piezoelectric elements that distort smoothly, on an Ångström scale, in response to applied voltages. The topographic image is represented by voltage applied to (deflection of) the z-piezo as controlled by feedback electronics to minimize the difference between the set tunnel-current and the actual tunnel-current. This keeps the tip-sample spacing approximately constant. The set-current is normally in a picoamp to nanoamp range. In the early days of STM it took some adjustment to realize that a single atom can rather easily pass a current of one nA, since it requires an electron hopping frequency through the atom of only f = $10^{-9}/(1.6 \times 10^{-19})$ = 6.25 GHz. As an atomic frequency this is small, as seen by comparison to the orbital frequency for the $n = 2, l = 1$ state of the electron in hydrogen that is 1.64×10^{15} Hz.

The tunneling tip is most frequently produced by etching a wire to a point, followed by various cleaning procedures. Often, however, simple methods for making tips are productive, such as simply cutting a 0.25 mm diameter platinum–iridium PtIr wire. The tip is normally rough on an atomic scale, but as long as it is clean it is likely to function. The tips can be selected for quality by imaging a reliable inert substrate such as cleaved $NbSe_2$ or graphene. Because of the very short range of the tunneling interaction between tip and sample, on the order of a Bohr radius, frequently only one atom contributes to the observed current. The interaction of tip and sample is similar to that of a covalent bond as described in Chapter 3.

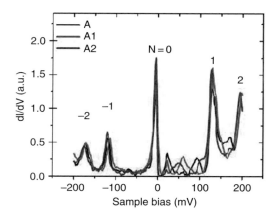

Fig. 6.5 Scanning tunneling spectroscopy of Landau levels in graphene sample at 10 T and 4.4 K. The three curves are from three separate locations on a graphene sample, indices are of the Landau level. (From Luican *et al.*, 2011. © 2011 by the American Physical Society).

An example of excellent atomic resolution STM work on graphene is that of Ugeda *et al.* (2010), who study single atomic vacancies induced in graphene by argon-ion bombardment. Each vacancy is found to be associated with a sharp peak in the density of states at the Fermi energy. The effect is discussed as indicating local magnetism.

The current in the STM experiment is influenced by the density of electronic states in the sample at the energy set by the bias voltage between tip and sample. The device thus allows measuring the density of electronic states in the sample, with energy resolution on the order of $5k_\text{B}T$, retaining the essentially high spatial resolution as described.

STM and the related spectroscopy have been used to study individual Co atoms deposited on graphene by Brar *et al.* (2011). These workers found that the valence state of the Co could be changed by the back-gate under the graphene mounted on a Si wafer. The effects on carrier mobility were discussed.

The STM energy spectroscopy is applicable to Landau level electrons in the quantum Hall effect, as there is no fundamental difficulty in making an STM operate in a high magnetic field. The interpretation of STM spectroscopy (STS) is basically simpler than inferring energy level positions from features observed in the longitudinal and transverse conductivities in a Hall geometry.

Figure 6.5 shows Landau levels in graphene as measured by scanning tunneling spectroscopy, see also Fig. 1.9(b).

6.4 Capacitance spectroscopy

The capacitance of a three-layer structure: *metal*–insulator–*graphene*–insulator–*metal* has a significant contribution from the changing density of states in the graphene. This is referred to as the "quantum capacitance"

$$C_\text{q} = e^2 D(E) \,, \tag{6.2}$$

where

$$D(E) = \mathrm{d}n/\mathrm{d}E \qquad (6.3)$$

at the Fermi energy E_F (Luryi, 1988; Eisenstein *et al.*, 1994; John *et al.*, 2004). It is closely related (Eisenstein *et al.*, 1994) to the electronic compressibility K, defined as

$$K^{-1} = n^2 \mathrm{d}E_\mathrm{F}/\mathrm{d}n, \qquad (6.4)$$

where n is the number of carriers per unit area and E_F is the Fermi energy or chemical potential.

The structure and its equivalent circuit are indicated in Fig. 6.6, following Luryi. The middle plate Q is assumed to allow partial penetration of electric field, so that the electric field due to charges on the top plate also has an effect on charge of the bottom plate. The quantum effect involved in the 2D middle plate, the quantum capacitance, is a consequence of the Pauli principle that requires an extra energy for filling such a 2D metal with electrons, this is, for example, the conical density of states in graphene.

A slightly more general description is given by John *et al.* (2004). These authors give an equation for the charge Q due to electrons and holes of charge q on the quantum plate of density of states g(E) (written in the more general case that the quantum plate may have an energy gap E_G):

$$Q = q \int \mathrm{g}(E) \left[\mathrm{f}(E + E_\mathrm{G}/2 + qV_\mathrm{a}) - \mathrm{f}(E + E_\mathrm{G}/2 - qV_\mathrm{a})\right]\mathrm{d}E. \qquad (6.5)$$

Here f(E) is the Fermi function and the integration over energy E is from 0 to ∞ with V_a the applied voltage. With this definition of Q, the quantum capacitance is

$$C_\mathrm{Q} = \partial Q / \partial V_a. \qquad (6.6)$$

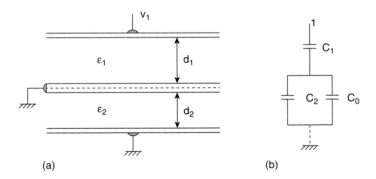

(a)   (b)

Fig. 6.6 (a) Schematic illustration of a three-plate capacitor where the middle plate represents a two-dimensional metal, such as graphene. (b) Equivalent circuit for the capacitance measured at the top (labeled V$_1$). (Reprinted with permission from Luryi, 1988. © 1988 AIP).

Devices of this sort using graphene have been measured by Ponomarenko *et al.* (2010), following earlier work by Giannazzo *et al.* (2009), Xia, *et al.* (2009), see also Droscher *et al.* (2010).

In the work of Ponomarenko *et al.* (2010), an extremely thin 10 nm Al_2O_3 insulator was provided between the graphene (mounted on an oxidized high resistivity silicon wafer) and an upper aluminum electrode. The graphene monolayer was exfoliated in the Scotch tape method and was provided with Ti/Au electrodes for use in the capacitance measurements. It was determined that the graphene retained high mobility after the application of the oxide and Al upper electrode, and that the Al top gate allowed variation of the Fermi energy in the graphene by as much as ± 0.5 eV. The devices with upper Al electrodes were found to have Dirac point (neutral point) close to zero gate-voltage, indicating native carrier concentration $n \leq 10^{12}$ cm^{-2}. The raw data were capacitance in units μF/cm^2 plotted vs. V_G (top gate voltage) in range ± 3 V. The capacitance was assumed to arise as

$$1/C = 1/C_{\text{ox}} + 1/C_Q \tag{6.7}$$

and the oxide capacitance was identified by a fit to the capacitance curve that showed a marked minimum around 0.32 μF/cm^2 at $V_G = 0$ and an asymptotic value at high voltage about 0.47 μF/cm^2. The expected behavior was observed to be fit by only two parameters, the Fermi velocity in the graphene and the oxide capacitance, and accepted value $v_F = 1.15 \times 10^6$ m/s was found. With this fit established, the data were plotted as shown in Fig. 6.7 in the conceptually appealing format of quantum capacitance (proportional to density of states) vs. E_F. This was done by determining the carrier concentration n for a given gate voltage. The data curve is cut off near zero

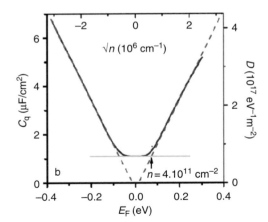

Fig. 6.7 C_Q as a function of Fermi energy (bottom) and $n^{1/2}$ (top) at 10 K and zero magnetic field. The right scale plots the density of states $D = C_Q/e^2$ in units of 10^{13}/eV cm^2. (From Ponomarenko *et al.*, 2010. © 2010 by the American Physical Society).

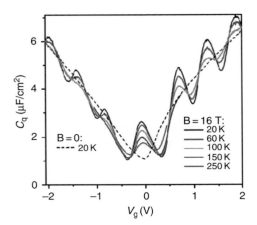

Fig. 6.8 Landau levels observed in capacitance spectroscopy. C_Q as a function of back gate voltage at several temperatures and magnetic fields zero and 16 T. The zero index Landau level is seen clearly at temperatures up to 250 K. (From Ponomarenko *et al*., 2010. © 2010 by the American Physical Society).

Fermi energy, as apparently, the sample has a minimum carrier concentration, due to "electrostatic puddling" around 4×10^{11} cm^{-2}.

Having carefully established the capacitance spectroscopy method, Ponomarenko *et al*. have applied it to provide an additional determination of the Landau level structure in graphene at 16 T and several temperatures ranging from 20 K to 250 K, as shown in Fig. 6.8. The anomalous zero-energy Landau level is clearly shown in these data.

6.5   Inverse compressibility with scanning single electron transistor (SSET)

The "inverse compressibility" [eqn (6.4)] is an unfamiliar quantity, yet closely related to the electronic density of states. It can be measured using scanning single electron transistor (SSET) microscopy. This technique was developed by Yoo *et al*. (1997) and now applied to graphene. The SSET device is capable of mapping static electric fields and charges on a surface with spatial resolution about 100 nm and charge sensitivity a small fraction of an electron charge. The device operates at low temperature and can be called a scanning electrometer.

The single electron transistor (SET) sensor device is mounted at the end of a sharp glass tip of diameter around 100 nm. Following Yoo *et al*. (1997), the SET measures the local surface potential V_s (closely related to the chemical potential μ). The surface potential V_s arises from the work function difference between surface and tip and also from charged centers distributed across and below the surface.

In more detail, the SET consists of a source and drain, connected by an "island" sensing electrode of such small size and capacitance that adding a single electron charge requires a charging energy $e^2/2C$ exceeding the thermal energy $k_{\mathrm{B}}T$. Here C is the total capacitance of the island with respect to its surroundings, including source, drain and surface. If the net charge Q on the island is $e/2$, or an odd multiple of $e/2$, the transistor is "on," while with integer net charge on the island the transistor is "off." Being off means that the Coulomb blockade occurs, electrons cannot be exchanged with source and drain because of the Coulomb energy barrier $e^2/2C$. The net charge Q on the island arises in part from an external electric field, E that induces on the island a charge density $\sigma = \varepsilon_0\, E$ in Coulombs/m^2, with ε_0 the permittivity of space. The electric field between SET "island" and the surface is essentially the potential difference between surface and SET island, divided by the spacing. The authors estimate that a single electron directly under the tip, if the tip-surface spacing is 25 nm, will induce up to 0.1 electron charge on the SET island. Under these conditions, that require cryogenic temperature, the source-drain current, under fixed source-drain bias, is approximately

$$I_{\mathrm{SET}} = A\,\sin(2\pi Q/e) = A\,\sin[2\pi C_{\mathrm{s}}(V_{\mathrm{B}} + V_{\mathrm{S}})/e]\,, \tag{6.8}$$

although full SET theory gives a triangle-wave in place of the sine-wave. This SET current is oscillatory in Q with period one unit e, and in the measurements 0.01 e is detectable. The island-charge Q can be written as $C_{\mathrm{s}}(V_{\mathrm{B}} + V_{\mathrm{S}})$, where C_{s} is the tip-surface capacitance that is primarily set by the spacing, and normally held constant in scanning. V_{B} and V_{S}, respectively, are an applied tip-sample bias and the surface potential of interest. One mode of operation is to scan the tip at fixed height and use a feedback circuit adjusting the applied potential V_{B} to maintain constant SET current. An application of this method to study quantum Hall states is given by Yacoby *et al.* (1999).

In adapting the scanning SSET electrometer to study graphene, it was natural for Martin *et al.* (2009) to use sinusoidal modulation of the carrier concentration n by use of the back-gate. These authors note that while the SSET scanning electrometer measures and maps the local electrostatic potential V_{S} as described above, this is equivalent to measuring and mapping the local chemical potential μ. This follows because, in equilibrium, any change in μ is compensated by a corresponding change in V_{S}. The graphene measurement then is of μ and, by modulating the back-potential, the quantity $\mathrm{d}\mu/\mathrm{d}n$. The quantity $\mathrm{d}\mu/\mathrm{d}n$ is inversely proportional to the local electronic compressibility and directly related to the local density of states.

The modern version of the technique as practiced by Martin *et al.* (2009) can map the chemical potential directly or obtain it by integrating the inverse compressibility $\mathrm{d}\mu/\mathrm{d}n$, directly related to the density of states. Color gray scale images of $\mathrm{d}\mu/\mathrm{d}n$ expressed in units of 10^{-10} meV cm^2 are given by Martin *et al.* (2009) and in the most recent paper based on the scanning SET device of Feldman *et al.* (2012). An example of such data is a "Fan diagram," relevant for the quantum Hall effect that is a grey scale image of the inverse compressibility (at a fixed sample location) vs. magnetic field B and carrier density n. We will return to discovery of fractional quantum Hall effects, using these methods, in monolayer graphene, in Section 8.4.

In the modern use of the scanning SET for graphene, the scanning capability is typically used to find and discriminate between ideal and defective regions in a sample, rather than to image surface charge. Indications of local variations in Landau levels were shown in Fig. 6.5, by Luican *et al.* (2011), while a conventional Hall-geometry measurement of Landau levels averages over the whole sample. The ability to exclude data from defective portions of a sample is valuable.

The data in Fig. 1.9(a) are due to Martin *et al.* (2009) using a scanning single electron transistor SSET, showing the Landau levels in monolayer graphene. The measured quantity is $d\mu/dn$ where μ is the chemical potential and n is the density of carriers.

This chapter has highlighted some of the experimental techniques used in graphene research, but is not an exhaustive survey. We have pointed out several lesser-known methods that have been applied to graphene.

7
Mechanical and physical properties of graphene

The mechanical behavior of graphene depends upon the size of the sample, as noted at the beginning of Chapter 1. If the sample has a linear dimension ten to one hundred times its thickness it appears to be extremely rigid. Of course this is an unavailable linear scale for ordinary observation, and on any conventional scale graphene is extremely soft. In micrometer sizes, of linear length at least 10^4 times its thickness, it readily conforms to any surface that it is placed upon, as suggested by Fig. 1.1. Electrically, graphene is highly conductive and capable of current density far in excess of any metal. Its covalent character allows the single layer to be atomically smooth and continuous, features unattainable in metals that break into islands when reduced in thickness. We begin our discussion with mechanical properties.

7.1   Experimental aspects of 2D graphene crystals

Single crystals of graphene have been found by Bunch $et\ al.$ (2007) to behave as would be expected for an elastic solid whose Young's modulus Y is nearly one TPa but whose thickness is 0.34 nm. [Nuances on these values were discussed in Section 2.7.1, following eqn (2.63).] Even though the material is inherently among the strongest known, because of the extreme thinness, its rigidity $\kappa = Yt^3$ is severely diminished. The frequency of oscillation of a cantilever (singly clamped beam) of thickness t and length L is predicted accurately by the conventional formula, as confirmed for graphene by Bunch $et\ al.$ (2007) with ρ the mass density,

$$\nu = 0.162\,(Y/\rho)^{1/2}\,t/$$

(7.1)

In case of a doubly clamped beam (the form used in most of their experiments) the prefactor 0.162 is replaced by 1.03.

This experimentally verified formula for frequency of oscillation can be put into the form

$$\nu = (2\pi)^{-1}\,(K^*/m)^{1/2}$$

(7.1a)

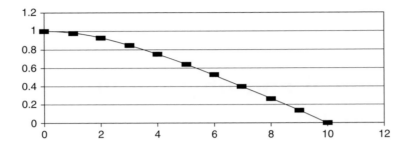

Fig. 7.1 Sketch of deflection (vertical deflection multiplied by factor 5) of 66 μm long graphene cantilever under uniform gravitational acceleration $9.8\,\mathrm{m/s^2}$. This deflection would apply also to a square of side L. Assumptions are $Y = 1\mathrm{TPa}$, density $\rho = 2270\,\mathrm{kg/m^3}$, and functional form for displacement $\zeta(x) = qx^2\left(x^2 - 4Lx + 6L^2\right)/24YI)$, (Landau and Lifshitz, 1959, p. 91) where $q = \rho gtw$ (t and w, respectively, the thickness and width of the cantilever) and $I = t^3 w/12$. (Courtesy Ankita Shah).

to deduce an effective spring constant K^* for a slab, of dimensions L, w, t, clamped at one end. As discussed below [eqn (7.17) and Section 7.4], K^* is proportional to Ywt^3/L^3. As a conventional measure of rigidity of a sample of graphene with $t = 0.34$ nm, we can ask for the side L' of a square of this material such that if one picks it up with a tweezers, the force of gravity would make its opposite side droop by $L'/10$. This size is on the order of 50 μm = 0.05 mm. This small size suggests, indeed, a very soft material that must always be supported except in applications on a micrometer or smaller scale.

This value is confirmed by the result from Landau and Lifshitz (1959) p 91 that the displacement $\zeta(L)$ of the end of a horizontal cantilever of length L under uniform gravitational force g is

$$\xi(L) = 1.5\ g\rho L^4/(t^2 Y), \tag{7.2}$$

where Y is Young's modulus, t the thickness, g the gravitational acceleration, and ρ the density. Evaluating this formula using $L = 66$ μ gives a deflection of 6.5 μ. A sketch of this deflection is shown in the Fig. 7.1, where the vertical scale is enhanced by a factor of 5.

While the shortness of the section to droop by 10% under gravity indicates the softness of the graphene sheet, if one considers that the length of the sheet is more than 10^5 times its thickness, 0.34 nm, this brings to mind its rigidity when considered on a small scale.

The fractional deformation rises as the cube of the flake size. On this basis, single layers of dimension larger than 0.1 mm would droop in gravity to behave more like saran wrap than a playing card. The thickness and not the inherent Young's modulus, is the origin of the limp nature of larger samples of graphene. The stiffness of graphene comes back into view, when we recognize that the vibrational frequency of the 0.05 mm square, if clamped on one side, would be ~424 Hz, using the formula above with parameters $Y = 1\text{TPa}$, $\rho = 2270$ kg/m^3.

This prediction is supported by experiment, in the measured frequency, 70.5 MHz, for a 1.1 μm long doubly clamped graphene beam (Bunch *et al.*, 2007), mentioned in Section 7.4. Bunch *et al.* (2007) adjusted this measured value to 5.4 MHz in consideration of strain due to the mounting, and then to 0.849 MHz to convert from a doubly clamped beam to a cantilever. That conversion is 6.35, compared to a factor 4 that one would get simply by changing the length by half. Then extrapolation to a length 50 μm yields 411 Hz, in good agreement.

Discussions of measurements on larger size graphene sheets leave implicit that measurement would have to be in good vacuum, to avoid the viscous damping of air, and even the effect of molecular fluctuations, on the movement of the sheet. The sheet is invisible to the eye. The effect of van der Waals attraction to the mounting is neglected.

A sample of graphene is seen to conform to the shape of a substrate, see Fig. 1.1, from the van der Waals attraction. The force per unit area can be estimated, between planar graphene and a planar substrate spaced by height h, as $A/6\pi h^3$ with the Hamaker constant $A \sim 10^{-19}$ J. It appears that a force comparable to the gravitational force would require a spacing h on the order of 10 μm, so that at 1 μm spacing the van der Waals force would be 1000 times larger than the gravitational force. It appears reasonable that the membrane would conform to its substrate on these grounds. To summarize these comments it seems that the word "sheet" is more appropriate than "crystal" or "flake" for the single extracted planes, if the size 0.1 mm or larger. At the scale of 50 micrometers, solid state physics is accurate to estimate the electronic bands, because a sample of dimension 50 microns is about 10^{10} unit cells. The band theory of solid state physics will still be useful for sheets of much smaller extent, even down to 100 atoms on a side, or about 14 nm.

Thermally grown (3D) graphite crystals are commonly millimeters in diameter and perhaps 0.1 mm in thickness. Such a single crystal flake ("Kish" graphite) grows in carbon-saturated molten iron in the processing of steel. It is not believed that single graphene sheets can grow directly, under these conditions, without a 3D support. It is still possible that graphene might be grown from a linear array of Fe or Ni nanoparticle catalysts, in analogy to the successful growth of carbon nanotubes in a chemical vapor deposition apparatus. We have seen earlier (Section 5.2.2) that graphene grows readily on a Cu or Ni surface with inflow of carbon-bearing gas, such as methane, at temperature near 1000°C. This 2D crystal grows spontaneously with no relation between its axes and those of the Cu support, and a discussion of the nucleation and other aspects is given by Chen *et al.* (2012).

7.1.1 Classical (extrinsic) origin of ripples and wrinkles in monolayer graphene

Graphene is robust with respect to bending: the structure does not fracture but easily bends. This is clear from the existence of carbon nanotubes, a family of hollow cylinders, with diameters in the low nm range, equivalent to the result of bending graphene around various directions, preserving six-fold bonding. ("Bending" of graphene into spheres "Buckyballs" requires the addition of five-fold rings.) The diameter of a carbon nanotube is

$$d = (a/\pi) \left(n^2 + m^2 + nm\right)^{1/2}.$$

(5.10)

Here n, m are integers describing the axis vector about which the graphene is rolled in concept. Here $a = 246$ pm is the lattice constant of graphene (but, see Section 7.1.3, this is slightly temperature dependent, reflecting an anomalous negative coefficient of expansion, due to flexural phonon modes, to be discussed). The two characteristic nanotube types are "armchair" ($m = n$), that are metallic and "zigzag ($m = 0$) that are semiconducting. It is suggested that nanotubes can be "unzipped" with a current pulse to lay flat the corresponding "GNR" (graphene nanoribbon). (In such a process strain energy would be released at the expense of lost bonding energy, the edge atoms would no longer have strong covalent bonding partners.) The nomenclature extends to graphene nanoribbons, with edges of the two limiting types. A well-studied nanotube is the (10, 10) tube with diameter $(a/\pi) \, 300^{1/2} = 1.35$ nm reported [see the excellent paper of Krishnan *et al.* (1998)] to have Young's modulus 1.25 TPa.

The literature speaks of the "Euler buckling instability"; see for example Golubovic *et al.* (1998) and references therein. As we have seen, graphene is stable in a great variety of rolled and rippling configurations, and "buckling" in the context of graphene does not mean a fracture but rather a smooth elastic distortion into a sinusoidal configuration. Graphene resists compression up to a critical strain about 1.25%, [according to Kumar *et al.* (2010), see Fig. 7.11 and text following eqn (7.20)] but much before that point it does not fracture but simply goes into one of the many oscillatory buckling shapes extending into the direction perpendicular to the graphene plane.[1]

The fundamental "buckled" state is a sinusoidal wave identical in configuration to the "flexural phonon" shown in Fig. 7.9 below. The wavelengths of sinusoidal variations that are observed vary from 2.4 nm [for Graphene grown on Ru(0001), Vazquez de Parga *et al.* (2008)] to micrometers. In the epitaxial growth on Ru, the graphene adjusts its ripple wavelength so that 11 carbon hexagons (each 0.246 nm) will match almost exactly 10 Ru–Ru interatomic distances (0.27 nm). (A slightly different observation is that of Martoccia *et al.* (2008.) The case of Fe(110) is discussed by Vinogradov *et al.* (2012). The ripples in these extrinsic cases are the result of confinement of the graphene layer by its boundary conditions.

The range of (substrate- or strain-induced) ripple wavelengths is characterized as 2 nm to 5 μm, dependent upon substrate, by Mao *et al.* (2011), using graphene exfoliated by sonication in an organic solvent (see Section 5.1.2, text following Fig. 5.4).

These workers actually observed a ripple wavelength of 0.64 nm, about three unit cells of the hexagonal lattice, using transmission electron microscopy (TEM) on a

---

[1]Such buckling is inhibited by an attractive force to the support, as occurs in growing graphene on SiC.

free-standing graphene bilayer affected by boundary conditions. In another case, a wavelength 0.36 nm was found in a folded monolayer near the folding edge. These authors find evidence that, in a structure with ripples, the bending of the structure in the perpendicular direction is made more difficult. (See the discussion of Gaussian curvature, in this regard, in Sections 2.7.2 and 2.7.3.) A buckling amplitude of 0.12 nm was measured, so that the curvature of the sheet, if the period is only 0.36 nm, is large.

The ripples that are observed are analogs of familiar macroscopic effects with nearly inextensible flexible sheets, discussed in Chapter 2, Section 2.7.3. Curtains hung above a window are seen to be *puckered* (compressed) at the attachment point, with waves spreading apart downward. A rubber sheet if pulled in tension will, at some point, break into a pattern of ripples along the direction of tension. A sheet if sheared in two dimensions by its mounting geometry will develop wrinkles or ripples in the direction between the two shearing supports. All of these phenomena appear in graphene, seen in the experimental work of Bao *et al.* (2009).

Questions have been raised whether the unstrained graphene sheet spontaneously develops wrinkles, and this largely theoretical topic is discussed in Section 7.2. In summary, however, the answer is almost certainly no that the observed ripples are the result of nearly unavoidable strain in any mounting and the nearly unavoidable decoration of the graphene with adsorbates that may locally weaken bonds and allow release of compressional strain in out-of-plane deformation.

Even small wavelength ripples preserve the honeycomb lattice by small distortions of bond angles, but not of bond lengths. An STM image (K. Xu *et al.*, 2009) is shown in Fig. 7.2 illustrating slightly distorted bonding of carbon atoms on the top of a short wavelength ripple.

The ripples shown in Fig. 7.3 are observed by atomic force microscope AFM on graphene single- and multi-layers exfoliated from graphite and laid across trenches in

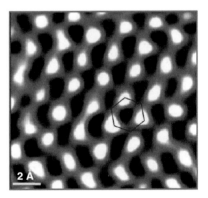

Fig. 7.2 STM image on top of wrinkle (of height 2.5 nm and width 14 nm) inscribed hexagon has side 142 pm, called "3 for 6 pattern." (Reprinted with permission from Xu *et al.*, 2009. © 2009 American Chemical Society).

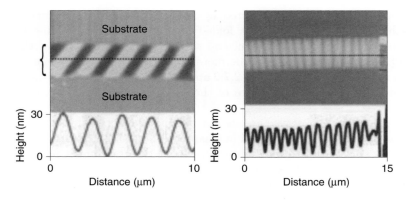

Fig. 7.3 Classical ripples due to mounting strain for graphene draped across a trench. (From Bao *et al.*, 2009, by permission from Macmillan Publishers Ltd., © 2009).

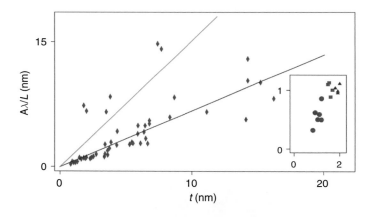

Fig. 7.4 Analysis of classical ripples due to mounting strain for graphene draped across a trench. Inset amplifies region near zero thickness. (From Bao *et al.*, 2009, by permission from Macmillan Publishers Ltd., © 2009).

SiO_2/Si substrates. As mentioned above, it is found that the single graphene layer ripples are similar to ripples observed in much larger systems. Classical behavior is retained in the limit of small thickness t, of a family of membranes, up to 20 nm in thickness, as seen in Fig. 7.4, due to Bao *et al.* (2009).

Taking $0 < x < L$ as the direction across the trench, y along the trench, the sinusoidal ripple is described as

$$\zeta(y) = A \sin(2\pi y/\lambda). \tag{7.3}$$

The classical theory, as applied to graphene by Cerda and Mahadevan (2003), is adapted by Bao *et al.* (2009). Bao *et al.* use relations between the wavelength λ, ripple amplitude A, thickness t, Poisson ratio ν (see Eqs. 2.59–2.73), and impressed longitudinal tensile strain γ. These are

$$\lambda^4/(tL)^2 = 4\pi^2/\left[3(1 - \nu^2)\gamma\right] \tag{7.4}$$

and

$$A^2/\nu tL = \left\{16\gamma/\left[3\pi^2\left(1 - \nu^2\right)\right]\right\}^{1/2}. \tag{7.4a}$$

Combining these two equations, to eliminate the strain variable γ, Bao *et al.* (2009) find two different expressions directly compared with the data (upper and lower solid curves in Fig. 7.4):

$$A\lambda/t = \left\{8/\left[3(1 + \nu^2)\right]\right\}^{1/2} t \tag{7.5}$$

(upper curve, based on assumption of shear strain), and

$$A\lambda/t = \left\{8\nu/\left[3(1 - \nu^2)\right]\right\}^{1/2} t \tag{7.6}$$

(based on assumption of tensile strain, lower curve).

It is seen in Fig. 7.4 that, of the 51 samples studied by Bao *et al.* (2009), with thicknesses t from 0.34 nm (single layer) to 18 nm, most lie on the lower line. This suggests that the ripples are due to tensile strain in the thin membranes under the measurement conditions. The two lines are drawn using the value $v = 0.165$ for the basal plane of graphite. The inset in Fig. 7.4 expands the scale at low thickness t. The data here are in three groups, the lowest point representing a graphene monolayer, the middle cluster of six points are bilayers, and the seven highest points assigned to trilayers.

Figure 7.5 shows a micrometer-scale image of draped graphene showing ripples with cusps or corners at their terminations. (The theoretical background for these classical effects was discussed in Section 2.7.) These features are classically expected responses of an inextensible flexible membrane to strains imposed by boundary conditions, and thus are not reasonably classified as unusual behaviors unique to graphene.

It is remarkable that these effects, clearly seen in Fig. 7.5, appear at one atomic thickness just as in macroscopic cases. The behavior is that of a sheet that cannot be stretched (inextensible) but can easily be bent. Both assumptions are reasonable under the conditions of the experiments, but at higher stress indeed the graphene becomes remarkably extensible [elastic, 15% to 25% tensile strain is tolerated (C. Lee *et al.*

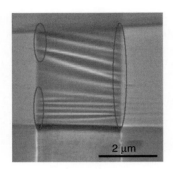

Fig. 7.5 Micrometer scale image of graphene draped across a trench. The wrinkles are classically expected responses of an inextensible flexible membrane to boundary conditions. They may be temperature-dependent by virtue of the anomalously negative temperature coefficient of graphene, vs. the positive coefficients of supporting structures. The circles show the cusps, regions of high elastic strain, but not plastic deformation, at the ends of the ripples. In the literature of this subject the word wrinkle appears (yet the strains in these cases are not irreversible) and the word cusp is replaced by "conical singularity, CS". (From Pereira *et al.*, 2010. © 2010 by the American Physical Society).

(2008)], yet less so than a rubber sheet (that may accept 100% elastic strain) and also has small inherent resistances to bending and to compression (see Figs. 2.15 and 2.16 and the related text).

Figure 7.6 shows development of ripples under strain imposed by thermal expansion of the mounting with change in temperature.

Apart from the observed slipping of the film, the relative compression between the second and third panels in Fig. 7.6, at coefficient of expansion $-7 \times 10^{-6}\,\mathrm{K}^{-1}$ is 0.21%, the authors state that ripples always appear after the anneal, thus under compressive strain, and may or may not be accompanied by buckling of the membrane. [This value of the coefficient of expansion measured by Bao *et al.* (2009) has been confirmed by Chen *et al.* (2009), who find $-7.4 \times 10^{-6}\,\mathrm{K}^{-1}$ and by Yoon *et al.* (2011), who find $-8 \times 10^{-6}\,\mathrm{K}^{-1}$ using Raman spectroscopy.]

The data in Fig. 7.6 show that the film is stable at 600 K and shows no evidence of ripples unless called for by the boundary conditions, as in the classical cases.

7.1.2 Stability of graphene in supported samples up to 30 inches

We have seen in Chapter 5 that 30-inch monolayer supported films of high quality graphene have been used to make panels for touch screen devices. Such large useful 2D samples may be surprising in view of a theoretical literature that suggests instability. We now turn to bridge the theory to experiment.

As we have just seen, the data of Bao *et al.* (2009) are quite straightforward, and leave no need to posit "inherent" ripples or wrinkles in the films.

Fig. 7.6 Demonstration of ripples resulting from strain induced by mounting. Sample shown by SEM picture (initially at 300 K in left upper panel) in situ annealing to 600 K (center upper panel), erasing ripples, with ripples appearing upon cooling to 300 K accompanied by drooping of the membrane downward into the trench. At 600 K the sample is under tension, since the temperature coefficient of expansion of graphene is large and negative, while that of Si is positive. The authors state that in the annealing stage, the taut film slips along the Si support, so that when it is cooled again, it is under compressive stress. Then it buckles downward and develops ripples, as suggested in the lower two schematic panels. It is important to note the evident stability of the film across the trench. This is an extended crystal free of ripples that is stable at 600 K. If the film is 3 μm long, and the cell length is 0.246 nm, the film is a single crystal containing 12 million unit cells along one direction. (From Bao *et al.*, 2009, by permission from Macmillan Publishers Ltd., © 2009).

Following Castro Neto *et al.* (2009), the number of flexural mode phonons per unit area is given by

$$N_{\mathrm{ph}} = (2\pi)^{-1} \int k \, \mathrm{d}k \left[\exp(ak^2) - 1 \right]^{-1} \tag{7.7a}$$

where the integral over wave-vector k is from 0 to ∞, the inverse bracket factor is the Planck or Bose–Einstein occupation function f_{BE} and

$$\alpha = \hbar(\kappa/\sigma)^{1/2}/(k_B T). \tag{7.7b}$$

In this expression $\kappa = Yt^3$ (with Y Young's modulus and t the film thickness) is a bending stiffness about 1 eV and σ the mass per unit area (about $7.5 \times 10^{-7}\mathrm{kg/m^2}$).

The Planck function, the thermal occupancy of the mode at frequency ν,

$$f_{\mathrm{BE}} = 1/[\exp(h\nu/k_B T) - 1] \tag{7.7c}$$

approaches $\exp(-h\nu/k_\mathrm{B} T)$ at high energy and approaches $k_\mathrm{B} T/h\nu$ at low energy, where $k_\mathrm{B} T/h\nu$ would be the number of phonons in the mode at frequency ν. The average energy in that mode at high temperature would then be $k_{\rm B}T$. This equation is integrated over all k to find the number of phonons per square meter. Classical theory predicts that for a solid at high temperature the thermal energy is $3Nk_{\rm B}T$, where N is the number of atoms, each with three modes of distortion potential energy and three modes of kinetic energy.

At small wave-vector k, the integrand of (7.7a) approaches $1/\alpha k$, but note that the corresponding integral for the lattice energy per unit area, is convergent. That is, by taking the phonon energy as $E = a'k^2$, the integral is

$$U_{\rm ph} = (2\pi)^{-1} \int a'\, k^3 dk [\exp(ak^2) - 1]^{-1}. \qquad (7.8)$$

Since the integrand approaches k as k goes to zero, the integral is convergent for any sample size L.

In the integral for $N_{\rm ph}$ (Eq, 7.7a), the integrand diverges as k approaches zero, so that the number of phonons per unit area is infinite for an infinite size sample. (This does not mean that the vibrational energy per unit area diverges.) Adopting a cutoff on small k: $k_c = 2\pi/L$, for sample size L, the result becomes

$$N_{\rm ph} = 2\pi/L_{\rm T}^2 \, \ln(L/L_{\rm T}), \qquad (7.7d)$$

where

$$L_T = 2\pi \, \hbar^{1/2} \, (\kappa/\sigma)^{1/4}/(k_{\rm B}T)^{1/2}. \qquad (7.9)$$

This estimate may appear as alarming because the numbers for 300 K give $L_{\rm T}$ about 0.3 nm, around one lattice constant. Castro Neto *et al.* (2009) state that this means that "free-floating graphene should always crumple at room temperature due to thermal fluctuations associated with flexural phonons." They go on to suggest that "renormalization" of bending by anharmonicity may be needed, although the "bare" parameters κ and σ of their analysis are confirmed by experiments. Crumpling of graphene at room temperature, of course, does not occur, and we have seen that the thermal energy per unit area is finite for an infinite sample. To seek reconciliation we need to examine more carefully the physical implication of the small value of $L_{\rm T}$ that appears to be correct, and to see why it is in fact not alarming.

First a few sentences may be useful on the nature of the predicted "crumpled" phase.

Crumpling requires large displacements of much of the sample, so that it fills a volume rather than lying flat with modulations. Crumpling in the literature is also associated with partial melting, loss of local order, weakening the membrane, but the crumpled phase is of low density, viewed in three dimensions. A review of transitions between flat and crumpled phases is given by Radzihovsky (2004).

Recall also the earlier mentioned prediction of Peliti and Leibler (1985) of a coherence length, with a the lattice constant,

$$\xi = a \, \exp\left(4\pi\kappa/3k_{\rm B}T\right) \qquad (7.10)$$

setting the size of a sample before "crumpling" will occur. This size [eqn (7.10)] at 300 K, as we saw earlier, is very large indeed, but it falls to around 4.7 micrometers at 4900 K, where the recent simulations suggest (see Chapter 8) that graphene will disintegrate.

So perhaps the "crumpling," rather, suggests that the large unsupported floppy sheet will have large displacements, when its size L increases, but the local properties in the sheet are stable.

Assuming the analysis leading to the Castro Neto estimate $L_T = 0.3$ nm is correct, it seems reasonable to examine the practical consequences of a small value of L_T. We choose a "large" sample, a platelet 100 μm on a side. For a 0.1 mm $=$ 100 μm size sample, the density of phonons per unit area is predicted here as $N_{ph} = 2\pi/L_T^2 \ln(L/L_T) = 7 \times 10^{19} \ln(3.3 \times 10^5) = 8.87 \times 10^{20}$ m^{-2}. Following the Castro Neto estimate, the platelet thus contains 8.87×10^{12} phonons. In the platelet, we can see that the energy range of the phonons starts at about 1.46×10^{-12} eV. To find this, we assume the minimum phonon energy occurs for $k_c = 2\pi/L$, with $L = 10^{-4}$ m, so $q = k = 6.3 \times 10^4$ m^{-1}. If we assume $E = a'k^2$, with a' a constant, for the flexural mode, and use the data of Nicklow *et al.* (1972) that gives (their Figure 6) the flexural phonon frequency as 2 THz at $q = 4.73 \times 10^9$ m^{-1}, then the phonon frequency at $k = 6.3 \times 10^4$ m^{-1} is 2 THz $\times (6.3 \times 10^4/4.73 \times 10^9)^2 = 355$ Hz. The energy for this phonon is thus 1.46×10^{-12} eV. If we assume that all phonons have this energy, the thermal vibrational energy of the platelet is only $8.87 \times 10^{12} \times 1.46 \times 10^{-12}$ eV $= 12.95$ eV. The maximum flexural frequency is given by Nicklow *et al.* (1972) measurements as about 14 THz. If all phonons were taken at this value, a large overestimate, the thermal energy of the platelet would be 511 GeV. (The correct value, as we will see shortly, is about 4 GeV.)

A realistic estimate for the thermal energy at 300 K for the assumed 0.1 mm platelet can be made from the calculation of Mounet and Marzari (2005), whose calculated specific heat curves are shown in Fig. 7.7. (There is no difficulty in calculating the thermal energy, and indeed the calculations [below] accurately predict the properties of the flexural phonon modes.) The dashed curve in Fig. 7.7, for graphene, when extrapolated to 300 K, gives an area under the curve of 84 234 J/kg from $T = 0$ to $T = 300$ K. Considering the mass of the assumed platelet as 7.7×10^{-15} kg, the thermal energy is 6.5×10^{-10} J $= 4.05$ GeV. (The simplest estimate of the internal energy at 300 K, based on $3k_B T$ per atom, gives 29.8 GeV for the platelet.)

We can then estimate the average phonon energy by dividing the platelet thermal energy by the number of phonons estimated from the Castro Neto formula. This is $4.05 \times 10^9/8.87 \times 10^{12}$ eV $= 0.46$ meV.

The conclusion is that the platelet is stable even with the quadratic dispersion of the flexural phonons near zero wavevector. It is also suggested [see Kumar *et al.* (2010), and Fig. 7.12 below] that the underlying assumption

$$\omega = a'k^2 + b\,|k| \quad (b = 0) \tag{7.11}$$

can be altered in the presence of strain to allow b nonzero that would remove the divergence for the infinite sample.

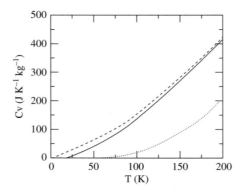

Fig. 7.7 Calculated specific heat for graphene and graphite (dashed and solid curves, respectively) and for diamond (dotted curve). The graphene specific heat is larger at low temperature because it has a larger density of the low frequency flexural phonons. Diamond has a lower specific heat at low temperature because it has no soft phonon modes. (From Mounet and Marzari, 2005. © 2005 by the American Physical Society).

7.1.3   Phonon dispersion in graphene

There is a high quality literature on the vibrational spectrum of graphite, and much of it is applicable to graphene. A good review is offered by Wirtz and Rubio (2004).

The graphene layers in graphite are separated by 0.34 nm and are weakly coupled by van der Waals forces, so the properties of graphene closely resemble the in-plane properties of graphite. Neutron diffraction data of Nicklow *et al.* (1972) for graphite, together with theoretical curves *for graphene* (solid lines) of Mounet and Marzari (2005), are shown in Figure 7.8. Very similar phonon spectra from inelastic x-rays for graphite are given by Mohr *et al.* (2007). The lowest curves in Fig. 7.8, labeled ZA, are the transverse acoustic or flexural modes. Nicklow *et al.* in their 1972 paper explicitly state that the ZA modes (TA_{perp} in their notation) require both linear and quadratic terms (see their Figure 6) to fit the graphite data, stated (in Nicklow eqn 2) [with reference to Komatsu (1958)] as

$$\nu^2 \propto Aq^2 + Bq^4. \tag{7.12}$$

This form is discussed by Zabel (2001), who identifies the coefficient A with the shear interaction between layers in graphite. Specifically, $A = C_{44}/\rho$, the shear coefficient divided by the mass density. This implies $A = 0$ for an isolated layer, but it is not entirely clear that this is true. $A = 0$ is also argued by Saito *et al.* 1998 on the basis of the D_{6h} point group symmetry of graphene, but the possibility of a dependence $\nu \propto A\,|q|$ is mentioned.

The Mounet and Marzari (2005) calculations for graphene seem to require something more than a pure quadratic term, and a dependence on $q^{3/2}$ as q goes to zero is derived for the flexural mode of the graphene monolayer by Mariani and von Oppen (2008).

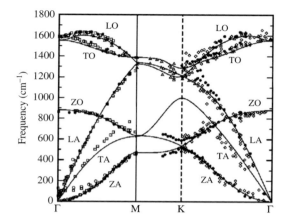

Fig. 7.8 Phonon dispersions (solid lines) for graphene, with data points measured in graphite. The lowest curves, marked ZA are the "flexural phonon" branches and appear to have a small linear component for small wavevector (near the point Γ). (Data points are from Nicklow *et al.*, 1972, very similar data are in Mohr *et al.*, 2007). (From Mounet and Marzari, 2005. © 2005 by the American Physical Society).

The work of Mounet and Marzari (2005) is a comprehensive comparison of graphite, graphene and diamond carried out using a combination of density-functional theory (DFT) total-energy calculations (Hohenberg and Kohn, 1964; Kohn and Sham, 1965) and density-functional perturbation theory lattice dynamics in the generalized gradient approximation. This *ab initio* calculation well predicts features including the lattice parameter of graphite and its temperature dependence. The assumption in such work, to determine phonon properties, is the Born–Oppenheimer approximation: one assumes that the electronic structure follows the small variations in atomic positions that are the lattice vibrations. These motions are relatively slow, and the electronic structure easily follows the changes. (For this reason, in calculation of electron scattering leading to resistivity, an excited phonon has the same effect as a static distortion.) The theory correctly predicts a negative temperature coefficient of expansion of graphite, with a most negative value near 250 K. For graphene it predicts a larger negative coefficient of expansion than that of graphite, continuing negative up to 2300 K. (See Fig. 7.10.) The value near room temperature, in the work of Mounet and Marzari (2005) for the in-plane temperature coefficient of expansion of graphene, is about $-5.7 \times 10^{-6}\,\mathrm{K}^{-1}$, close to the recently measured value $-7 \times 10^{-6}\,\mathrm{K}^{-1}$ found by Bao *et al.* (2009). Schelling and Keblinski (2003) consider that the thermal contraction along the graphene planes is directly a consequence of growing amplitude of the out-of-plane flexural phonon, and that the coefficient is larger in graphene than for graphite because that flexural motion is less constrained. As the plane bends into an arc, or becomes corrugated in the flexural mode, the linear distance between its ends diminishes. The same effect is discussed in terms of entropy gain by Kwon *et al.* (2004). In neither of these accounts is a change of in-plane C-C bonding discussed.

Fig. 7.9 Flexural transverse acoustic mode (ZA) that appears in graphite and in graphene as a consequence of the nearly 2D atomic arrangement. This specific mode eigenvector is at $k = (2\pi/a)\,(0, 0.1, 0)$ that is 30% of the way from Γ to M (see Fig. 1.2, where M corresponds to $k = (2\pi/3a)$). This mode is also called the bending mode, and its importance in graphite and other layered compounds is also called the "membrane effect." (From Mounet and Marzari, 2005. © 2005 by the American Physical Society).

The wavelength of the illustrated flexural mode is $\lambda = 2\pi/k = 10a$, roughly consistent with Fig. 7.9, where about 10 benzene rings separate the maxima. The frequency of this mode in graphene is about 40 cm^{-1} (= 5 meV) [see Figure 10 of Mounet and Marzari (2005)] and the inclusion of at least 25 nearest neighbors was needed in the calculation to make the frequency converge. At 300 K ($k_{\mathrm{B}}T = 26\,\mathrm{meV}$), the Bose occupation factor $f = \exp(h\nu/k_{\mathrm{B}}\,T) - 1]^{-1}$ for this mode is thus 4.8, but will become 0.146 at 30 K. The occupation of these modes means that distortions of this type are present in the graphene at room temperature, but much less so at 30 K, and corresponding scattering of charge carriers will be sharply reduced. A further prediction of Mounet and Marzari (2005) is a monotonically decreasing lattice parameter with increasing temperature, a feature not found in the limited simulation of Zakharchenko *et al.* (2009).

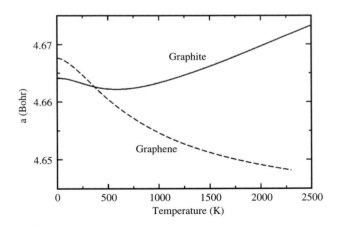

Fig. 7.10 Thermal expansion and thermal contraction in graphite and graphene, respectively. The latter is due to flexural modes. (From Mounet and Marzari, 2005. © 2005 by the American Physical Society).

The latter authors also disagree that the temperature coefficient of graphene remains negative up to 2300 K, as found by Mounet and Marzari, and their value for this coefficient, $-4.8 \times 10^{-6} \, \mathrm{K}^{-1}$ up to 300 K, is less negative than either the prediction of Mounet and Marzari (2005) or the experiments.

The DFT theory is seen to reasonably predict properties of graphene, including the frequency of the flexural phonon, and indicates that graphene is stable. However, as pointed out by Schelling and Keblinski (2003), the mesh size in a calculation limits the actual length of the sample that is being modeled, it is not clear what sample size is implied in the work of Mounet and Marzari (2005). There is also a question as to a criterion for instability. If a thermal mode amplitude larger than a lattice constant indicates instability, this criterion is easily reached in practical sample sizes at room temperature. For example, if the square of side L were clamped on opposite sides, the thermal amplitude at the fundamental frequency is shown [eqn (7.19)] to be approximately:

$$y_{\mathrm{th}} = \left(kT/32Yt^3\right)^{1/2} L, \qquad (7.13)$$

where Y is Young's modulus and t is the thickness. At room temperature for a mono-layer platelet of side 0.1mm this amplitude is 181 nm, many times larger than a lattice constant. If this is evaluated at $L = 1$ m, the thermal amplitude is 1.81 mm. Since this rms value is proportional to L, it can reach any amplitude at large enough L. This describes a breakdown of crystalline order of a particular sort. Unit cells are no longer at the positions $\mathbf{R} = n\mathbf{a} + m\mathbf{b} + l\mathbf{c}$, but the large scale on which it occurs makes it irrelevant for most presently envisioned applications. This situation in principle is no different from the situation with plates or beams in the world of civil engineering, but the effects appear sooner because of small bending energy, due to the small thickness t. Of course for a three dimensional sample the thermal vibration amplitude is small and independent of sample size, as discussed near eqn 2.37a, b.

If the amplitude of distortion for crumpling has to be comparable to the sample size L, then graphene is stable (but carbon nanotubes are not). But to suggest that a graphene sample crumples at $L = 0.3$ nm is clearly incorrect. It harder to decide if, has been suggested, graphene's bending parameter $\kappa \sim 1$ eV, corresponding to Young's modulus ~ 1 TPa, requires renormalization. (The success, see Section 7.4 below, in fitting observed oscillations in cantilevers and beams of Bunch *et al.* (2007) using the common values of the essential parameters, suggests that these values are correctly normalized.)

To return to the flexural mode shown above for graphene at $k = (2\pi/a)$ (0,0.1,0) where the energy is approximately 5 meV, and the dispersion, following the data of Nicklow *et al.* 1972 is of the form

$$\nu^2 \propto Aq^2 + Bq^4. \qquad (7.12)$$

The wave shown in Fig. 7.9 is of the type present along the conventional bending beam of mechanical engineering,

$$h(x,t) = A \, \exp\left(-ikx + i\,\omega t\right) \qquad (7.14)$$

where the dispersion relation is $\omega = (YI/\rho A)^{1/2} k^2$ and Y (also denoted E) is Young's modulus. Here $I = wt^3/12$ is the second moment of the area $A = wt$ of the beam of dimensions width w and thickness t, and ρ is the volume mass density. In a doubly clamped beam of length L, this corresponds to a resonant frequency (Carr *et al.*,1999)

$$\omega = \left(4.73/L^2\right)\left(YI/\rho A\right)^{1/2}. \tag{7.15}$$

To check this with the mode description given above, one can ask for the spacing between nodes (doubly clamped cantilever length L) that corresponds to the frequency of the 5 meV ZA mode that is 7.5×10^{12} rad/sec. (The ZA mode exhibits a flexing motion.) Using parameters $t = 0.34$ nm, density 2270 kg/m^3 $Y = 10^{12}$ Pa, the length between nodes is 1.121 nm. Looking at the $k = (2\pi/a)\,(0, 0.1, 0)$ mode in the diagram, this would correspond to half the wavelength (a standing wave has half the wavelength of the corresponding running wave) and thus would be $5a$ or 5×0.246 nm $= 1.23$ nm, directly comparable to the 1.121 nm. So this is a consistent picture. This engineering picture of the graphene as a doubly clamped beam was extended by Bunch *et al.* (2007) to include imposed tension T. Their formula for the frequency is

$$f_0 = \left\{\left[A\left(E/\rho\right)^{1/2} t/L^2\right]^2 + A^2 0.57T/\rho L^2 wt\right\}^{1/2} \tag{7.16}$$

where $E = Y$ is Young's modulus, T the tension applied in Newtons, and ρ the density. Here, t, w, and L are the dimensions of the beam and the clamping coefficient A is 1.03 for a doubly clamped beam and 0.162 for cantilevers. (These values of A that differ slightly from the values of Carr *et al.* (1999), come from Ekinci and Roukes (2005). The appearance of the second term in eqn (7.16) with tensile strain T is discussed by Mariani and von Oppen (2010).

The value of E was taken as 1 TPa. (Bunch *et al.*, 2007). Measurements of vibrational frequencies of graphene sheets of various thicknesses suspended across a trench (doubly clamped beam) were well fit by this formula, sometimes requiring assumption of a tension T.

The thermal amplitude of such a cantilever mode can be estimated. Vibration described in the two formulas can be put in the form $\omega = (K^*/M^*)^{1/2}$ where the effective spring constant for a force applied at the midpoint of the doubly clamped beam is (Shivaraman *et al.* 2009)

$$K^* = 32Ywt^3/L^3. \tag{7.17}$$

Following Bunch *et al.* (2007), the rms thermal amplitude in the fundamental mode is $y_{\mathrm{rms}} = (k_{\mathrm{B}}T/K^*)^{1/2}$. For a particular cantilever of thickness 5 nm and estimated $K^* = 0.7$ N/m, Bunch *et al.* 2007 found the rms thermal amplitude at room temperature to be 76 pm. If we apply this formula to a square, so that $w = L$, we find the thermal amplitude of the fundamental flexural mode scales linearly with the size L:

$$y_{\mathrm{rms}} = (k_{\mathrm{B}}T/K^*)^{1/2} = \left(k_{\mathrm{B}}T/32Yt^3\right)^{1/2} L. \tag{7.18}$$

If we go to zero temperature, the zero point vibration can be estimated by replacing $k_B T$ by the zero point energy. This energy is $hf/2 \approx 0.723 \times 10^{-12}$ eV. The zero point vibration is then 9.5×10^{-4} nm.

This picture does not lead to a "crumpling transition" for a square sample at large L for any finite temperature, because the rms thermal amplitude is always a small fraction of the sample dimension L (although it could be many lattice constants):

$$y_{\mathrm{rms}} \Big/ L = \left(k_B T / 32 Y t^3 \right)^{1/2}. \tag{7.19}$$

If the sublimation point for graphene is 3900 K, this this thermal amplitude is about 0.65% of the length. At $T = 0$ the system recovers crystalline order. It is hard to see how this conclusion will be affected by small changes in stiffness (Y, Young's modulus) that may occur with possible renormalization effects, to be mentioned.

Crumpling would mean an appearance like a crumpled sheet of paper, where fluctuation displacements have the same size scale as the sheet. While there are systems where such transitions have been observed (Radzihovsky, 2004), it does not seem that graphene is in this category.

A pattern of thermal standing wave vibrations at frequencies 1 MHz to 1 THz, depending on the size L of the mounting, is set up by the boundaries of any graphene sheet. The mode pattern (the thermal version of a Chladni nodal pattern, most familiar in the analysis of vibrations of the body of a strongly bowed violin), could be construed as an origin of "corrugations". This appears to be what has been done in the work of Fasolino *et al.* (2007).

A calculation related to the flexural phonon dispersion in absence of strain is offered in the work of Kumar *et al.* (2010). These workers find a small linear contribution b to the flexural mode frequency dependence on k, so that the integral for the phonon occupation will be of the form

$$N_{\mathrm{ph}} = (2\pi)^{-1} \int k \, \mathrm{d}k \left[\exp \left(ak^2 + b\,|k| \right) - 1 \right]^{-1}, \tag{7.20}$$

that is no longer divergent as $k \to 0$. Kumar *et al.* (2010) suggest that this linear term leads to an intrinsic strength of the layer against buckling under compressive stress. (The classical theory of buckling was mentioned in Chapter 2, see eqns (2.89a, b). This strength is experimentally confirmed by the flatness of graphene under compressive strain -0.8% when grown on SiC, where the graphene experiences an attractive force.) The physical origin of the linear term of Kumar *et al.* (2010) is a restoring force against bending the sheet coming from interaction among the π orbitals as suggested in Fig. 7.11. The authors state that the intrinsic buckling strength of the graphene comes from curvature-induced mixing among the π orbitals.

Kumar *et al.* give values of the elastic parameters Y, μ, and λ of 20.2, 8.8 and 3.0, respectively (in units of eV/A^2) and Poisson ratio $\nu = 0.16$. These values, not greatly different from earlier values, and incorporating the small linear dispersion term, are the basis for their estimate of the buckling strength of graphene.

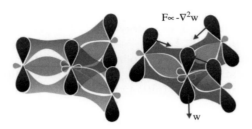

Fig. 7.11 Sketch showing role of p_z orbitals in maintaining planarity of graphene. (From Kumar *et al.*, 2010. © 2010 by the American Physical Society).

A recent article by Si *et al.* (2012) (see also Marianetti and Yevick, 2010) finds that the tensile strength of pure graphene can be increased by up to 17% by charge doping. The 17% is beyond a critical tensile strain known to be about 15%. The article suggests that the effect has to do with stiffening of the K_1 phonon mode (the highest frequency mode at point K, see inset to Fig. 7.12) by electronic doping.

A different prediction for a change in the flexural dispersion has been offered by Mariani and von Oppen (2008). These authors find a critical wavevector q_c below which the frequency is proportional to $q^{3/2}$, with an enhanced stiffness

$$k = k_0(1 + q_c^2/q^2)^{1/2} \qquad (7.21)$$

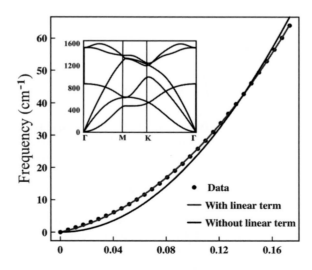

Fig. 7.12 Flexural mode dispersion for graphene as calculated with and without a linear term, compared with data points. (From Kumar *et al.*, 2010. © 2010 by the American Physical Society).

that is said to stiffen the membrane at low q. (A similar q dependence of ω and stiffening of κ was later found by Zakharchenko *et al.* (2010).

The expression of Mariani and von Oppen (2008) obtained at small q is

$$\omega = (\kappa_0 q_c/\rho)^{1/2} q^{3/2} \tag{7.22}$$

The cutoff wavevector is given as

$$q_c = (3K_0 k_B T/8\pi\kappa_0^2)^{1/2} \tag{7.23}$$

that appears to be about $2.5 \times 10^9 \, \text{m}^{-1}$ for graphene at 300 K, taking K_0, (described as an in-plane Young's modulus), about $20 \, \text{eV/A}^2 \approx 320 \, \text{N/m}$. [A similar expression for a critical q was later found by Braghin and Hasselmann (2010)].

The corresponding frequency at the q_c of Mariani and von Oppen (2008) is about 0.45 THz. This point would lie in the Brillouin zone about 0.08 from Γ toward M (see Fig. 1.2a) and is close to the data of Nicklow *et al.* (1972), see their Figure 6. The data curve approaching $q = 0$ indeed appears with a shape consistent with $q^{3/2}$. In the analysis of Mariani *et al.*, $K_0 = 4\mu(\mu + \lambda)/(2\mu + \lambda)$, where $\mu = 4\lambda = 9\text{eV/A}^2$ are Lame coefficients, and the stiffness κ_0 is taken as 1 eV. Since this dispersion is close to what is calculated by Mounet and Marzari, that well fits neutron scattering data [and also x-ray scattering data of Mohr *et al.* (2007)] it would be consistent with values of specific heat that they have calculated. The interest of Mariani and von Oppen (2008) is in the resistivity resulting from electron scattering by the flexural modes. They predict a resistivity of the form $\rho \propto T^{5/2} \, \ln(T)$. They attribute the ln T term to renormalization of the flexural mode dispersion due to coupling between bending and stretching degrees of freedom.

An additional calculation of resistivity has been described by Castro *et al.* (2010). These authors do not find a renormalization of the flexural phonon modes at zero strain, but find that tensile strain u induces a linear dispersion according to

$$\omega^2 = (\kappa_0/\rho)q^4 + u\nu_L^2 q^2, \tag{7.24}$$

where

$$\nu_L^2 = (2u + \lambda)/\rho = (2.1 \times 10^4 \, \text{m/s})^2. \tag{7.25}$$

They find a contribution to the resistivity proportional to T^2, in conflict with the prediction of Mariani and von Oppen. Their paper mentions qualitative agreement of their high temperature resistivity with "classical theory assuming elastic scattering by static ... ripples" discussed earlier by Katsnelson and Geim (2008). We discuss in Section 7.6, doubt about assumptions of "intrinsic ripples" and more that such ripples can be "quenched" or "frozen."

It is not clear what experiment would be suitable to check the modified dispersion relation proposed by Mariani and von Oppen (2008). Conceivably it could have an effect on the temperature dependence of the resonant frequency of a graphene cantilever, of the sort that were carried out by Bunch *et al.* The freely suspended cantilever, vs. the doubly clamped beam, is free of strain effects that can have a large effect in the latter case. The renormalization [found by Mariani and von Oppen (2008)] stiffens the membrane at low wavevector $q < q_c$, that becomes more prominent as the size L is increased, and thus offers little support for the idea of a crumpling transition.

7.1.4   Experimental evidence of nanoscale roughness

The best evidence for static "roughness" or "corrugations" in the single graphene layer comes from the careful studies of Meyer *et al.* (2007), and Kirilenko *et al.* (2011) of diffraction patterns observed in a transmission electron microscope. In these experiments diffraction spots arise as an image of the reciprocal lattice of the atomic array. [See also Geringer *et al.* (2009)]. The reciprocal lattice for graphene is hexagonal. In the case of a single layer, the reciprocal lattice spots are replaced by "diffraction rods," as indicated in Fig. 7.13 below. (In Fig. 7.13 the rods are tilted, to represent an array of crystal platelets with a spread of normal directions.) The formation of "diffraction rods" can be visualized in the following way. Imagine a 3-dimensional atomic array AAA of stacked identical layers, for which the reciprocal lattice is a hexagonal array of points. If the imagined 3D atomic array is expanded, say in the vertical direction, then the 3D diffraction spots will collapse vertically leaving the array of rods. The diffraction condition is the intersection of the rods with the Ewald sphere, representing the monochromatic radiation. For 60 keV electrons, the radius of the Ewald sphere is $k = 2\pi(2mE)^{1/2}/h = 1.26 \times 10^{12} \mathrm{m}^{-1}$, much larger than the reciprocal lattice basis vector $G = 0.52\,\text{Å}^{-1}$. This sets the radius of the diffraction rods from the origin (spacing of the rods) (Kirilenko *et al.*, 2011, Figure 2), that is given as $0.47\,\text{Å}^{-1}$ in Figure 3f of Meyer *et al.* (2007). The Ewald sphere, with its large radius, approximates a horizontal plane, intersecting the rods as indicated in Fig. 7.13. The diameter,

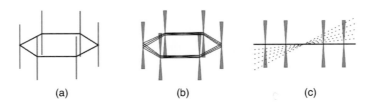

(a)   (b)   (c)

Fig. 7.13 Aspects of TEM diffraction from graphene. (a) Reciprocal lattice for perfect graphene. The rods extending vertically from the hexagonal points are a consequence of the 2D nature of graphene. (b) Tilted diffraction rods from assumed randomly tilted platelets of graphene. The rods are spread into cones of increasing width with increasing displacement in reciprocal space. (c) Tilting of Ewald sphere (of large radius, represented by dotted lines) reveals systematically broadened intersections with viewing surface at increasing angles of tilt. (After Meyer *et al.*, 2007).

in k-space, of the sharp diffraction spots seen by Meyer (in their Figure 3g) is 1.0×10^8/m, a value that the authors attribute to the instrumental line width. The diffracted intensity $I(k)$ from a sample of diameter w is of the form

$$I(k) = [\sin(\pi wk)/\pi k]^2, \tag{7.26}$$

with full-width at half-maximum near $\pi wk = \pi/2$. This width nominally implies diffracting regions of width w at least 5 nm. Since the observed width is known to be instrumental, the actual width of the diffracting regions is larger, perhaps 50 nm. The sharp hexagonal pattern indicates a single crystal lattice, with all carbon atoms spaced equally by the basis vectors of Fig. 1.2(b). (Diffraction rings would be obtained if rotational order were lacking.) It is to be noted that thermal motion, in this situation, does not broaden diffraction spots, but merely weakens them by the Debye–Waller factor, $\exp[-<(\mathbf{k} \cdot \mathbf{u})^2>]$, where u represents an isotropic random thermal displacement.

Following Meyer *et al.* (2007), the 60 keV TEM beam illuminated a diameter of 250 nm of the sample, and the sharp diffraction spots were imaged with a CCD camera for various angles of tilt, about a horizontal axis normal to the beam, at angles 0°, 14° and 26°.

The broadening of the spots increased with tilt, and was greater for spots further from the tilt axis, as shown in Fig. 7.14. This implies a spread of angles among the originating regions from the normal around ± 5°. Meyer *et al.* (2007) conclude that their suspended single layer graphene sample exhibits static deformations of up to 1 nm in the normal direction, with domain sizes L less than 25 nm.

The surface normal varied randomly from position to position, and fell inside cones of angles between 8° and 11°. The TEM does not directly allow the vertical displacement to be determined, but analysis of the diffraction pattern suggested that the vertical displacement associated with the isotropic "waviness" was at most 1 nm. The

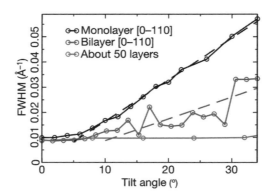

Fig. 7.14 Width of diffraction spots vs. tilt angle. The upper two curves extrapolated to zero tilt angle indicate inherent diffraction spot widths less than 0.001 A^{-1}. (From Meyer *et al.*, 2007, by permission from Macmillan Publishers Ltd., © 2007).

effect was seen in many samples, was similar at different locations on a sample, the effect was diminished for bilayer films, and absent for films of many layers. None of these samples were cleaned by current heating that leaves the strong chance that weakly adsorbed molecules from the ambient atmosphere or handling, capable of locally warping single layer graphene, produced the effects.

It was specifically concluded by Meyer *et al.* (2007) that the corrugations were *static* because the effect of the corrugations were directly seen in TEM images with atomic resolution at 300 keV, "because otherwise changes during the TEM exposure would lead to blurring and disappearance" of the effects.

A second tool for observing graphene is low-energy electron diffraction, LEED: the geometry is shown in Fig. 7.15.

Much broader diffraction spots were observed in LEED at 42 eV by Knox *et al.* (2008), whose micro-mechanically cleaved graphenes were supported on SiO_2 grown on Si. This conventional support is now known to have a roughening effect on the graphene (see Fig. 5.1), a larger effect than chemical adsorbates that likely were also present. The authors estimate from their LEED-spot-width for the monolayer that an angular spread of about 6° might be present in the surface normal distribution due to "undulations."

The FWHM (full width at half maximum) of the monolayer graphene central spot is about 10^{10} m^{-1} (right-hand panel in Fig. 7.16), compared to 10^8 m^{-1} in the data of Meyer *et al.* (2007). The smaller diffraction spot width in LEED observed for graphite (lowest trace) confirms that the monolayer spot width 10^{10} m^{-1} exceeds the instrumental width.

The larger diffraction spot diameter for single layer graphene seems to arise, as Knox *et al.* (2008) state, from extrinsic undulations arising from charge and positional disorder in the amorphous quartz layer grown on the Si substrate. (See Fig. 7.17.) This suggests that all measurements of graphene monolayers supported on the traditional oxidized Si substrate thereby exhibit large extrinsic surface roughness. (STM evidence of this was shown in Fig. 5.1.) While these effects are not, in our opinion, intrinsic to graphene, they obviously are a practical matter in much of the work on graphene. (It appears that a hexagonal boron nitride substrate is smoother, but the use of h-BN is more difficult experimentally.)

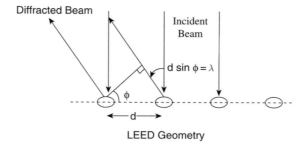

Fig. 7.15 Sketch of incident- and diffracted- electron beams in a LEED measurement. The diffraction condition is d sinφ = nλ, where d is the atom spacing. (Courtesy Ankita Shah).

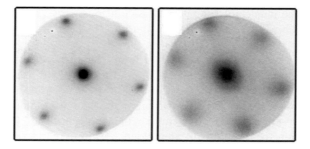

Fig. 7.16 LEED patterns of silica-mounted micro-mechanically cleaved graphene at 42 eV. The patterns (left and right respectively) are from exfoliated graphene bilayer and monolayer laid on quartz thermally grown on Si wafer. (From Knox *et al.*, 2008. © 2008 by the American Physical Society).

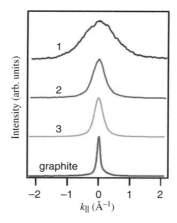

Fig. 7.17 Width of the central intensity maximum of LEED pattern at 42 eV, top to bottom, exfoliated graphene monolayer, bilayer, trilayer and graphite samples, respectively, all laid on quartz on Si wafer. (From Knox *et al.*, 2008. © 2008 by the American Physical Society).

As a complication to this literature, Kirilenko *et al.* (2011) state that their method, similar to that of Meyer *et al.* (2007), does not discriminate between time varying and static strain. It is believed that AFM and STM see static strain only. It appears that the work of Fasolino *et al.* (2007) actually is a calculation of oscillating strains, as a function of temperature, rather than static strains. Thermally excited strains mean lattice vibrations that are time varying. (The confusing and likely misleading phrase "static thermally excited ripples" or even "frozen ripples" (Katsnelson, 2012, Section 10.2) comes from a suggestion of Katsnelson and Geim (2008) that thermal vibration corrugation patterns might be "quenched" in putting the sample down onto a substrate. The frequencies of the thermal oscillations as we have seen above are in the THz range or above, likely too fast to be quenched by drooping the film onto a silicon wafer.)

The TEM work of Kirilenko *et al.* (2011) found for suspended monolayers an rms roughness of 1.7 A, a wavelength about 10 nm and a typical tilt angle about 6°. These results are similar to those earlier reported by Meyer *et al.* (2007). The technique of Kirilenko *et al.* allowed temperature variation of the sample, reduction from 300 K to 150 K. It was surprisingly found that the corrugation amplitude *increased* at 150 K vs. 300 K. [The chance of temperature dependent strain being introduced, as emphasized in the work of Bao *et al.* (2009), and Yoon *et al.* (2011) seems possible, and is not mentioned by Kirilenko *et al.* (2011)] The effect of reducing temperature on the free-standing (i.e., supported on a microscope grid) graphene was to increase the observed rms corrugation above the 1.7A value, and also to increase the wavelength of the corrugations from 10 nm at 300 K to 18 nm at 150 K. The changes were repeatable on graphene suspended across trenches of variable size L, between 0.5 and 5 μm. The authors conclude that the clear increase in the measured corrugation with decreasing temperature is not indicative of a process involving flexural phonons. No particular steps were taken to "clean" the surface of the graphene films, and in related work (Gass *et al.*, 2008) it has been discovered that hydrocarbon contamination of graphene is not uncommon in experiments of this sort. It was found by Bolotin *et al.* (2008) see Section 7.1.5, that heating the graphene by passing a current through it greatly improves the quality of electrical measurements, clearly by releasing adsorbed molecules. It is also reported that graphene films treated chemically with photoresist can have mass per unit area from two to six times the carbon mass. Gass *et al.* (2008) comment: "It remains unclear whether the presence of surface contamination, likely to be hydrocarbons, plays a part in the stabilization of the 2D structure, causing, or being associated with small deformations of the film. . . ." It thus seems likely that the ripples arise from surface adsorbates that can be present in the laboratory air, even if the sample is not treated with photoresist. These small ripple effects probably would not be seen in the work of Bao *et al.* (2009). It is well known that graphene can be doped by adsorbing, for example, water or ammonia on its surface.

A specific example of a chemisorbed atom that is known (Hossain *et al.*, 2012) to desorb at 260°C, is oxygen, first mentioned in Section 2.4, see Fig. 2.12. The oxygen in this study was introduced in atomic form, so it is not likely that this particular adsorbate is common in suspended graphene samples. However, the behavior carefully studied in this case may occur in more likely adsorbates such as OH. As shown in Fig. 2.12, the oxygen finds a chemically uniform binding state. This state was modeled by Hossain *et al.* (2012), using density functional theory (DFT) calculations in the local density approximation (LDA). Figure 7.18 shows their result applicable to oxygen adsorbed to graphene with strain −0.8% (compression).

The motivation of the calculation of Hossain *et al.* was to model the STM image of the oxygen atom (Fig. 2.12) taking into account the known compressive strain existing in graphene grown on SiC. Epitaxial graphene thermally grown on SiC (0001) is compressively strained because the graphene expands and the SiC contracts during cooling. This compressive strain is not released through graphene buckling or warping because

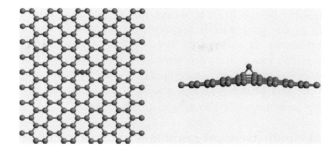

Fig. 7.18 Static long range distortion induced by adsorbate, assuming negative (compressive) strain in the graphene. Density functional theory DFT calculations of oxygen adatom on epitaxial graphene. The calculation is for –0.8% strained graphene. The oxygen atom (center of the left panel) is seen in right panel to lie above the centerpoint of a C–C bond. The right panel reveals the adatom and also the upward bowing (buckling) of the graphene (under 0.8 % compression) that surrounds the adatom. This effect enlarges considerably the size of the adatom as seen in STM, as was shown in Fig. 2.12. (From Hossain *et al.*, 2012, by permission from Macmillan Publishers Ltd., © 2012).

of a critical compressive strain release barrier at -1.25 % (Kumar *et al.*, 2010) arising from π-π orbital interactions within the graphene lattice, or as a consequence of the van der Waals attraction of the graphene to the SiC. The DFT calculations of Hossain *et al.* (2012) reveal that the oxygen adatoms perturb the planar π-π interactions of the sp^2-bonded graphene lattice by adding sp^3 bonding and thereby act as strain release centers that allow buckling to locally occur. Oxygen chemisorption induces topographic distortion of the epitaxial graphene in the region surrounding the bonding site, increasing the height of the STM image of the oxygen adatom from about 0.5 nm at zero compression to about 1.3 nm at -0.8% compression.

It has been shown by Ferralis *et al.* (2008) that growth of graphene on the (0001) surface of 6H-SiC leads to a graphene film under uniform (hydrostatic) compression at room temperature of 0.8%. As commented earlier, the flat conformation of the graphene at such strain is likely the result of van der Waals attraction to the substrate. A free graphene plane buckles at much smaller compression, as suggested in Fig. 2.16.

Under the modeled conditions it appears that the vertical amplitude of the buckling is modest, about 0.5 Å (1.65 Å including the oxygen itself) and the diameter of the buckled region is 13 Å. It is reasonable to assume that the locally induced buckling around adatoms will be increased as the compressive strain is increased, with 1.25% being the limit calculated by Kumar *et al.* (2010) for elastic compression of pristine graphene before buckling.

One can speculate that such an effect could account for the observation of Kirilenko *et al.* (2011) of larger undulations at lower temperature, assuming that lower temperature in the typical sample mount for graphene would be accompanied by larger compressive strain. In a related theoretical calculation Giannopoulos (2012) has found that sinusoidal buckling occurs in pure, defect-free graphene nanoribbons, under compression. Buckling of the graphene is regarded as reasonably the expected mode of failure under compression, in view of its low bending stiffness. It is accepted in this literature that graphene has a non-zero threshold for distortion (buckling or bowing from the plane) under compression, as explicitly shown by Kumar *et al.* (2010). This analysis is consistent with stability of free sheets of 2D graphene against spontaneous distortion, contrary to some early theoretical predictions.

7.1.5 Electrical conductivity of graphene in experiment

Electrical conductivity of graphene is greatly improved by modest surface cleaning by annealing (Moser *et al.*, 2007) as revealed in the work of Bolotin *et al.* (2008). (Many early published data were likely affected by the surface adsorbates, as was later learned, are easily removed by modest heating.)

The change in conductance vs. carrier density is from near linear to sub-linear, respectively, before and after annealing. The conductivity minimum is greatly narrowed by annealing, which suggests that before annealing local potential fluctuations

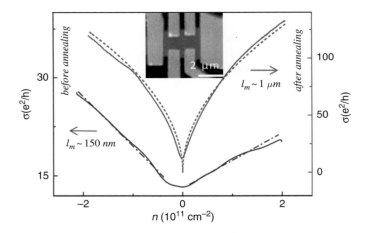

Fig. 7.19 Electrical conductance of graphene, as improved by surface cleaning. Lower curve, left scale shows conductance data at 40 K for suspended graphene sample in 4-terminal measurement configuration (upper inset). The data are characterized by mean free path 150 nm, and show minimum that is broadened and shifted to the left. Upper curve (right, **note wider scale range**) shows conductance again at 40 K after current annealing to modest temperature (estimated as 400°C) in the cryostat. The mean free path is now estimated as 1 μm (comparable with device size, see inset) and the dashed curve is a theoretical estimate assuming ballistic propagation of carriers. The mobility in the annealed state is estimated as 17 m^2/Vs at 40 K, with n = 2 × 10^{11} cm^{-2} gate-induced charge density. The only effect of such modest annealing is to allow desorption of weakly bound surface species that are seen to play a large role. (From Bolotin *et al.*, 2008. © 2008 by the American Physical Society).

shifted the local Fermi level positions. This could be described as generating locally negative and locally positive conducting "puddles". Even though the impurities must be weakly bound to be released at 400°C, they evidently induce potential fluctuations in the graphene, shifting and broadening the minimum of the curve. The related static potential fluctuations also evidently considerably reduce the mobility, approximately by a factor of 6.6.

A similar observation of electrical change after annealing in vacuum (Wang *et al.*, 2009) is shown in Fig. 7.20. A large change in the doping of the film is evident in the large shift of the conductivity minimum voltage from 40 V to about 8 V.

Desorption of weak adsorbates, shown in Fig. 7.19 and Fig. 7.20 (whose influence was largely unrecognized in pioneering work), strongly reduces the resistivity of the film. The resistivity reduction comes directly from removal of scattering centers, leaving only to speculation possible secondary effects related to reduction of adsorbate-related Angstrom scale ripples.

The observed conductivity shown in Fig. 7.19 has two distinct regions in carrier density. At large densities the conductivity of graphene is metallic (i.e., decreasing with increasing T), while at low carrier densities in "ultra-clean" samples the conductivity shows pronounced nonmetallic T-dependence, i.e., conductivity decreasing with T. Values of mobility $\mu = \sigma/en$ are larger in the metallic range, and these values are

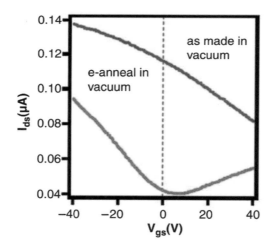

Fig. 7.20 Source-drain current at 1 mV source-drain voltage in 125 nm wide graphene nanoribbon before and after annealing in vacuum with high current, measured to raise temperature to $\sim 300°$C. The Dirac point moved from beyond 40 V to ~ 8V. The device is mounted on Si/SiO$_2$ with 300 nm oxide thickness. AFM measurements on the graphene nanoribbon indicated its thickness (height) decreased from about 1.5 nm to about 1.0 nm, attributed to loss, during the heating, of residues of PmPV polymer coating used in the fabrication process and known to produce p-doping of graphene. The authors also cite physisorbed oxygen as source of p-doping. (From Wang *et al.*, 2009, with permission from AAAS).

usually extracted at carrier density $n > 0.5 \times 10^{11}$ cm^{-2}. The metallic mobility in the clean sample at 240 K, near room temperature, is 12 m^2/Vs, higher than in conventional materials of any type (but see Fig. 2.2).

The conductivity dependence $\sigma(n)$ on carrier density n is roughly linear before annealing and is sublinear after annealing. In the "dirty" case the linearity suggests scattering by charged impurities. For screened Coulomb potential scattering the mean free time is $\tau \propto k_F$ where k_F is the Fermi wave-vector. In this regime, following Bolotin *et al.* (2008), the conductivity is

$$\sigma = (2e^2/\hbar)k_F \nu_F \tau, \tag{7.27}$$

proportional to carrier density, n, with τ the mean free time. The mean free path is

$$l = \sigma h/2e^2 k_F \sim 150\,\text{nm} \tag{7.28}$$

in the "dirty" case.

After annealing, the "ultraclean" device appears to have ballistic carrier motion. This can be deduced from the match between the data and dashed curve in Fig. 7.19. The dashed curve is obtained taking the ballistic conductivity

$$\sigma_{\text{ball}}(n) = 4e^2/h\,N = 4e^2 W k_F/\pi \, \alpha \, n^{1/2}. \tag{7.29}$$

Here N is the number of longitudinal quantum channels that, for width W and wavevector k_F, is $W k_F/\pi$. The fit to the annealed sample data measured at 40 K (upper curves in Figure, dashed fitting curve) is obtained with $W = 1.3$ μm, close to the sample dimension. With rising temperature, it is found that the conductivity only slightly declines in the temperature range from 40 K to 240 K, where the mobility is still 12 m^2/Vs. This exceeds the highest known mobility in semiconductors, 7.7 m^2/Vs, in InSb. The resistivity in cleaned samples rises proportionally to T as might be expected for phonon scattering.

The match of the ballistic model to data means that the sample at 40 K is an ideal quantum conductor. There is nothing anomalous here, no reason to mention an electronic "stiff membrane" with intrinsic "ripples." The reader should be wary of a theoretical literature that posits as intrinsic, effects probably not present in suitably cleaned samples (Morozov *et al.*, 2006; Abedpour *et al.*, 2007; Gibertini *et al.*, 2010; Katsnelson, 2012; and see Sections 4.6, 10.2), that bear some relation to the early but incorrect notion of an intrinsic "minimum conductivity" for graphene.[2]

With ideal ballistic carrier motion, as achieved in the data of Fig. 7.19, the mobility is ill-defined, because there is no diffusive drift velocity. If the data were forced to fit

---

[2]While we are discussing in Fig. 7.19 perfect electrical conduction with bias, so that the Fermi energy is above or below the Dirac neutrality point, it is now believed that the minimum conductivity at low temperature is in fact zero in pure, suitably cleaned and shielded samples, at zero T and zero bias (see Fig. 1.6 above). The observed minimum, in commonly processed samples, has turned out, to be an extrinsic response of the material to stray electric fields (Ponomarenko *et al.*, 2011).

as $\sigma = ne\mu$, the artificial μ would scale as $n^{-1/2,}$ that is non-physical. In following work, a detailed analysis of data at temperatures above 40 K, presumably not in the ballistic regime, was carried out, showing $\mu \propto 1/T^2$. A discussion of the limit to the conductivity in graphene arising from electron-electron interactions has been given by Rosenstein *et al.* (2013).

Ballistic transport has also been observed at 300 K in graphene "encapsulated" between boron nitride layers by Mayorov *et al.* (2011), see also Ponomarenko *et al.* (2011).

In related work on cleaned graphene, Castro *et al.* (2010) have extracted the temperature dependence of the mobility at temperatures between 5 K and 200 K. The mobility was extracted, with some difficulty, from 2-terminal measurements, creating the necessity of estimating and correcting for contact resistances, a problem not present in the data of Bolotin *et al.* (2008).

Inverse mobility data shown in Fig. 7.21 are from three similarly processed samples, cleaned by current annealing in a manner not closely described but similar to that used by Bolotin *et al.* (2008). The emphasis of Castro *et al.* is on the possible effects of tensile strain (that has never been a measured quantity) on the resistivity and mobility, with the basic assumption that tension changes the dispersion relation for the flexural phonon, $\omega = ak^2 + b|k|$, to increase the value of b. (A discussion of the result of tensile strain on resistivity is given by Mariani and von Oppen 2010).

The three curves in Fig. 7.21 have been analyzed using a formula

$$1/\mu = (1/\mu)_{T\to 0} + \gamma T^2 \tag{7.29}$$

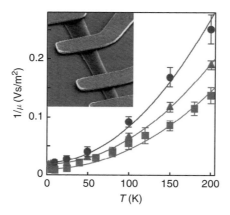

Fig. 7.21 Quadratic dependence of inverse mobility on temperature, showing very high mobility values at low T. Inset shows 2-point resistance geometry, in sample (darker near vertical stripe) of width 1 μm. The sample was cleaned by current annealing, following methods of earlier workers. (From Castro *et al.*, 2010. © 2010 by the American Physical Society).

where

$$\gamma = (Dk_{\mathrm{B}}/\kappa\nu_{\mathrm{F}})^2 (64\,\pi e\hbar)^{-1}\ln(k_{\mathrm{B}}T/\hbar\omega_{\mathrm{c}}). \qquad (7.30)$$

Here ω_{c} is a low-energy, long-wavelength cutoff that was adjusted to model different states of assumed strain in the measured samples. In this formula $\kappa = Yt^3 \approx 1\mathrm{eV}$ is the bending rigidity, $\nu_{\mathrm{F}} \approx 10^6\mathrm{m/s}$ is the Fermi velocity, and D is a derived deformation potential, related to the screened electron-phonon scattering potential $g = g_0/\epsilon(k_{\mathrm{F}})$ $\approx g_0/8 \approx 3$ eV, where $\varepsilon(k_{\mathrm{F}})$ is the static dielectric function. The parameter $g_0 \sim 20$–30 eV is the bare deformation potential, and the nearest neighbor distance, a, is 1.4 A. The expression for D is

$$D = [g^2/2 + (\beta\hbar\nu_{\mathrm{F}}/4a)^2]^{1/2}. \qquad (7.31)$$

The parameter β is defined as

$$\beta = -\partial\log(t)/\partial\log(a), \qquad (7.32)$$

estimated to be in the range 2 to 3, with t $\sim$3 eV the hopping of electrons between the nearest neighbor carbon π orbitals.

The authors attribute the factor-of-2 sample-to-sample inconsistency in their data to changes in the state of sample strain, accommodated in their analysis by changes in the cutoff parameter ω_{c}. For this purpose the needed change in cutoff ω_{c} would be a factor 7.4 that seems rather large. While it is not discussed by the authors, it seems possible that the discrepancy between samples might alternatively arise by changes in the annealing details, not described, between samples that may retain different densities of adsorbates. Plausibly adsorbates, possibly electrically charged, may locally change the dielectric function $\epsilon(k_{\mathrm{F}})$ that enters, squared, in the T^2 coefficient γ, through the expression for D. Credence is lent to the analysis of Castro *et al.* by the good indication of T^2 dependence of $1/\mu$ in the data. The conclusion that strain is central seems weaker. The strong dependence of the transport on adsorbates, shown in the work of Bolotin *et al.* (2008), does not directly indicate a role for strain.

Cleaning of graphene was discussed by Ishigami *et al.* (2007). These authors find that heating to 400°C in an atmosphere of hydrogen and argon removes residues of photoresist that cannot be removed by standard solvents and glacial acetic acid. Using STM on micro-mechanically cleaved graphene supported by quartz on Si, the authors found, after the heating procedure that the atomic lattice of graphene was visible. After annealing the authors concluded that the corrugation was smoother and arose solely from the curvature of the underlying quartz.

We return to the topic of molecular adsorbates as sources of electron scattering in graphene, and as the source of the "corrugations" that are a practical feature of real samples. An indication that adsorbates have a strong effect on the electrical conductivity is given by the application of graphene, through changes in its conductivity, as a detector of single molecules. Schedin *et al.* (2007) found that electrons are extracted (holes are introduced) by typical atmospheric exposure.

In a detailed study of adsorption of NO_2 on SiC-grown graphene at 20 K, Zhou *et al.* (2008) found that the Fermi level can be changed from + 0.4 eV (above the Dirac point, typical for graphene grown on SiC) to about −0.4eV. Charge is directly transferred, and leaves the molecular ion on the surface as a Coulomb scatterer. It is found that the NO_2 can be completely removed by heating to 400°C. The change to the local bonding of the graphene may be slight, but, as in the case of the oxygen, it may locally weaken bonding such that the adsorption center could act as a center of "buckling," if the sample were under compressive strain, as it is believed be in the case of growth on SiC. These authors find that a metal-insulator transition is caused by the hole doping, moving the Fermi level from conducting states to a gap, an artifact of growth of graphene on SiC (Zhou *et al.*, 2007).

In a similar doping experiment, McChesney *et al.* (2010) found that strong electron doping of graphene on SiC can be achieved by depositing monolayer amounts of K or Ca in an ultra high vacuum chamber.

7.2   Theoretical approaches to "intrinsic corrugations"

There remains literature arguing the presence of intrinsic corrugations in graphene, although the present experimental situation suggests that all observed static effects are better explained as extrinsic, coming primarily from mounting strain and as effects of adsorbates that weakly distort local bonding. The literature of oscillating corrugations has early roots going back many years. The work of Abraham and Nelson (1990) (see their eqn 1.10), [drawing on earlier work by Nelson and Peliti (1987) and Aronovitz and Lubensky (1988), see also Paczuski *et al.* (1988)], predicts that the mean square thermal fluctuation, h_{rms}^2, of the layer of length L and lattice constant a is given as

$$<h^2> = h_{rms}^2 = (k_B T/2\pi) \int_{1/L}^{1/a} \frac{qdq}{q^4}/k(q) \sim TL^{2\varsigma}. \qquad (7.33)$$

Here, following Abraham and Nelson, $\varsigma = 1$, and the limits on the integration are $1/L$ and $1/a$, respectively. This means that an effective rms thickness of the fluctuating sheet is proportional to L. (In agreement, the contribution to this result from the fundamental mode was described in eqns. 2.96 and 7.18, based on bending-beam theory and measurements on cantilevers and doubly clamped beams of graphene.)

According to Le Doussal and Radzihovsky (1992) the exponent ς is reduced from 1 to 0.59 by consideration of coupling between in-plane and out-of-plane motions.

These are random out-of-plane fluctuations with frequencies not discussed, but in the range of the phonon frequencies of the system. These fluctuations, in the nature of lattice vibrations, do not present static patterns of distortion. The spectrum of these fluctuations has been of interest in regard to the observed corrugations, even though the experimental corrugations were stated as static and the predicted fluctuations are of high-frequency.

The first simulation analysis of such vibrations on graphene was reported by Fasolino *et al.* (2007), who performed Monte Carlo simulations of carbon atom arrays

Fig. 7.22 Representative configuration of the $N = 8\,640$ atom sample at 300 K. Arrows embedded in Fig. are about 80Å long. Typical height of fluctuation h is stated as about 0.7 Å. (From Fasolino *et al.*, 2007, by permission from Macmillan Publishers Ltd., © 2007).

(typically 8640 atoms) using an accurate long-range "reactive bond order potential for carbon" (Los *et al.*, 2005; Ghiringelli *et al.*, 2005). The simulations extended to 3500 K, showing no evidence of "melting" and in fact no evidence of any defect to the hexagonal lattice. The height fluctuations were calculated and showed a broad distribution of heights h_{av} with typical value 0.07 nm for a sample of 8640 atoms at room temperature, using periodic boundary conditions. The authors stated that "ripples spontaneously appear owing to thermal fluctuations with a size distribution peaked around 80 Å which is compatible with the experimental findings (50–100 Å)". A picture of one typical configuration of the total sample (about 93 atoms on a side) is shown in Fig. 7.22. The height distribution h in the figure indicates that the local normal vector **n** to the plane varies. The further analysis of the data is in terms of the in-plane vector **u**, the out-of-plane displacement h, and the normal vector **n** that has in-plane component $-\nabla h/[1 + (\nabla h)^2]$ indicating a tilt angle θ. In the harmonic approximation, the bending h and stretching u modes are decoupled. It is natural to express the height fluctuations h in terms of the Fourier components h_q.

$$<h^2> = \sum <|h_q|^2> \; \alpha(k_B T/\kappa)L^2, \tag{7.34}$$

where

$$<|h_q|^2> = (k_B T N/\kappa S_0 q^4). \tag{7.35}$$

In this expression N is the number of atoms and $S_0 = L_x L_y/N$ is the area per atom in the simulation. The "correlation function for the normals" (or "normal-normal correlation function"), $G(q)$ for graphene, is the central result of their calculation. This is defined as

$$G(q) = <|n_q|^2> = q^2 <|h_q|^2> \tag{7.36}$$

In the simplest harmonic approximation the result is

$$G_0(q)/N = (k_B T/\kappa S_0 q^2). \tag{7.37}$$

In more detail, the angle θ between the local normal $\mathbf{n}$ and the average perpendicular to the plane is defined by

$$\cos \theta = 1/[1 + (\nabla h)^2] \approx 1 - \tfrac{1}{2}<\theta^2> \tag{7.38}$$

so that

$$<\theta^2> = <(\nabla h)^2> = (k_{\mathrm{B}}T/4\pi^2) \int \frac{q dq}{q^2}/\kappa(q). \tag{7.39}$$

This integral is convergent (only) if κ is renormalized to the form $\kappa_{\mathrm{R}}(q) \sim (k_{\mathrm{B}}TK_0)^{1/2} q^{-1}$ (Nelson and Peliti, 1987, eqn 14b).

Thus the implication, in the harmonic case that the mean square angle between the normal vectors diverges in a logarithmic fashion as the sample size L becomes large (Abraham and Nelson, 1990): is removed by renormalization. The renormalization is based on interaction of the bending and stretching motions, described by Nelson and collaborators, see also Le Doussal and Radzihovsky, 1992, mentioned above.

The effect of renormalization (Nelson and Peliti, 1987) is to greatly reduce the height fluctuations below $<h^2> \propto (k_{\mathrm{B}}T/\kappa) \, L^2$, and to change the sample-size-dependence of h_{rms}, from linear to $L^{0.6}$, approximately. Nevertheless, the fluctuations are anomalously large (because the film is 2D) and can exceed the interatomic distance for large samples. On this basis, Fasolino et al. (2007) state that their theory (of simulated lattice vibrations) predicts "an intrinsic tendency toward ripple formation." At the same time the amplitude of the transverse fluctuations, proportional to $L^{0.6}$, remains much smaller than the sample size L and preserves the long range order of the normals so that the sample can be considered approximately flat and not crumpled. In their original work the calculated $\mathrm{G}(q)/N$ for the 8640 atom sample showed a peak in the vicinity of $q = 0.08$ A^{-1}, corresponding to a length on the order of 80 Å. This peak feature is absent in the more recent version of such calculations.

A more recent version of the spectrum $\mathrm{G}(q)/N$ is given by Zakharchenko et al., 2010. The lower three curves in Fig. 7.23 reflect the reduction in fluctuations from the renormalization. The paper of Zakharchenko et al. (2010) goes on to predict the dispersion of the flexural modes, as a consequence of renormalization, as $\omega \propto q^{1.6}$, taking $\kappa_{\mathrm{R}}(q) \sim q^{-\eta}$, with $\eta = 0.82$.

Radial distribution functions from the early work of Fasolino et al. (2007) are shown in Fig. 7.24, at 300 K and at 3500 K. (From this work the typical height fluctuation at 300 K is about 0.07 nm = 0.7Å). It appears that their 2D model predicts that the sample remains intact at 3500 K, with a broadened radial distribution.

The title of the Fasolino et al. (2007) paper is "Intrinsic Ripples in Graphene", and references the work of Meyer et al. (2007) described in Section 7.1.4. Figure 7.24 shows at 300 K bond length variations from 1.54 Å to 1.31 Å, anomalously large, on the order of 15%. In view of the high Debye temperatures of graphene (see Section 7.2 below), 300K is near the vibrational ground state of the system. In Fig. 3 of their paper (reproduced in Katsnelson (2012) as Fig. 10.1) it is shown that in the modeling, one of the bonds, to nearest-neighbor carbon atoms is frequently much shorter than the

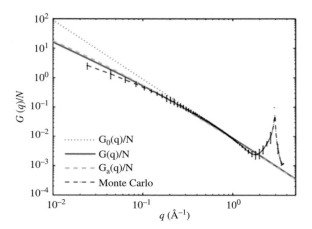

Fig. 7.23 Comparison of Monte-Carlo height-height correlation function (lower dashed curve with data points) to unrenormalized (top, dotted) and SCSA (self-consistent screening approximation) solutions (middle two curves). In a previous version the Monte-Carlo data a "maximum of G(q) at $q \approx 0.08$ A^{-1} signals a preferred length scale of about 80 Å". (From Zakharchenko *et al.*, 2010. © 2010 by the American Physical Society).

other two. Such large local distortions in the ground state appear unlikely, simply from energetic grounds, and seem further questionable on their implications for graphite. It is more likely that the bonds adjust leaving the atoms undisturbed, as happens in the benzene molecule. Since in graphite the graphene planes are separated by 3.4 Å, the 0.7 Å vertical motion found by Fasolino *et al.* (2007) would largely not be impeded, and their predicted properties found would appear also in graphite. According to Krumhansl and Brooks (1953), the atomic motions of carbon atoms in graphite can be described by Debye temperatures $\theta_D \approx 2500K$ for in-plane motions and $\theta_D \approx 900K$ for out of plane motions.

These numbers suggest that, at 300 K, graphene is in its ground state configuration, and the enormous distortions predicted by the analysis of Fasolino *et al.* (2007) above (see also Katsnelson, 2012) would be found also in the ground state of graphite. This is counter to a large body of experimental evidence on graphite.

Fasolino *et al.* (2007) offer that "an anomalously broad distribution of first neighbor bond lengths (and) changes of bond conjugation are . . . the reason for the negative thermal expansion coefficient of expansion in graphene. . . ." This statement is counter to comments by Schelling and Keblinski (2003) and Kwon *et al.* (2004) that the negative expansion coefficient directly results from the flexural motion (the low frequency transverse acoustic or ZA mode), that shortens the length between the ends of the plane, and is un-related to in-plane bonding. (This modeled "anomalously broad" nearest neighbor distribution is repeated in the work of Zakharchenko *et al.* (2009), where it is related to "strong anharmonicity" in graphene.) Graphite (3D) is of course a refractory material with an extremely high sublimation temperature of 3900 K. The

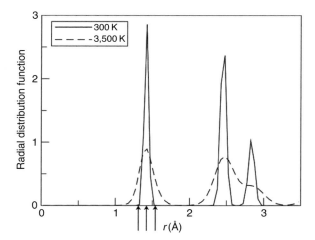

Fig. 7.24 Distribution of graphene (graphite) carbon–carbon distances in the modeling of Fasolino *et al.* (2007). The modeling of a 8640 carbon atom sample at 300 K and at 3500 K (dashed) shows an anomalously wide distribution of bond lengths centered at the bond length (intermediate arrow) 0.142 nm, close to 0.1399 observed for benzene. The left arrow marks double bond length 0.131 nm, the right arrow marks single bond length 0.154 nm. It is well known that in benzene the double bond position alternates with the single bond position, leaving for benzene a sharply defined bond length 0.1399 nm, so that all bond lengths are equal. The anomalously broad distribution here for 300 K, associated with the postulate of intrinsic graphene distortions, may be inconsistent with known x-ray diffraction properties of graphite. (From Fasolino *et al.*, 2007, by permission from Macmillan Publishers Ltd., © 2007).

robust nature of graphene is confirmed by the Fasolino *et al.* (2007) simulation, with no tendency for point defect formation seen in the simulation. No pre-melting anomalies (formation of vacancies, topological defects, etc.) were found up to 3500 K. Nor were any defects seen in simulations up to 2200 K with imposed deformation up to 10% (Zakharchenko *et al.*, 2009). In later work, a melting phenomenon was found near 4900 K (Zakharchenko *et al.*, 2011) to be discussed in Chapter 8. The cohesive energy of graphene is extremely high, 7.37 eV/atom in graphite, and defect formation energies are also large (Carlsson and Scheffler, 2006).

Realistic work on divacancy defects in carbon nanotubes and graphene is reported by Amorim *et al.* (2007). These authors find that divacancies of forms "585" (two pentagons side-by-side with an octagon) and "555777" (three pentagons with three heptagons) form in nanotubes and graphene (see Fig. 7.25.) These authors project that at large radii (going toward graphene) the "555777" defect is more stable. The estimated formation energies in graphene are 7.8 eV ("585") and 7.0 eV ("555777").

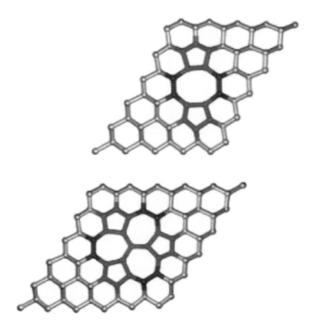

Fig. 7.25 Model of 585 defect (upper) and 555777 defect (lower). (Reprinted with permission from Amorim *et al.*, 2007. © 2007 American Chemical Society).

This can perhaps be related to the pioneering work of Nelson and Peliti (1987), in connection with the possibility of crumpling. These authors state that "finite temperature crumpling transitions are likely in crystalline membranes," and go on to estimate the occurrence of dislocations that would trigger the crumpling. According to Nelson and Peliti (1987) (their eqn 20), a correlation length, suggesting the size of a crystal at crumpling, is:

$$\xi_{\mathrm{T}} \approx a(R_c/a)^{\beta}, \quad \text{with} \quad \beta = K_0 a^2/16\pi k_B T. \tag{7.40}$$

Here $K_0 \approx 20\mathrm{eV}/\mathrm{\AA}^2$ and $R_{\mathrm{c}} \approx 60\,a$, with a the lattice constant, is a critical size of a flat plate against buckling by formation of dislocations. According to Nelson and Peliti (1987), a membrane with local crystalline order will appear fluid on length scales exceeding ξ_{T}. At room temperature, with the parameter values appropriate to graphene, this is astronomically large, but, at 3900 K, it is much smaller. Taking a =142 pm, one finds the exponent as $\beta = 30.9$ *at* 300 K but $\beta = 2.376$ *at* 3900 K. Taking $R_c/a = 60$, the value ξ_T/a at 300 K is about 9×10^{54}, but at 3900 K it is 16,783, or sample size 2.4 μm. This seems to be rough analysis, however, and has not been applied in the literature to graphene to the author's knowledge. Dislocations in graphene are not easily observable. However point defects that might generate dislocations have been reported (Gass *et al.*, 2008; Jeong *et al.*, 2008; and Hashimoto *et al.*, 2008). Energy estimates for dislocation cores are in the range 9.2 eV to 15.6 eV.

A simple defect in the hexagonal lattice where the center two hexagons become pentagons, coupled by upper and lower heptagons has been considered by Los *et al.* (2005). This "5-77-5 defect" can also be viewed as the result of rotating a bond between two atoms by 90 degrees.

The idea of this analysis is that the dislocation is a "pre-melting defect" as mentioned above. That such defects were not seen in early simulations could be taken as an indication of the difficulty of simulation, or that crumpling does not occur in graphene below the graphite sublimation temperature, 3900 K. We return to this in Section 8.6.

A different, adsorbate-based, possibility for the origin of the small static corrugations seen by Meyer *et al.* is given by Thompson-Flagg *et al.* (2009). They argue that an oscillating system, as calculated by Fasolino *et al.*, would on average leave the atoms in their lattice positions and would thus produce an unchanged electron diffraction pattern, contrary to observation. Thompson-Flagg *et al.* have looked for a mechanism of *static* distortions based on adsorbates, and one that would not be strongly temperature dependent. In simulation at $T = 0$ of a 10 nm square sample of graphene, Thompson-Flagg *et al.* have calculated the effect on atomic positions resulting from small initial displacements of atoms on the edge of the sample. As shown in Fig. 7.26, the small initial displacements do produce waves, but these waves do not penetrate into the interior of the sample. The decay length of the surface ripples was 3 Å compared with the 100 Å length of the sample. A single square was modeled, with initially all atoms on regular lattice sites, unless a perturbed initial boundary was chosen, and the system was allowed to relax toward its steady state. Similar results have been obtained by other workers.

This result shows that graphene has no inherent tendency to form corrugations, since such corrugations, if introduced at a boundary, die off exponentially going to the interior. While the Thompson-Flagg modeling is for $T = 0$, in view of the extremely high Debye temperatures, 900 K and 2500 K, respectively, expected for transverse and in-plane motions (Krumhansl and Brooks, 1953) the Thompson-Flagg prediction should still be relevant at 300 K and below, where localized distortions have been reported. On this basis, the observed distortions seem unlikely to have an intrinsic origin. ["Intrinsic ripples", in a contrary view, are elaborately supported by Katsnelson (2012), and references therein, including discussion of "frozen ripples."]

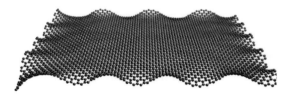

Fig. 7.26 Modeling to show that ripples do not propagate into a graphene sheet from irregularities at its boundary. (Reprinted with permission from Thompson-Flagg *et al.*, 2009. © IOP Publishing 2009).

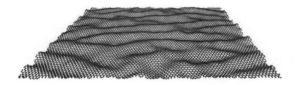

Fig. 7.27 200 Å square of graphene as modeled at $T = 0$ with assumed 20% of carbon atoms with two slightly enlongated bonds (by 4%), simulating effect of OH adsorbate. The static ripples shown are similar to static ripples inferred from TEM experiments. (Reprinted with permission from Thompson-Flagg *et al.*, 2009. © IOP Publishing 2009).

Going beyond this, Thompson-Flagg *et al.* (2009) have assumed that the occasionally observed static ripples must reflect distributed defects introduced onto the interior atoms themselves, since distortions cannot propagate inward from the boundaries. They thus introduced bond deformations at random positions, producing the $T = 0$ ripple pattern shown in Fig. 7.27. The nature of the distortion is an increased length, from 1.42 Å to 1.48 Å (about 4%), of two bonds on affected carbon sites. The density of affected sites is 20% of the sample, and the distortion models the effect of an OH molecule adsorbing to graphene, following the work of Xu *et al.* (2007). The energy of the OH attachment is about 0.17 eV. The modeling of the static "buckles" was done with all atoms initially in undisturbed lattice positions, with the perturbed bond set in place, and the system was allowed to relax in steps. The density of perturbed bonds was varied, and it was found, for 20% coverage that the average ripple wavelength is about 55 Å. while the rms amplitude is about 1.2 Å. The peak amplitude is about six times larger than the rms amplitude, and would thus be about 0.72 nm, fairly close to the estimate of Meyer *et al.* (2007). This amplitude is about 10 times the amplitude predicted by Fasolino *et al.* (2007), and is static, as was stated by Meyer *et al.*(2007). (The corrugations of Fig. 7.22, in contrast, are associated with anomalous bond length fluctuations on the order of 2%, modeled to occur spontaneously. See Fig. 7.24 and following text.)

Ishigami *et al.* (2007) have measured surface undulations on graphene deposited on quartz on Si, and have, after annealing by passage of current, distinguished between ripples induced by the substrate and those induced by adsorbates. They found that the undulations decreased after the heat treatment, consistent with the mechanism of adsorbate-induced buckling.

7.3   Impermeable even to helium

It has been observed (Bunch *et al.*, 2008) and theoretically confirmed (Leenaerts *et al.*, 2008) that a single atomic layer of graphene is impermeable to helium gas. The theory has also found that even if the graphene layer has defects, rather large defects are required to allow any penetration.

In the experiments of Bunch *et al.* (2008) helium, nitrogen, and air were trapped in a micro-chamber 4.75μm × 4.75μm × 380 nm, etched into SiO_2 on a Si wafer.

The small drumhead membrane withstands a full atmosphere pressure difference. The membrane's degree of extension with overpressure (or of downward deflection in case of a negative relative pressure) are monitored with an atomic force microscope. It is basically found that the sealed membrane will hold a pressure difference for a time on the order of 24 hours, but that the loss rate is not through the graphene but through the walls of the micro-chamber. The raw data, in more detail, give rates of loss of helium, nitrogen and air, respectively, of 3x 10^5, 2 × 10^5 and 2 × 10^3 atoms per second. These rates, for each gas, were independent of membrane thickness from 1 atomic layer up to 75 atomic layers. The authors interpret the raw rate in terms of the transmission probability per helium atom impact, dN/dt (2d/Nv) $\leq 10^{-11}$, but state that "in all likelihood,... the true permeability is orders of magnitude lower than the bound given."

The theoretical work indicates that the barrier for penetration of helium through the center of a six-fold (benzene) ring in graphene is about 18.8 eV, with a barrier thickness about 1.43 Å. The authors model a helium atom colliding with the center of a ring, and find that the reflection has occurred before any relaxation of the membrane develops. This is shown in the Fig. 7.28.

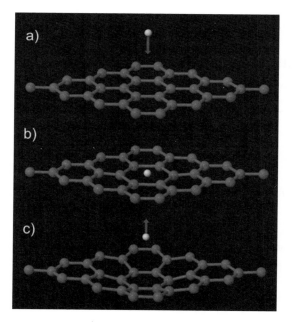

Fig. 7.28 Modeling the reflection of a helium atom of 18.6 eV from a graphene surface. (a) The atom approaches the perfect graphene layer. (b) The helium atom comes to rest. Note that the distortion of the graphene layer is very small at this moment. (c) The He atom is reflected and the surface starts to relax. (Reprinted with permission from Leenaerts *et al.*, 2008. © 2008 AIP).

The conventional WKB tunneling estimate for the transmission of a particle of mass 4 AMU through a barrier of height 18.8 eV and width 1.43 Å is $\sim \exp(-546)$, thus entirely negligible. The fullerene literature suggests that penetration does occur as an activated process at about 80 kCal/mole (or 3.47 eV) (Saunders *et al.*, 1993).

Saunders *et al.* (1993) conclude that the observed passage of helium through fullerene cage walls occurs by a mechanism where "one or more bonds reversibly break to open a 'window' allowing atoms to pass in or out." The rate for this mechanism for single layer graphene must fall below the limit observed by Bunch *et al.* (2008). The conclusion is indeed that graphene is impermeable to helium and other gases. Devices can be constructed using gas pressure to vary resonant frequencies, for example.

In a related development, Nair *et al.* (2012) discovered large permeability of water through graphene-based membranes, while still blocking gases including He. These Graphite Oxide membranes were composed of parallel-aligned, 1μm-sized, platelets of graphite oxide, with vertical spacings $d \sim 1$ nm between platelets. The thickness of the membranes is about 1 μm. The membrane blocks gases but transmits water, and the mechanism is described as an array of two dimensional capillaries between the platelets that becomes clogged with a monolayer of water. The water rapidly diffuses in the capillary network, but gases are blocked by the platelets and diffuse only slowly through the water contained in the network of capillaries.

7.4   Nanoelectromechanical resonators

Resonators made by suspending graphene layers, ranging in thickness from 0.3 nm to 45 nm over trenches in an oxidized silicon surface, have revealed resonant frequencies in the 1 MHz to 170 MHz range, with Q values from 20 to 850. The fundamental frequency was found, for doubly clamped beams, to be well fitted by the classical expression

$$f_0 = \left\{ \left[A(E/\rho)^{1/2} t/L^2 \right]^2 + A^2 0.57 \boldsymbol{T}/\rho L^2 wt \right\}^{1/2} \tag{7.41}$$

where E is Young's modulus, $\boldsymbol{T}$ the tension applied in Newtons, ρ the density; t, w, and L are the dimensions of the beam and the clamping coefficient A is 1.03 for a doubly clamped beam and 0.162 for cantilevers. The value of E was taken as 1 TPa. (Bunch *et al.*, 2007). The second term in this equation, linear in tension $\boldsymbol{T}$, is discussed by Mariani and von Oppen (2010), and the measurements seem to confirm the latter analysis. The success of the experimental fits to eqn 7.41 of Bunch *et al.* (2007) finding actual renormalized values of graphene parameters suggests that the renormalization effects are not excessive. The experiments of Bunch *et al.* (2007) are summarized in Figs. 7.29 and 7.30.

The graphene sheets in this study were mechanically exfoliated and draped across pre-existing trenches in oxidized Si wafers. This process does not involve solvent touching the graphene, and the measurements do not involve a large electric field from a

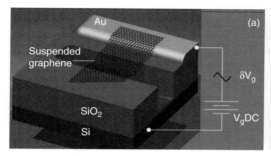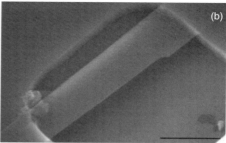

Fig. 7.29 (a) Graphene electro-mechanical resonator. Sketch of the mounting of graphene suspended across a trench in an oxidized silicon surface. The electrical contact is shown as evaporated gold that allows bias voltage and variable excitation frequency to be introduced. (b) Graphene as doubly clamped beam resonator. Micrograph of suspended graphene across trench in silicon dioxide surface. (From Bunch *et al.*, 2007, with permission from AAAS).

gate. The chance for tension T appearing in the graphene during the process of draping the film across the trench is mentioned by the authors.

Bunch *et al.* (2007), following Ekinci *et al.* (2005), define an effective spring constant for given mode as $\kappa_{\text{eff}} = m_{\text{eff}}\,\omega_0^2$ with $m_{\text{eff}} = 0.735Lwtp$. This gives $\kappa_{\text{eff}} = 30.75\,Ywt^3/L^3$ for a doubly clamped beam, close to the value given in eqn 7.17 and by Shivaraman *et al.* (2009). As mentioned before, this successful analysis, if extrapolated to a hypothetical 1 meter strain-free square of graphene at 300 K, predicts a thermal amplitude around 2 mm.

A related study of single layer resonators has been reported by Chen *et al.* (2009). In their work the gate voltage was used to vary the tension in the graphene. Their fabrication process involved chemical steps, the trench was removed after the graphene was deposited, and high current annealing was used to partially remove chemical residues from the graphene. By measuring the frequency of the device, and its changes with annealing, the authors were able to detect excess mass on the graphene. Chemically processed graphene layers in the work of Chen *et al.* (2009) routinely exhibited 2D mass densities two to six times larger than for pure graphene, $7.4 \times 10^{-19}\,\text{g}/\mu m^2$! It appears that the photoresist material PMMA poly(methyl-methacrylate) is adherent and difficult to remove from graphene. It was also found that the chemical residue was accompanied by built-in strain, with typical values 4×10^{-5} to 2×10^{-4}.

7.5 Metal-insulator Mott–Anderson transition in ultrapure screened graphene

The recent measurements of Ponomarenko *et al.* (2011) show, at low temperature, a conventional metal-insulator transition in ultra-pure electrostatically screened graphene, when the Fermi level is moved to the Dirac point. Their data (their

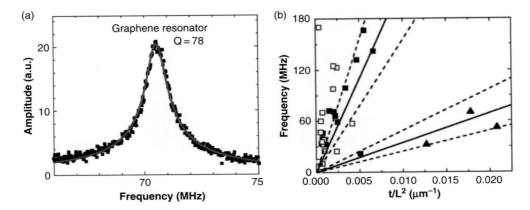

Fig. 7.30 (a) Amplitude curve for single-layer resonator (doubly clamped beam), with dimensions 0.3 nm thickness, 1.1 μm length and 1.93 μm width. With the correct density and $Y = E = 1\text{TPa}$, the resonant frequency with zero tension should be 5.4 MHz, while the observed value is 70.5 MHz. Using eqn (7.41), a tension $T = 13\text{nN}$ will give the observed frequency. The corresponding strain can be deduced from the relation $\Delta L/L = T/(YA) = \tau$ that yields tensile strain $\tau = 2.2 \times 10^{-5}$. This is a small strain, but its role in determining the resonant frequency is seen to be large. (b) Measured oscillation frequencies for 33 graphene electro-mechanical resonators ranging in thickness from monolayer to 75 nm. Square symbols are doubly clamped beams, open squares have thickness less than 7 nm, solid squares have thickness larger than 7 nm. The solid line is the prediction of eqn (7.41) for zero tension, taking Young's modulus as 1TPa and the density as 2200 kg/m³. The two dashed lines are calculated with Young's modulus values 2 TPa (upper) and 0.5 TPa (lower). Among doubly clamped beams, many points lie above the theory curve, and the expectation is that extraneous tension T is present from the mounting process. Points below the theory lines can arise from extra mass due to adsorbates. The triangle symbols and related theory lines are from singly clamped cantilever devices. In that case tension T is not a possibility. (From Bunch *et al.*, 2007, with permission from AAAS).

Figure 2b, inset) shows resistivity more than 500 kΩ/square at 10 K. This resolves the mystery of the "minimum metallic conductivity", thus revealed to be an (extremely subtle) experimental artifact. The electric field effect of inducing carriers, a defining feature of graphene as explained in the first works (Novoselov *et al.*, 2004; Zhang *et al.*, 2005, see Fig. 1.4) makes the local Dirac point, in an experimental sample, dependent on the local electric field. It is inevitable that electrostatic fields from charged impurities in the surroundings will broaden the Dirac point, leaving a distribution of conductive localities with positive and negative charge densities. This means that observing the true zero of carrier concentration is extremely difficult, and until the work of Ponomarenko *et al.* (2011) the expected Mott–Anderson transition at the Dirac point had never been convincingly observed.

The original idea of Mott (1949) (see also Mott, 1968) was of a metallic transition in a 3D array of hydrogenic atoms at increasing concentration, such that the electrons, once freed, screen away the electrostatic binding of the ionic centers. Transitions observed in semiconductors in fact are more influenced by disorder, as was explained by Anderson (1958), in important work that has been refined and reviewed by several authors, including Evers and Mirlin (2008). The related work of Abrahams *et al.* (1979), directly applicable to graphene, concludes that in 2D "there is no true metallic behavior; the conductance (of a sample of size L) crosses over smoothly from logarithmic or slower to exponential decrease with L." Thus, there is (in a disordered 2D system) no universal minimum metallic conductivity, and in all cases the conductance $g(L \to \infty) = 0$. This theoretical work offers no exception for conical bands. All of this work means that at low enough carrier concentration and low temperature an insulator must appear. The maximum value of resistivity more recently measured [see Fig. 1.6, due to Ponomarenko *et al.* (2011)] is in fact strongly temperature dependent, and reaches $33k\Omega$ (about $5.12h/4e^2$) at 10 K. (Their Figure 2b inset shows points above 500kΩ.) Although the title of the paper speaks of "Tunable metal-insulator transition in double-layer graphene heterostructures," this article is about the intrinsic metal-insulator Mott–Anderson transition in pure monolayer graphene when shielded from electric fields.

Measurements were normally made to show the insulating regime in the lower graphene layer (always found to have the higher mobility because of its encapsulation in BN) with fixed V_t, to establish a nearly constant carrier concentration n_c in the upper screening layer. It was found that the mobility of the exposed upper layer gradually decreased, while that of the encapsulated lower layer was stable. The screening concentration n_c was varied and found to have a large effect on the resistivity of the measured layer, believed due to screening of electrostatic fields away from the measured layer. The measurements were made with variation of the back-gate voltage V_b to vary the carrier concentration in the lower studied layer. It was found that the neutral point resistivity ρ^{NP} of the lower layer with the $d = 4$ nm spacer could easily be driven into the megaohm range below 4.2 K by maintaining a high carrier concentration in the control screening layer. The insulating phase appeared below 70 K, becoming more pronounced at lower temperature. The same effects were present if the roles of measured and screening layers were reversed, but higher resistivities approaching the insulating region were found when the fully encapsulated higher mobility layer was used as the sample.

As shown in Fig. 7.31(a), the tunable double-layer graphene heterostructure is conceived to introduce screening into the usual field-effect transistor geometry. The structure incorporates h-BN that has an inherently low random charged-impurity density. In more detail, the measured sample film is placed above an initial screening graphene layer, lying between the measured upper film and the biasing substrate, to reduce random electrostatic fluctuations. The tuning is achieved by the back gate to reach the Dirac point, as in all such measurements, see Fig. 1.4.

Fig. 1.6 shows a rapidly rising resistance with falling temperature, below 50 K, only when the carrier concentration in the measured film is less than about 10^{10} cm^{-2}. The data showing localization are obtained with a spacing 12 nm between the measured

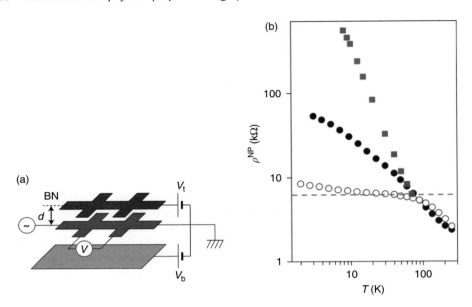

Fig. 7.31 (a) Double graphene layer experimental and measurement geometry used by Ponomarenko *et al.* (2011). Each Hall-configured sheet of monolayer graphene rests on an h-BN single crystal, the lower 20–30 nm thick layer placed on the underlying oxidized Si substrate, and the upper of thicknesses $d = 4$nm, 12 nm and 16 nm, such that no electron tunneling is possible between the two graphene layers. Separate Hall coefficient and resistivity measurements are made of the sample (lower layer), whose carrier concentration is controlled by the back-gate voltage V_b, and the upper screening layer. A bias V_t applied between the two graphene layers induces opposite charge densities, but does not result in current between the layers. The Hall voltage is measured in each graphene layer with vertical magnetic field, not shown, to monitor the concentration in each layer, at variable temperatures typically below 70 K. (b) Measurements of peak neutral point resistivity ρ^{NP} vs. T in double graphene layer experimental and measurement geometry shown in previous figure. Open circles and filled circles, respectively, are for $d = 12$ nm device with screening concentrations $n_C = 0$ and $n_C = 3 \times 10^{11} cm^{-2}$. (The resistivity curves for this device, when screened, were shown in Fig. 1.6.) Open squares show ρ^{NP} vs. T for $d = 4$nm device with screening concentration $3 \times 10^{11} cm^{-2}$. (From Ponomarenko *et al.*, 2011, by permission from Macmillan Publishers Ltd., © 2011).

and screening graphene films, and a high concentration 3×10^{11} in the screening film. The metal-insulator transition is not observed without the screening layer, and indeed a large carrier concentration in that layer is needed. The authors state that even using hexagonal boron nitride as encapsulating layers, but without the screening layer, "electrostatic puddles" are still induced in the measured layer. However, they are broader in extent and shallower in carrier density (than those in the traditional SiO_2/Si mounting) but still have carrier density above 10^{10} cm^{-2}. This unfortunately still leaves an inhomogeneous measured layer, where portions are conducting, local Fermi energy moved away from the Dirac point where the metal-insulator transition can occur. In such samples the resistivity is still determined by "inter-puddle ballistic transport with $\rho \sim h/4e^2$". We will return to models for this in Chapter 8, on anomalous properties of graphene, Section 8.2.

7.6 Absence of "intrinsic ripples" and "minimum conductivity" in graphene

From an experimental point of view graphene is extraordinarily susceptible to foreign atoms or molecules on its surface, for it truly exists only as surface. It is difficult to "clean" that surface in many experiments. The second basic property of graphene that it can be electrostatically doped, makes it susceptible to random local electrical doping in the presence of random local electric fields. It has turned out that the early substrate of choice, amorphous SiO_2, has surface roughness and charge centers. The graphene adheres to the uneven surface profile by van der Waals forces (see Fig. 5.1) and locally acquires positive and negative dopings by the action of the random local electric fields. It is thus not possible, on SiO_2, to maintain a graphene sample uniformly at the Dirac point: what one has instead is a distribution of positively and negatively conductive regions. The conical density of states means that rather large shifts in local Fermi energy are needed to maintain small surface charges, as required by externally generated electric fields. This effect is undoubtedly the reason for the minimum conductance usually seen (Figs. 1.4 to 1.6) unless extraordinary steps are taken (Section 7.5). Such steps minimize the "electrostatic puddles" mentioned by use of non-polar substrates and electrostatic screening, steps typically unavailable in applications. However, it is now clear that graphene in its intrinsic state has zero conductance at zero temperature as is required in a two dimensional metal by the work of Abrahams *et al.* (1979). The large body of theory devoted to "intrinsic corrugations" seems to have been in response to experiments, whose sophistication has increased, largely removing evidence for any intrinsic effect. The experimental reports of corrugations (See Section 7.1.4, as well as Figs. 5.1, 7.14, 7.16) have in many cases been retracted (e.g., Morozov *et al.*, 2006, see "note in proof") or superseded as better samples or better mounting surfaces were obtained. These details have frequently not been noted by theorists. For example, Gibertini *et al.* (2010) cite the paper of Morozov *et al.* (2006) (where the note in proof stated that improved samples did not show the corrugations) as support for "intrinsic ripples". The paper went on to base a complicated analysis, not on experimental data but on an artificial graphene surface, using results of an earlier simulation (Los *et al.*, 2009) said to model "ripples generated by thermal fluctuations." The height fluctuations in this modeled surface (given as $h_{\mathrm{rms}} \approx 9$ Å in Los *et al.,* 2009) exceeded by more than a factor of 10 that in the earlier modeling shown in Fig. 7.22 (0.7Å) for the same temperature. (There is no indication in the experimental reports that the heights of observed local surface irregularities, around 1.7 Å rms, on length scales 10 to 25 nm, and likely due to a combination of adsorbed atoms and sample strain, increase with the overall sample size.) In both cases the topograph is represented as static, although in any thermal origin the structure would be oscillating. It is hard to see how two sophisticated estimates for local thermal motions in a small sample of graphene could differ by a factor of 10. A simple estimate for thermal motion $h_{\rm rms}$ can be made based on eqn (7.18) above,

$$h_{\rm rms} = (k_{\rm B}T/K^*)^{1/2} = (k_{\rm B}T/32Yt^3)^{1/2}L, \qquad (7.42)$$

for a square of side L clamped on opposite sides. (Changing the boundary conditions to clamping on four sides, as treated in eqn (2.91) above would further reduce the estimate.) This estimate $h_{\rm rms}$ is backed by experimental work of Ekinci *et al.* (2005); Bunch *et al.* (2007); and Shivaraman *et al.* (2009). For $L = 10$ nm this gives a root-mean-square thermal oscillation $h_{\rm rms} = 0.182\,A$, smaller than the experimental estimates that, further, call for a static effect. (A suitable static effect was discussed above in connection with Fig. 7.27.) The frequency of the oscillation can be estimated from eqn (2.91) as

$$\omega = [Yt^2/12\,\rho(1-\nu^2)]^{1/2}\pi^2[2/L^2] \qquad (7.43)$$

that, taking $Y = 10^{12}\,Pa$, $t = 0.34$ nm, $\rho = 2270\,{\rm kg/m}^3$, $L = 10$ nm, and Poisson ratio $\nu = 0.165$, is $\omega = 4.1 \times 10^{11}$ rad/s $= 6.55\,GHz$.

It thus seems clear that thermal oscillations are not the origin of localized static surface irregularities, to the extent that these are still reported in improved experimental conditions. There remains no evidence for "intrinsic ripples," apart from obviously expected lattice vibrations that were treated accurately, e.g., by Krumhansl and Brooks (1953).

Gibertini *et al.* (2010) use their theoretical/modeling constructs to search for "electrostatic puddles." These might perhaps arise from the assumed "intrinsic corrugations," to support the failed notion of a "minimum conductivity." This concept was earlier indicated, see Figs. 1.4, 1.5, but is now disproven. As has been explained in Section 7.5, the "puddles," and related minimum conductivity, are not intrinsic, as they are removed by mounting graphene on h-BN substrates, and providing electrostatic shielding [see Figs. 7.6, 7.31(a), (b)]. The fundamentally expected Mott–Anderson transition (Abrahams *et al.*, 1979) has been observed. The graphene literature is left with theoretical papers justifying an intrinsic minimum conductivity of order e^2/h in graphene, all disproven by the results indicated in Section 7.5. The minimum conductivity of pure graphene at low temperature at the Dirac point is zero.

These experimentally supported clarifications of the pioneering literature, fortunately, make graphene less anomalous than was earlier imagined. Yet, there still remain more subtle anomalous features of graphene, treated in Chapter 8.

8
Anomalous properties of graphene

As we have seen, the mechanical properties of graphene are dominated by the large tensile strength and small resistance to bending, arising, respectively, from the strong covalent bonding and vanishingly small (one atom) thickness. The common formulas for bending beams quantitatively describe the behavior, on the micrometer scale, using conventional values of Young's modulus, etc. The tendency to wrinkle under conditions of strain is a well-known property of "inextensible" soft membranes, with many classical examples. The electronic properties of graphene might be considered (favorably) anomalous, but are adequately described by the usual tight binding theory of solids, based on Schrödinger's equation that reduces, with careful algebra, to the Dirac-like Hamiltonian in eqns (1.1–1.4). The main differences arise from the linear dependence, near the Fermi energy, of the band-edge energies on wavevector, first noted in 1947 and always fully encompassed by the Schrödinger-equation-based tight-binding methods of solid state physics. Beyond this, the lack of backscattering and Klein tunneling phenomena, related to the dual sublattice and spinor nature of the electron wavefunction, can reasonably be called anomalous and deserve more discussion in this chapter.

Anomalous properties, that are still not fully understood, include the high temperature disintegration of graphene, our first topic.

8.1   Sublimation of graphite and "melting" of graphene

Graphite is stated as sublimating at 3900 K, which plausibly means that graphene layers detach from the graphite crystal. This implies that the melting or disintegration temperature of graphene itself is higher than 3900 K and we thus expect that graphene layers exhibit local order at 3900 K. Melting in a strictly 2D situation is similar to the familiar 3D cases: a loss of local order at nearly constant density. In the case of graphene (one atom thick) in real three-dimensional space, the film will disintegrate into fragments moving to infinity leaving no remnant. Important progress has been reported by Zakharchenko *et al.* (2011) on atomistic modeling of such disintegration process of the graphene layer. Briefly, these authors, confirming the original report of Ito and Nakamura (2007), suggest that graphene "melts" at about 4900 K into a space-filling anomalous "liquid" of carbon strings, in the nature of polymer strands. Part of the nomenclature problem, suggesting a "liquid," rather than molecular fragments departing ballistically to infinity, may come from the simulation strategem of a fixed volume with periodic boundary conditions, so that the final density is finite rather than zero. The extensive literature of 2D melting cannot fully explain this, because that literature deals with matter confined to a plane, while the disintegrating graphene easily explores the third direction. The further literature of "tethered membranes," with predictions of "crumpling," does not emphasize the rigid covalent bonding that exists in graphene, resulting in its large Young's modulus. However, the idea of "local crumpling" does seem to play a role in the disintegration of graphene, assuming the validity of the recent simulations.

Briefly, it seems that the loss of local order, via planar defects like five- and seven-membered rings (see Fig. 7.25), first appears in regions that are "locally crumpled," i.e., displaced from the initial plane. On the other hand, micrometer-size samples probably do not "crumple" in the sense of undergoing displacements on the scale of L, following formulas 2.50 and 2.51 in the earlier text, even at 4900 K. (See text above eqn 2.51.) Below 4900 K, it may be that the classical elastic treatments, as of bending beams and sheets or plates, are a good guide, since the local order of the elastic membrane holds up to near 4900 K, where a proliferation of defects first appears in the simulations. In Chapter 2 we presented a rough estimate that a *1 meter square* of graphene, hypothetically unsupported in thermal equilibrium at 300 K, would have a flexural vibration amplitude near 2 mm, about 0.2% of the length L and that at 3900 K the amplitude would increase to about 7 mm.[1]

The classical fits to the data found by Bunch *et al.* (2007) would predict thermal amplitudes falling off with frequency and inverse wavelength, note that an example of a short wavelength flexural mode is shown in Fig. 7.9.

The first report to reveal the essential nature of the disintegration of graphene was that of Ito and Nakamura (2007), one of whose original figures, at simulation temperature 5800 K, is entitled "Creation of Chain-like Carbon Molecule by Graphene Melting." Their work was in connection with the use of graphite as a first-wall material in a fusion reactor. It was found that the disintegration products are linear carbon chain fragments, rather than atoms. Ito and Nakamua (2007) note that "almost all six-membered cyclic structure" remains in nearby preserved regions. [These linear chain molecules are evident in Fig. 8.1(b)]. Carbon-chain fragments had earlier been identified as prominent in the liquid phase resulting from melting fullerenes, in a molecular dynamics study (Kim and Tomanek, 1994).

In Fig. 8.1(a) and (b) are shown recent extended simulations performed by Dr. Atsushi Ito and Dr. Hiroaki Nakamura, using the method of Ito and Nakamura (2007). These two images are simulations of a 40 nm × 40 nm square of graphene at 5400 K, with time elapsed as 100 ps in the first and 500 ps in the second. The simulation thus includes 64 000 atoms, the space simulated is a 40 nm cube and periodic boundary conditions are used so that atoms emitted from the top reappear at the bottom. The authors note that almost all of the emitted carbon is in "carbon-chain" forms.

A separate line of work relating to the disintegration of graphene is indicated in Fig. 8.2. Calculations (Kowaki *et al.*, 2007) of the melting temperature of carbon nanotubes as a function of radius shows a smooth curve extrapolating to about 5800 K at infinite radius. The relation between the indices n, m and the carbon nanotube diameter were given in eqn (5.10).

---

[1] This assumes one could monitor the invisible film, assumed floating freely in gravity-free vacuum, still in temperature equilibrium.

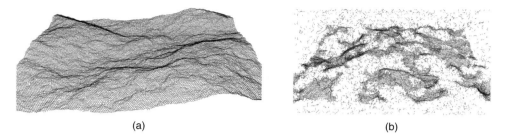

(a)   (b)

Fig. 8.1 (a) 40 nm by 40 nm square of graphene melting at temperature 5400 K simulated in 40 nm cube with periodic boundary condition. Snapshot after 100 ps. Step time of 5×10^{-18} s, using Brenner potential. (b) Disintegration of graphene 40 nm by 40 nm square of graphene melting at temperature 5400 K simulated in 40 nm cube with periodic boundary condition. Snapshot after 500 ps. Step time of 5×10^{-18} s, using Brenner potential. (Courtesy of Atsushi Ito).

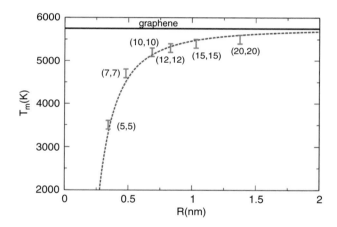

Fig. 8.2 The radius-dependence of the melting temperature T_m of single-wall carbon nanotubes as estimated from the temperature dependences of the radial distribution functions, mean-square deviations and atomic configurations. The upper curve represents a separate simulation on a graphene sheet. (Reprinted with permission from Kowaki *et al.*, 2007. © IOP Publishing 2007).

An earlier theoretical literature on liquid carbon, at 5000 K and above, was refined and compared to experimental measurements of radial distribution functions of quench-condensed amorphous carbon. Reasonable agreement was achieved and some results are now shown in Fig. 8.3. Tetrahedral bonding dominates in liquid carbon and in amorphous carbon, while the decay product of graphene disintegration seems to involve chain-fragments with double- and triple-bonds as in linear molecules of

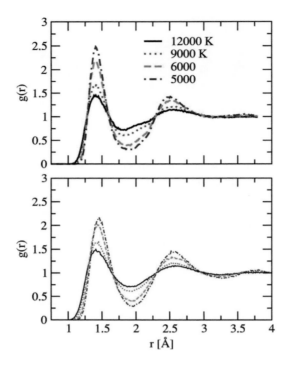

Fig. 8.3 (upper) Comparison of radial distribution functions for liquid carbon at (top to bottom, in first peak) 12 000 K, 9000 K, 6000 K and 5000 K, using potential referred to as LCBOPII (see also Brenner *et al.*, 2002). The density assumed is 2.9 g/cc. (lower) The lower panel is from work of Harada *et al.* (2005) using a 64 atom simulation with local density approximation LDA in the Density Functional Molecular Dynamics DFMD method, regarded as more accurate but requiring more computer time. These results are consistent with the earlier work, Marks (2000). (From Ghiringhelli *et al.*, 2005. © 2005 by the American Physical Society).

carbon. For example, Marks (2000), in their Figure 3, show a pair distribution function $g(r)$ for 2.9 g/cc liquid carbon, simulated using *ab initio* methods, with results well-describing experimentally studied (condensed) amorphous carbon. Experimental measurements find amorphous carbon ("amorphous diamond") at density 2.9 g/cc. The liquid in the Marks (2000) figure was equilibrated at 5000 K in the simulation, where it exhibited high diffusivity, as expected for a liquid phase. This line of work validates the methods more recently used to simulate the melting of graphene that seems essential, since experiment in this area is probably impossible.

Experimental measurements find amorphous carbon ("amorphous diamond") at density 2.9 g/cc. See for example McKenzie *et al.* (1991) Figure 7, who plot $G(r)$ for undoped vacuum-arc-deposited carbon that resembles the computed radial distribution function for liquid carbon.

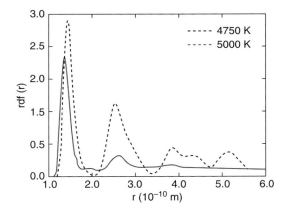

Fig. 8.4 Radial distribution functions for graphene above (solid curve) and below (dashed line) modeled melting transition. The transition is seen to reduce the nearest-neighbor distance (attributed by the authors to chain formation) and to wash out further structure typical of the crystalline phase. Comparison of the melted phase, here, to liquid carbon seen in earlier figures, makes clear that typical distributions observed, as time progresses toward disintegration, do not resemble those of conventional liquid carbon. (Reprinted with permission from Zakharchenko *et al.*, 2011. © IOP Publishing 2011).

The temperature where graphene disintegrates seems difficult to assess experimentally. The simulations generally have been started at a temperature high enough that disintegration is expected and show change in the atomic positions as time progresses. A recent simulation study of Zakharchenko *et al.* (2011) shows radial distributions of decay products (see Fig. 8.4) with peaks at smaller radial distances than in liquid/amorphous carbon, suggesting bonding similar to that in linear carbon molecules, in agreement with the earlier work of Ito and Nakamura (2007).

The simulations were done with fixed number N of atoms assuming pressure zero and kinetic energy set by the temperature. The article is not clear on the use of periodic boundary conditions. At one point in the article it is stated that ring defects occur at temperatures as low as 3900 K. Some information on this is given in our Fig. 2.10, relating to the melting criterion in 2D.

Figure 8.5 depicts a snapshot of graphene at $T = 5000$ K, during the melting process, in a Monte Carlo simulation. Eroded void regions display sparse linear chains, while large parts of the sample are intact with hexagonal order, as earlier found by Ito and Nakamura. This figure depicts a transient: a snapshot in a disintegration process that is complete on a time scale of approximately 500 ps.

In Fig. 8.6 are shown details of the melting process with passage of time, as revealed by numbers of bonds of several types. Hexagons give way to chains of various lengths, as shown.

We earlier showed in Fig. 2.10 of Section 2.4.1.2 the corresponding simulated temperature dependence of a Lindemann-differential-vibrational-amplitude melting

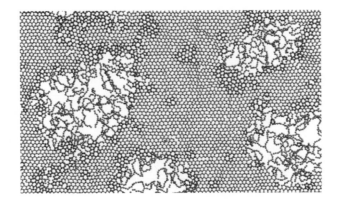

Fig. 8.5 Snapshot of graphene at $T = 5000$ K during the melting process in a Monte Carlo simulation using the LCBOPII potential. Eroded void regions display sparse linear chains, while large parts of the sample are intact with hexagonal order. Boundary layers surrounding the voids are marked to show high concentration of pentagons and heptagons. (Reprinted with permission from Zakharchenko *et al.*, 2011. © IOP Publishing 2011).

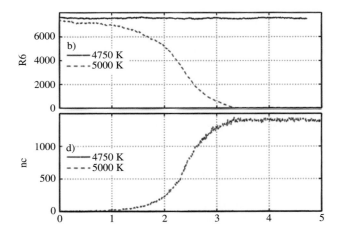

Fig. 8.6 Indication that graphene is stable at 4750 K (uppermost curve), but disintegrates at 5000 K. Number of hexagons (R6, upper panel) and number of chains (nc, lower panel) in simulation of 16 128 atom sample at temperatures 4750 K (upper curve in upper panel; lower curve, essentially zero, in lower panel). Number of chains is counted as number of cases with more than three connected two-fold-coordinated atoms. At 5000 K, melting occurs after about 2.5×10^7 Monte Carlo steps (MC), starting from flat graphene layer of 16 128 atoms. (Reprinted with permission from Zakharchenko *et al.*, 2011. © IOP Publishing 2011).

criterion for graphene, indicating melting at 4900 K. [In that figure, the lowest curve (circles) is based on 3-nearest-neighbor distances, the intermediate curve (crosses) based on 12 neighbors and upper curve (diamonds) based on 9 nearest neighbors.]

These authors state "the molten state forms a three-dimensional network of entangled chains rather than a simple liquid," and speak also of "the molten phase," that is only suggestive in a disintegration. From the point of view of the earlier literature on the breakup of 2D layers, it does seem reasonable to describe the disintegration as initiated by "crumpling". In the original work of Ito and Nakamura (2007) (Fig. 8.1) it appears that large vertical waves on the graphene appear before any loss of local order. The local order initially erodes in small areas of the distorted plane with larger local strains, making the loss of local order inhomogeneous across the area of the initial graphene sample. The local order does seem to decay following the suggestions of Nelson (1982) of generation of dislocations (see Section 2.4.1.2). The products of the distortion-initiated disintegration are molecular chain fragments, not atoms and, in a transient sense, the result of the disintegration is a disordered "liquid phase" of lower density than the well-studied liquid phase of 2.9 g/cc carbon. The simulations suggest that a single graphene layer heated to 5000 K will disintegrate and disappear into surrounding vacuum on a sub-nanosecond time scale. One might say that the disintegration originates in a "local crumpling," but not "membrane crumpling" in the sense of displacements on the scale L of the sample itself, as was much earlier predicted by Nelson and collaborators, leading to eqns (2.50) and (2.51), above. Equation (2.50) leads to a coherence length 4.0 µm at 4900 K, a scale much larger than that of the simulated samples, where L is on a nanometer scale.

If this estimate is correct, then one can say that graphene does not "crumple" but rather disintegrates inhomogeneously starting from local melting in regions of out-of-plane displacement.

The behavior as simulated does not indicate a strong role for the considerations of the HLMPW theorem (Mermin, 1968), that applies to a strictly 2D melting, whereas the actual disintegration occurs into 3D.

8.2 Electron and hole puddles, electrostatic doping and the "minimum conductivity"

As was clearly explained in the pioneering work (Novoselov *et al.*, 2004), graphene is quite unusual, exhibiting an "Electric Field Effect in Atomically Thin Carbon Films." Conventional metal films have a screening length on the order of 1 nm that shields the metal bulk from electric field and the usual case is that bulk metal carrier concentrations are large compared to any surface charge that can be induced by the field effect. The key observation of Novoloselov *et al.* (2004) and Zhang *et al.* (2005) was that, in graphene, carrier concentrations n up to $10^{13}/\text{cm}^2$ could be induced by applying a gate voltage V_G (see Figs. 1.4, 1.5), following a simple formula $n = \varepsilon\varepsilon_0 V_G/de$, where d is the thickness of a dielectric of permittivity ε. In a frequent case of graphene placed on quartz grown on doped Si, 100 V applied across a 300 nm quartz layer induces $7 \times 10^{12} \text{ cm}^{-2}$. Because of the linear density of states spectrum and the zero DOS at the Dirac point, this gives a large change in the Fermi energy, on the order of 0.28 eV. In more detail, in 2D the density of states $g(E)$ for graphene can be taken as

$$g(E) = g_0 \, |E|/A', \tag{8.1}$$

where $A' = 5.18$ Ångström^2 is the area of the graphene unit cell and $g_0 = 0.09/(\text{eV}^2$ unit cell) (Giovannetti *et al.*, 2008). To accommodate N total states per unit area, we have

$$N = \int g(E)\mathrm{d}E, \tag{8.2}$$

so that

$$N = E_\text{F}^2 g_0/2A', \tag{8.3}$$

to give $E_\text{F} = 0.28$ eV for the example. This useful electric field effect unique to graphene has a corollary that stray electric field can locally modulate the Fermi energy, along with the local charge density, especially when E_F is near the Dirac point. It is estimated by Giovannetti *et al.* (2008) that a transfer of 0.01 electron per unit cell shifts the Fermi energy by 0.47 eV. This makes it difficult to prepare a homogeneous sample near the Dirac point and, in this regard, graphene placed on hexagonal BN is more ideal than graphene placed on SiO$_2$.

The difficulty entailed by this sensitivity in the accurate measurement of the graphene conductivity near the Dirac point was not immediately recognized. This practical result of a perceived minimum conductivity is important experimentally, but can now be regarded as an artifact, not representing a fundamental property of graphene. The metal-insulator transition in ultra-pure graphene was reported by Ponomarenko *et al.* (2011), removing a mystery, but leaving a now confusing literature devoted to an explaining, as intrinsic, an effect that only appears in non-ideal samples. Of course, the effects remain important in device applications using typical samples. The non-ideal effects were investigated by Zhang *et al.* (2009) and by Martin *et al.* (2009). A random resistor network model of the (extrinsic) minimum conductivity in graphene is given by Cheianov *et al.* (2007).

In the low density limit, near the Dirac point, where the carrier concentration becomes smaller than the charged impurity density, the non-ideal system breaks up into puddles of electrons and holes, where a duality in two dimensions guarantees that locally transport occurs either through the hole channel or the electron channel.

Percolation resistance arising from these electron and hole puddles is important in understanding the experimentally observed minimum conductivity in real samples mounted on electrostatically "noisy" substrates. Important contributions to understanding the puddles have been made by Adam *et al.* (2007), Galitski *et al.* (2007), Hwang *et al.* (2007), Rossi and Das Sarma (2008) and Tan *et al.* (2007). The situation was addressed and reviewed by Blake *et al.* (2009), before the work of Ponomarenko *et al.* (2011).

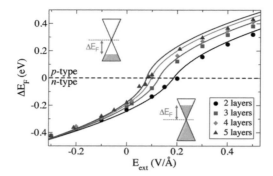

Fig. 8.7 Electrostatic doping of graphene in a specific structure, with a field-electrode of copper separated by 2 to 5 monolayers of h-BN. This electrostatic doping effect illustrates the origin of the troublesome disordered puddles in common circumstances where stray electrostatic fields exist. It also is the basis for field-effect transistors where source and drain electrodes are electrostatically charged. (Reprinted with permission from Bokdam *et al.*, 2011. © 2011 American Chemical Society).

The basic effect arises because the monolayer cannot shield an external electric field and the charge induced by the field, through the usual relation $E = \sigma/\varepsilon_0$, where σ the surface charge density and ε_0 the permittivity, shifts the graphene Fermi level. A recent calculation of this shift, due to Bokdam *et al.* (2011) (see eqn. 9.7 below), is shown in Fig. 8.7.

8.3 Giant non-locality in transport

A remarkable non-local potential gradient, appearing only at substantial magnetic field, has been measured in graphene monolayers by Abanin *et al.* (2011). The effect is larger in more perfect samples mounted on BN, but is present also in the usual samples on SiO_2.

If one imagines a Hall-bar geometry, [see Fig. 8.8 (a)] with terminals numbered, from one end, as 1–6, with 1 and 4 carrying current (slightly different from Fig. 2.5), a longitudinal resistance is defined as

$$R_{2,3} = V_{2,3}/I_{1,4}. \tag{8.4}$$

If the width of the Hall-bar is w and the spacing between potential terminals is L, then the longitudinal resistivity is

$$\rho_{xx} = (w/L)\, R_{2,3}. \tag{8.5}$$

In contrast, a *non-local resistance* is exemplified [see Fig. 8.8(c)] by

$$R_{NL} = V_{3,5}/I_{2,6}. \tag{8.6}$$

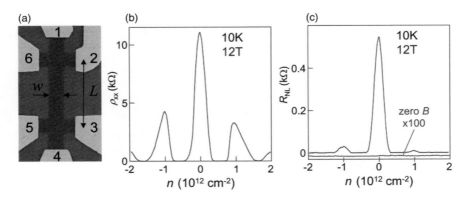

Fig. 8.8 Measurement geometry, showing local and non-local resistivity, at 10 K and 12 T, of graphene /SiO$_2$ (GSiO) device. (a) TEM of device, with $w = 1$ μm showing definition of L. (b) Vertical current flow, ρ_{xx} vs. carrier density n in perpendicular field 12 T. The features are from the anomalous quantum Hall effect, showing prominent peak at zero and at Landau level indices $\nu = \pm 4$ and ± 8. (c) Horizontal current flow through terminals 2, 6, measuring voltage at terminals 3, 5. Zero magnetic field (bottom curve, displaced and amplified) shows no non-local resistance. At 12 T, anomalous non-local resistance R_{NL} is similar to ρ_{xx}. R_{NL} was confirmed to be independent of drive current magnitude. (From Abanin *et al.*, 2011, with permission from AAAS).

The current $I_{2,6}$ now flows *across* the bar and one expects a voltage between these terminals 2, 6, but hardly any voltage across parallel terminals 3, 5 displaced by L along the bar. In a normal conductor, the current density between terminals 2 to 6 would be confined to the vicinity of those contacts and the usual construction of equipotentials and current flow lines would predict a very small voltage to appear between terminals 3 and 5. This expectation of locality is quantified as

$$R_{\text{NL}} = V_{3,5}/I_{2,6} \propto \rho_{\text{xx}} \exp(-\pi L/w) \qquad (8.7)$$

(van der Pauw, 1958). In the cited recent graphene measurements, the concentration of electric field between the current terminals is in effect defeated by the magnetic field: an anomalously large voltage persists across parallel-displaced terminals, but only when a magnetic field is applied. The authors suggest that a *spin current* diffuses in the longitudinal direction and that spin imbalance, diffusing away from region 2, 6, arrives near region 3, 5 where it generates a voltage gradient. (A spin current can be envisioned as a combination of a current of spin-up electrons in one direction with a current of spin-down electrons in the opposite direction. The resulting transport is of spin angular-momentum rather than of charge.) These aspects are shown in Figs. 8.8 and 8.9. The graphene bands near the Dirac neutral point, in magnetic field, break up into two pockets, of spin-down electrons and spin-up holes that move oppositely from terminals 2 to 6, under drive current. The Lorentz force $\mathbf{F} = q\,\mathbf{v}\mathbf{x}\mathbf{B}$

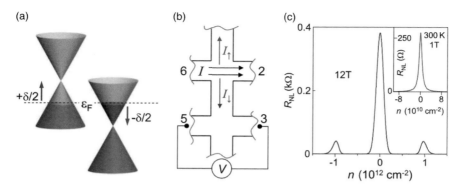

Fig. 8.9 Spin Hall effect (SHE) model mediated by spin diffusion for non-local resistivity at 10 K and 12 T, of graphene /SiO$_2$ (GSiO) device. (a) Zeeman splitting at charge neutrality produces two pockets, filled with spin-down electrons and spin-up holes. (b) Horizontal current flow $I_{2,6}$, produces net spin current but no charge current in the longitudinal (vertical) direction. (c) Model of Abanin *et al.* resembles data of Fig. 8.8(c). (From Abanin *et al.*, 2011, with permission from AAAS).

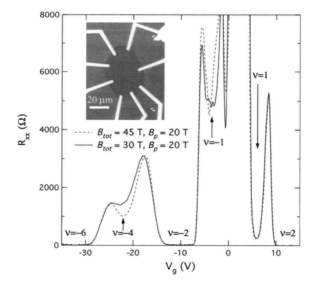

Fig. 8.10 Observation of plateaus in Hall resistance h/νe^2 of graphene sheet, at 4.2 K and 18 T, with $\nu = \pm 2, \pm 6$ and ± 10. Filling factor values ν are indicated and change of sign relates to electron vs. hole character of the levels. Inset shows device geometry for electrical measurement. Jiang *et al.* (2007) also measured absorption of infrared radiation confirming the levels shown in Fig. 8.11. Those data were previously shown in Fig. 6.2 of Section 6.2. (From Jiang *et al.*, 2007. © 2007 by the American Physical Society).

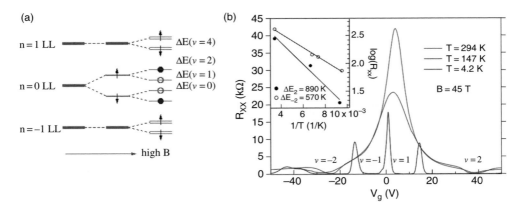

Fig. 8.11 (a) Schematic of splitting of graphene Landau levels at high magnetic field B. The arrows represent spin of the carriers and the open and closed circles represent different valleys in the graphene band structure, K and K'. (b) Splitting of graphene Landau levels measured at high magnetic field $B = 45$ T, at several temperatures. The zero level is clearly seen at room temperature. In this figure the neutral point is near zero gate voltage that controls the carrier density and the central peak that is split occurs at this Dirac point. (From Jiang *et al.*, 2007. © 2007 by the American Physical Society).

diverts electrons and holes in the same direction, since the product qv is the same for each, leading to no net transverse charge current, as measured. While this simple argument suggests spin-down particles (electrons) and spin-up particles (holes) deflect in the same direction, note that the spin of the hole resides on the missing electron and thus is deflected oppositely. A spin current is thus generated. Refinements, as given by Kane and Mele (2005) and Castro Neto (2011), confirm the conclusion of Abanin *et al.* (2011) that, in fact, the spin-up and spin-down particles are diverted oppositely, see also Kato *et al.* (2004), leading to a net spin current. This interpretation is outlined in Fig. 8.9, including modeling of the nonlocal resistance in Fig. 8.9(c), that is close to the observation. This version of the spin-Hall effect (SHE) does not involve spin-orbit coupling. The temperature- and field-dependences of the observed R_{NL} were measured (not shown) to help understand its origin. The Landau-level peaks for $\nu = \pm 4$ and ± 8, shown in Fig. 8.8(c) disappear above 70 K. This fact, in the view of the authors, indicates that their origin is in edge states of the quantum Hall effect that have a similar temperature dependence. Since nonlocal observation of the zero-index Landau level $\nu = 0$ persists to much higher temperature, even to 300 K at low magnetic field, the authors argue that its origin is not connected with the quantum Hall effect. This zero peak in R_{NL} has two temperature regimes: at low temperature it rises strongly, but, at high temperature, exhibits a slower decay. The authors associate the low temperature effect with a gap opening in the zero Landau level, while the high temperature non-locality arises as a bulk transport mechanism, possibly as sketched in Fig. 8.9. The non-local effects are 10 to 100 times larger in GBN samples (mounted on hexagonal boron nitride), that have higher mobility and, presumably, longer spin relaxation times. The authors further suggest a role for a bulk transport mechanism, as they have described, also in the low temperature peak at the Dirac point, consistent with earlier suggestions (Checkelsky *et al.*, 2009; Jiang *et al.*, 2007; Du, *et al.*, 2009; and Dean *et al.*, 2010). An intuitive discussion of these effects is given by Katsnelson (2012), see his Section 11.5. There are many aspects of these effects that remain to be clarified. For example, it is suggested that imbalance in occupation of the inequivalent valleys K, K' could contribute to the non-locality.

8.4   Anomalous integer and fractional quantum Hall effects

The quantum Hall effect has been reviewed in Chapter 2 following a summary in Chapter 1 of the features of the effect unique to graphene. The effect is characterized by Hall resistance R_{xy} exhibiting quantized plateaus in regions where $R_{xx} = 0$, with peaks in longitudinal resistance R_{xx} at Landau levels (see Figs. 2.3–2.7). The plateau values for graphene are given by (Zhang *et al.*, 2005)

$$R_{xy}^{-1} = \pm g_s \left(n + \frac{1}{2} \right) e^2 / h, \tag{8.8}$$

where $n = 0, 1, 2, \ldots$ and the degeneracy $g_s = 4$ for graphene, taking into account spin- and sublattice-degeneracies. The connection with the conventional filling factor is, for integer n,

$$\nu = g_s \left(n + \frac{1}{2} \right). \tag{8.9}$$

The filling factor is also given as $\nu = ne/hB$, where n is now the carrier density. The two anomalous features of the quantum Hall effect (QHE) in graphene are the level at zero energy and the square-root-dependence of the succeeding Landau level energies on magnetic field B, shown in eqn 1.5:

$$E_n = \pm \nu_F \left[2e\hbar B \left(n + \frac{1}{2} \pm \frac{1}{2} \right) \right]^{1/2} \quad \text{where } n = 0, 1, 2, \ldots \tag{8.10}$$

These features were shown in Figs. 1.8, 1.9 and also in Fig. 8.8(b). The zero Landau level, at the neutrality point [prominent in Fig. 1.9(a)], is shared by hole- and electron-like carriers. The unequal spacing of levels is a consequence of the linear energy dispersion near the Dirac point. The observation of the Landau levels has recently been reported in capacitance spectroscopy (Ponomarenko *et al.*, 2010) as shown in Fig. 6.8, as well as in scanning tunneling spectroscopy STS (Miller *et al.*, 2009) as shown in Fig. 1.9(b), and in infrared spectroscopy by Jiang *et al.* (2007).

Infrared spectroscopy has also been used to find asymmetry in the electron band structure of bilayer graphene by Li *et al.* (2009).

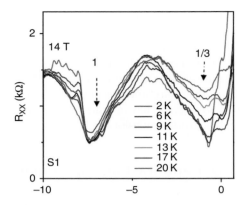

Fig. 8.12 Fractional quantum Hall effect. Graphene suspended across a trench with four terminal measurement. Longitudinal resistance, measured at 14 T and several temperatures, shows minimum corresponding to filling factors 1 and 1/3. (From Ghahari *et al.*, 2011. © 2011 by the American Physical Society).

The high quality of conduction in graphene permits observation of the quantum Hall effect at room temperature, first reported by Novoselov *et al.* (2007).

Beyond the quantum Hall states at $\nu = \pm 2$, ± 6 and ± 10 (first observed by Novoselov *et al.*, 2005 and Zhang *et al.*, 2005), broken-symmetry Integer quantum Hall states at $\nu = 0$, ± 1 and ± 4 were found by Zhang *et al.* (2006) and Jiang *et al.* (2007).

The fractional quantum Hall effect (FQHE) in graphene has been reported by Du *et al.* (2009) and Bolotin *et al.* (2009).

More details of the $\nu = 1/3$ fractional quantum Hall effect have been reported by Ghahari *et al.* (2011), using suspended graphene as shown in Fig. 8.12.

The longitudinal resistance for these states does not reach zero, but they are regarded nonetheless as quantum Hall states. The minimum value R_{xx} for $\nu = 1/3$ has a noticeable temperature dependence, fit as $R_{xx}{}^{\mathrm{min}} \sim \exp(\Delta E/k_B T)$ at fixed llarge field B. The authors find that the activation gap energy ΔE increases approximately as $\sqrt{B}$ (although a linear fit is almost as good) to about 25 K at 14 T.

The origin of the fractional quantum Hall Effect (originally discovered in silicon devices, see Tsui *et al.*, 1982 and D. C. Tsui, Nobel Lecture, 8 Dec., 1998) lies in repulsive electron–electron interactions, leading to particular forms of localization, in the state first described by Laughlin (1983) (see also Laughlin, 1998). Laughlin 1983 describes the 1/3 state as a fundamental state of 2D matter, comprised of electrons that condense at a particular density, 1/3 of a full Landau level. The states are capable of carrying electric current at little or no resistive loss and have a Hall conductance $1/3\ e^2/h$. Laughlin 1983 gives wavefunctions that are Gaussian $\exp(-\alpha r^2)$ in the radius r of electrons circling the magnetic field with conserved angular momentum (see text following Fig. 2.7). If this description applies to the state observed in graphene,

Ghahari *et al.* 2011 estimate that the activation energy would scale as $e^2/\kappa\varepsilon_0 l_B$ where κ is the permittivity and l_B is the magnetic length (cyclotron radius), that is proportional to $1/\sqrt{B}$. Namely, the magnetic length is

$$l = (\hbar/eB)^{1/2}\,, \tag{8.11}$$

(e.g., 6 nm at 18 T). This would agree with a $\sqrt{B}$ fit to the observed activation energy and the numerical value is reasonable taking permittivity as 5.2 and comparing to a detailed calculation of Apalkov and Chakraborty (2006). On the other hand, if the fit is really linear, then a model proposed by Dethlefsen *et al.* (2006) would be more appropriate.

Feldman *et al.* (2012) have successfully used the scanning single electron transistor (SSET, see Chapter 6) to find a large number of new fractional quantum Hall states in monolayer graphene. The measured quantity in this SSET work is the inverse compressibility $d\mu/dn$ (μ representing the Fermi energy), that is directly related to the density of states. The high quality of the data is related to their use of suspended graphene. By modulating the carrier density in the graphene and monitoring the resulting change in SET current they measure both the local chemical potential μ and the local value of $d\mu/dn$. They show (in their Fig. 1B) a color scale image of $d\mu/dn$ the ranges $B = 0$ to $B = 12$ T and carrier density $n = 0$ to $n = \pm\,3 \times 10^{11}$ cm^2, where 23 distinct integer and fractional quantum Hall states are revealed, in the inverse compressibility. Feldman *et al.* (2012) find that the scanning capability of the SET device is useful to assess how these many states are influenced by local disorder in the sample. It is clear that understanding all of these effects will take some time. In the words of the authors "observation of incompressible behavior at multiples of $\nu = 1/9$ indicates a substantial improvement in sample quality.". . "Graphene provides an especially rich platform in which to investigate correlated electronic states and their interplay with underlying symmetry."

8.5   Absence of backscattering, carrier mobility

Our text, in Chapter 1, including eqns. 1.1 to 1.4, described the absence of carrier backscattering in monolayer graphene as arising from the forbidden nature of reversing the pseudo-spin, essentially because the forward moving carrier is built on sublattice A and the reverse moving carrier is built on sublattice B, thus orthogonal electronic states. The states in question are centered on the inequivalent valleys K and K' (see Brillouin Zone in Fig. 1.2). These points were described by Semenoff (1984) as "right- and left-handed degeneracy points." Experimental evidence for no backscattering and a clear discussion were given by McEuen *et al.* (1999). These workers found that the mean free path in metallic carbon nanotubes was about 100 times longer than in similar quality carbon nanotubes that were semiconducting. They attributed this large difference to the change in the band structure from conical in the metallic case to parabolic in the semiconductor case. This is relevant to graphene, because metallic carbon nanotubes and graphene have similar bandstructures.

McEuen *et al.* (1999) analyze the important lack of backscattering that increases the carrier mobility in graphene. The starting point is the validity, near **K** and **K'**, of the 2D Dirac Hamiltonian, reached by the **k** · **p** approximation in tight binding theory (Yu and Cardona, 2010)

$$\hat{H} = b\,\hbar c^* \boldsymbol{\sigma} \bullet \mathbf{k} \tag{8.12}$$

(see eqn 1.1), where $b = 1$ (-1) for states above (below) those at point **K**. Features are shown in Fig. 8.13.

McEuen *et al.* (1999) write the wavefunction as

$$|\mathbf{k}> = (1/\sqrt{2})\,e^{i\mathbf{k}\cdot\mathbf{r}} \begin{pmatrix} -ib\,\exp(-i\theta_{\mathbf{k}}/2) \\ \exp(i\theta_{\mathbf{k}}/2) \end{pmatrix}. \tag{8.13}$$

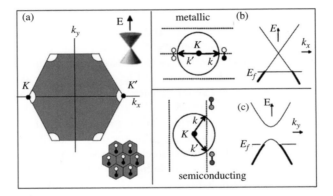

Fig. 8.13 Depiction of conduction and scattering in p-type metallic nanotube, applicable to p-type graphene. (a) Brillouin zone [see Figs. 1.2(a) and 1.3(b)] with shading indicating filled electron states. Fermi surface includes circular cone-sections of radius k at the degeneracy points K, K′ and their symmetry-induced counterparts. Upper inset in extended zone scheme collects together three Fermi arcs to form cones at K, K′. Lower inset indicates two carbon atoms per unit cell. (b) Fermi circle in metallic nanotube is shown centered in K degeneracy point. Arrows **k** and **k′** representing Fermi-energy holes, are associated, respectively, with bonding (white-black dumbbells) and anti-bonding (white-white dumbbells), indicating the nature of the underlying molecular states. The orthogonality of the **k** and **k′** basis states forbids backscattering transitions from **k** to **k′** (scattering angle π). (In Fig. 1.4 the same prohibition is described as requiring reversal of the pseudo-spin.) In (b) and (c), respectively, horizontal and vertical dashed lines represent allowed wavevectors specific to the assumption of a nanotube with size quantization. (c) Assumption of semiconducting nanotube, with energy gap and parabolic band dispersion, similar to bilayer graphene, with illustration of scattering angle $\theta_{\mathbf{k},\mathbf{k}'} < \pi$. In this case the molecular basis states are mixed (gray dumbbells) and not orthogonal, allowing substantially greater scattering rates. (From McEuen *et al.*, 1999. © 1999 by the American Physical Society).

where $\theta_{\mathbf{k}}$ is the angle that $\mathbf{k}$ makes with the y axis (see Fig. 8.13). This equation shows that the electrons are described by a two-component vector that gives the amplitude of the electronic wavefunction on the two sublattice atoms. This vector can be described as a "pseudo-spin," in analogy to the electron's real physical spin. The direction of this pseudo-spin determines the character of the underlying molecular orbital state (bonding or anti-bonding) as shown in Fig. 8.13. The wave function shows that the pseudo-spin is tied to the $\mathbf{k}$ vector, such that it always points along $\mathbf{k}$. This is completely analogous to the physical spin of a (nearly) massless neutrino that points along the direction of propagation. The states around $\mathbf{K}$ correspond to right-handed neutrinos (pseudo-spin parallel to $\mathbf{k}$) while those around $\mathbf{K}'$ are left-handed (pseudo-spin antiparallel to $\mathbf{k}$).

For the antiparticles (b = −1) this situation is reversed. Physically this pseudo-spin means that the character of the underlying molecular orbital state depends on the propagation direction. For example, a negative energy state near $\mathbf{K}$ with a positive $\mathbf{k}_{\mathbf{x}}$ is built from antibonding molecular orbitals, while the state with $-\mathbf{k}_{\mathbf{x}}$ is built from bonding orbitals. Using this basis, Ando *et al.* (1998a, 1998b) calculated the scattering probability between states $\mathbf{k}$ and $\mathbf{k}'$ in an assumed long range disorder potential with Fourier component $V(q)$.

The resulting, squared, matrix element for scattering was found to be

$$|<\mathbf{k}'|V(\mathbf{r})|\mathbf{k}>|^2 = |V(\mathbf{k}'-\mathbf{k})|^2 \cos^2\left[\theta_{\mathbf{k},\mathbf{k}'}/2\right] \tag{8.14}$$

where $\theta_{\mathbf{k},\mathbf{k}'}$ is the angle between the initial and scattered electron states. In the backscattering shown in Fig. 8.13(b), where $\theta_{\mathbf{k},\mathbf{k}'} = \boldsymbol{\pi}$, the matrix element (8.14) is zero, from the $\cos^2\left[\theta_{\mathbf{k},\mathbf{k}'}/2\right]$ term that describes the spinor overlap.

The case shown in Fig. 8.13(c) is quite different, because the scattering angle $\theta_{\mathbf{k},\mathbf{k}'} < \boldsymbol{\pi}$, so the scattering is only partially suppressed by the spinor overlap. McEuen *et al.* (1999) conclude that metallic nanotube, corresponding to graphene, is not sensitive to long-range disorder in its carrier mobility, while semiconducting nanotubes, with parabolic bands and conventional massive carriers, corresponding to bilayer graphene (see Section 4.8), will suffer mobility reduction with long-range disorder. Evidence is given, for metallic nanotubes that the effective scattering length is 8 μm (a value 3 μm had earlier been reported by Tans *et al.*, 1997), while McEuen *et al.* (1999) state that mean free paths in semiconducting nanotubes of similar quality are reduced by orders of magnitude. It has not been verified that a similar expected reduction appears in bilayer graphene, with massive electrons in parabolic bands. In fact, Mayorov *et al.* (2011) have reported very high mobility in suspended bilayer graphene that, however, may be associated with linear bands under certain circumstances. Further, Qiao *et al.* (2011) have recently predicted that, by applying suitable gating electrodes, carrier mean free paths of hundreds of microns could be expected in attainable-quality bilayer graphene. As mentioned in Section 4.8, transverse electric field can be used to control a small band gap in bilayer graphene.

The resistivity of graphene has a contribution from flexural phonons that was investigated by Mariani and Von Oppen (2008). A contribution to $\rho \sim T^{5/2} \ln T$ is reported.

8.6   Proposed nematic phase transition in bilayer graphene

We described in Section 4.6 the history of bilayer graphene, including recent work of Mayorov *et al.* (2011), identifying a transition, at low carrier density, to a "nematic" phase, supporting suggestions by Vafek *et al.* (2010) and Lemonik *et al.* (2010). A nematic phase is uniaxial, suggested by cigar-shaped molecules that are aligned parallel, but fluctuate randomly in position. (How something like this could appear in an electron system is indicated below.) A similar conclusion, based on careful experimental work, has been reached by Weitz *et al.* (2011). Graphene and bilayer graphene are operationally 2D free electron gases where most properties can be understood neglecting the Coulomb repulsion between electrons. The carrier density of graphene systems can be varied by inducing charge by gate electrodes and, of course, the temperature can be varied. At low density and low temperature indications of a transition have been noted. Electrons in metals and semiconductors are usually successfully understood as free non-interacting particles. On the other hand, at low kinetic energy, as attained at low temperature, we saw in Section 2.2, Fig. 2.1 that an insulating crystalline state of electrons appears on the surface of liquid helium. Ordered conducting states of electrons, the quantum Hall states, have also been discussed in Chapter 2 and the superconducting state of metals also is based on electron–electron interaction. It appears that nematic ordering of electrons, driven by their interaction, in bilayer graphene has likely been observed. The electronic nematic phase is likened to a molten phase of an anisotropic electron crystal (Fradkin *et al.*, 2007). The data of Mayorov *et al.* (2011) and Weitz *et al.* (2011) by no means specify the detailed nature of the electronic nematic phase, but do make fairly clear a reduction to two-fold symmetry.

Bilayer graphene is described most simply as a semimetal with Hamiltonian (Novoselov *et al.*, 2006; McCann and Fal'ko, 2006)

$$\hat{H}_0 = -\frac{\hbar^2}{2m} \begin{pmatrix} 0 & (k_x - ik_y)^2 \\ (k_x + ik_y)^2 & 0 \end{pmatrix} \qquad (8.15)$$

The latter authors however, find that the electronic state in bilayer graphene changes, as the concentration is lowered, from the parabolic bands, expected from this Hamiltonian (including nearest-neighbor intralayer and nearest-neighbor interlayer hopping) and observed by Ohta *et al.* (2006), to a band with three-fold symmetry. Adding two correction terms to the Hamiltonian, to include next-neighbor interlayer hopping, but ignoring effects of electron–electron interaction, McCann and Fal'ko find, at low carrier density, bands as suggested in Fig. 8.14. This change of the Fermi surface from a single circle to four separate closed curves is a topological transformation, referred to as a "Lifshitz transition."

The suggested nematic phase transition depends upon electron–electron interaction, neglected in the work of McCann and Fal'ko. The experimental results of Mayorov *et al.* (2011) on excellent high mobility bilayer samples are given in Fig. 8.14(b), contrasting results for bilayer graphene (upper) and single layer graphene (lower). The curves show the temperature dependence of the *width* of the conductance minimum (the raw data, not shown here, are of the type shown in Figs. 1.4 and 1.6). The circles

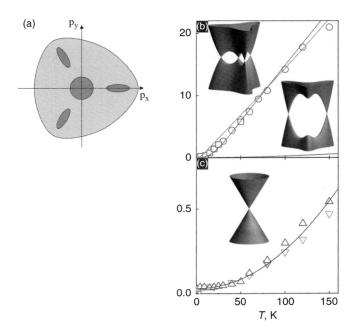

Fig. 8.14 (a) Two depictions of Fermi surface (Fermi line) in bilayer graphene. The larger, trigonally distorted, circle applies at high concentration. In the low concentration phase (estimated to appear around 10^{11} cm^{-2}), Fermi lines appear as enclosing (in darker shading) a central circle and three trigonally arranged ellipses. (Adapted from McCann and Fal'ko, 2006. © 2006 by the American Physical Society). (b) Width of the conductivity minimum (expressed as carrier concentration measured from Dirac point) at the neutral point in two (circles and squares) excellent suspended and cleaned bilayer graphene samples, vs. temperature. Solid lines are fits, see text. Upper inset is four-cone single electron band structure, lower inset is two-cone band structure for nematic phase, exhibiting biaxial symmetry. (c) Width of conductivity minimum for two monolayer graphene samples. Inset is band structure for monolayer graphene. (From Mayorov *et al.*, 2011, with permission from AAAS).

and squares are data from two bilayer graphene samples in two-terminal measurements. Two fitting curves are shown as solid lines. The lower curve, better fitting the low temperature data points, is derived from a theory assuming two cones in the band structure (lower inset), a uniaxial form associated with the suggested nematic phase, following Vafek *et al.* (2010) and Lemonik *et al.* (2010). This "reconstructed" fit gives an effective mass 0.028 m_e, an energy parameter $u = -6.3$ meV and a velocity $v_3 = 1.41 \times 10^5$ m/s. The curve that is uppermost in the lower temperature range (that assumes the four-cone bands) is obtained in the single electron approximation (upper inset) by McCann and Fal'ko for small carrier density. Mayorov *et al.* 2011 (in extensive supplementary material) give the width of the bilayer resistance peak as

$\Gamma(T) = 8m (2\ln 2)^{1/2} T/\pi\hbar^2$ and a similar formula, proportional to T^2, for monolayers. A theoretical perspective is offered by Kotov *et al.* (2012).

The interesting results of Mayorov *et al.* (2011) described above are somewhat puzzling, however, in view of the recent work of Ponomarenko *et al.* (2011), discussed in Section 7.5 above, where (single layer) graphene is concluded to undergo a metal-insulator transition starting around 80 K. Data are shown with resistivity exceeding 500 kΩ at 10 K. The conclusions of Mayorov *et al.* are based on the temperature dependence of the width and height of the resistivity peak (conductance minimum) at the Dirac point, data such as shown in Figs. 1.4 and 1.6. (Their conclusions are strengthened by consideration of Landau-level effects, not discussed here.) The paper of Mayorov *et al.* compares minimum conductivity in bilayer graphene, seen in their Figure 1A as about $(9 \text{ k}\Omega)^{-1}$ that is equivalent to 2.86 e^2/h (the text gives a value $20/\pi\, e^2/h$, to the theoretical value $24/\pi e^2/h$ of the "minimum conductivity" derived by Cserti *et al.* (2007) in their paper "Role of trigonal warping on the minimal conductivity of bilayer graphene."

Cserti *et al.* predict a "universal value" $24/\pi\, e^2/h$ that they present as six times larger than the "universal value" $4/\pi\, e^2/h$ established for single layer graphene, citing at least ten authors (their Refs. 8–18, see also Refs. 7 and 30) who have calculated minimal conductivity values on the order of e^2/h for monolayer graphene. It is pointed out by Ziegler (2006) that the theoretical minimum conductivity idea predated experiments on graphene (see Cserti *et al.* Ref. 7). The calculational method used by Cserti *et al.* for bilayer graphene is similar to that used for monolayer graphene. As improved experiments now show, there is no minimum conductivity for undoped monolayer graphene, i.e., the low temperature value is zero. The improved experimental evidence leaves this theoretical literature in some question. The ground state of neutral graphene is an insulator. It may be that the delocalized massless Fermions, the assumed starting point for the theoretical calculations, are in fact replaced by a basis of localized states, making the "universal" calculations irrelevant. The experiments of Mayorov *et al.*, supported by the early basic theory of Abrahams *et al.* (1979), indicate that the ground states of graphene and bilayer graphene will be insulating. This does not mean that interesting phases, such as a nematic, cannot exist before the localization occurs, but certainly further work is needed to clarify this situation. A recent report of charged skyrmion particles in bilayer graphene has been given by Lu and Herbut (2012) and a canted antiferromagnetic phase is described by Kharitonov (2012).

8.7 Klein tunneling, Dirac equation

A consequence of the Dirac Hamiltonian and spinor wave function that apply near the degeneracy point (see eqns 1.1–1.4), is unit probability of forward tunneling through a perpendicular potential barrier (angle of incidence $\alpha = 0$; Klein, 1929; Katsnelson *et al.*, 2006; Beenakker, 2008). (The effect is hard to observe because it occurs only very close to normal incidence.) The full features of this effect in graphene were observed recently by Young and Kim (2009). A review including these effects has been given by Das Sarma *et al.* (2011), following that of Castro Neto *et al.* (2009). As mentioned earlier, following Fig. 4.2, these effects were, to some degree, known in the literature of narrow bandgap semiconductors (Keldysh, 1964; Aronov and Pikus, 1967). The Klein tunneling effect is closely related to the lack of backscattering (scattering angle π), described in Section 8.5. The corollary of zero backscattering is unit probability of transmission. The mathematics of this effect was first given (Klein, 1929) in the context of electrons and positrons in vacuum, where no experiment has been possible. The electron of energy E approaches a barrier of height V_0 as shown in Fig. 8.15. In this figure the barrier that extends infinitely in the y-direction, is in the center, of width D, where the potential is raised. The electron on the left, with positive wavevector k and positive pseudo-spin ($\rightarrow$), matches, in energy and pseudo-spin, a positive hole of opposite wavevector q. The current is continuous in the barrier, where it is carried by a matching hole. The same transformation occurs between the barrier and the right hand electrode. This gives unit probability of transfer of the electron by Klein tunneling from left to right. As long as the interfaces are sharp and the barrier permits coherent ballistic motion, there is no limitation of unit transmission from increasing width and/or height of the barrier. In the original thought experiment of Klein (1929) in vacuum the hole is represented by a positron, with a threshold creation energy $mc^2 = 511\,\mathrm{Ke\,V}$. This fact precludes experimental realization, as a huge electric field would be needed to change the energy by 511 keV over a distance of the order of the Compton wavelength $\hbar/m_\mathrm{e}c = 0.039\,\mathrm{nm}$. Such a field is on the order of $1.3 \times 10^{16}\,\mathrm{V/m}$, or about 25,000 times the electric field at the electron in the hydrogen atom. In graphene there is no threshold energy and the unit probability of specular forward tunneling is surely expected. (In practice it has been difficult to achieve ballistic specular conditions providing the required precise normal incidence in barrier structures of a PNP form, so that subtlety in experimental design has been required,

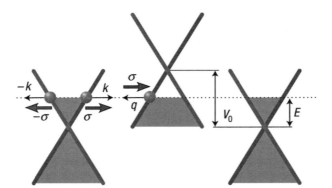

Fig. 8.15 Schematic of npn structure of graphene (extended out of the page) with current flow in the x-direction. The width of the barrier is taken as D and the equivalent barrier height is V_0. Isospin or pseudo-spin conservation provides perfect transmission precisely at normal incidence. (From Katsnelson *et al.*, 2006, by permission from Macmillan Publishers Ltd., © 2006).

focusing on predicted oscillatory features.) In more detail, following Katsnelson *et al.* (2006), the reflection coefficient r as a function of the angle of incidence α is

$$r = 2i \exp(i\alpha) \sin(q_x D)[\sin \alpha - ss' \sin \theta]$$
$$\{ss'[\exp(-iq_x D) \cos(\alpha + \theta) + \exp(iq_x D) \cos(\alpha - \theta)] - 2i \sin(q_x D)\}^{-1}. \qquad (8.16)$$

The corresponding transmission factor $T = 1 - |r^2| = |t^2|$, in the limit of a high barrier $|V_0| >> |E|$ is

$$T = |t^2| = \cos^2(\alpha)/[1 - \cos^2(q_x D) \sin^2(\alpha)]. \qquad (8.17)$$

These two equations also show that perfect transmission for $\alpha = 0$ reappears periodically at other angles controlled by

$$q_x D = n\pi, \quad \text{where} \quad n = 0, \pm 1, \pm 2, \dots \qquad (8.18)$$

In these equations

$$q_x = \left[(E - V_0)^2 \big/ \hbar^2 v_F^2 - k_y^2 \right]^{1/2}, \qquad (8.19)$$

and s and s' are ± 1 according to $s = \text{sgn} E$, $s' = \text{sgn}(E - V_0)$. In the bending of the trajectory as the particle enters the barrier, the refraction angle is $\theta = \tan^{-1}(k_y/q_x)$. In terms of the Fermi wavevector k_F, we have $k_x = k_F \cos(\alpha)$ and $k_y = k_F \sin(\alpha)$. These equations are obtained by matching coefficients of the $\rightarrow$ and $\leftarrow$ components of the wavefunction at the boundaries, $\pm D/2$ that represent PN junctions. (In their paper, Katsnelson *et al.*, 2006 derive the corresponding equations for bilayer graphene, a semimetal with parabolic bands. The results are conventional and do not include perfect transmission.)

The transmission is seen to remain near 1.0 for a small range of angles around $\alpha = 0$, this is a collimating action of the barrier on the motion of particles. It is seen from the reflection equation that the sign of r, the reflection amplitude, *reverses* as the angle passes through normal incidence, $\alpha = 0$. This feature is present in the transmission of the individual PN junctions that was analyzed carefully by Cheianov and Fal'ko 2006, who derive the transmission probability T of the single PN junction as

$$T = \exp[-\pi(k_F D) \sin^2(\alpha)] = \exp[-\pi \hbar v_F k_y^2/(eE)], \qquad (8.20)$$

with E the electric field at the junction. The second form of the equation is presented by Young and Kim (2009) and attributed to Cheianov and Fal'ko (2006). Strong collimation is present, as transmission off normal incidence is greatly reduced by the traditional tunneling barrier, often approximated in the WKB method as

$$T \sim \exp(-2\theta_{\text{WKB}}) \qquad (8.21)$$

where

$$\theta_{\mathrm{WKB}} = \hbar^{-1} \int p_x\left(x'\right) dx'. \tag{8.22}$$

Experimental approaches to observing the Klein tunneling effect have been reported by several authors, including Huard *et al.* (2007), Gorbachev *et al.* (2008), Ozyilmaz *et al.* (2008) and Stander *et al.* (2009), prior to the work of Young and Kim (2009). The problem has been that diffusive scattering remains in the devices and the characteristic effects sought in these works, near unit transmission near normal incidence and the Klein exponential collimation effect, are not distinguishable from a bulk resistance measurement that is sensitive only to the total transparency of the PN junction. A more intuitive analysis toward finding an experimentally recognizable signature for the Klein tunneling effects has been given by Shytov *et al.* (2008), whose predicated geometry is identical to the device structure of Young and Kim (2009) shown in Fig. 8.16(a). This is essentially the geometry of Fig. 8.15, with emphasis on making the barrier region as thin as possible, to further the ballistic fraction of motion in the device. Shytov *et al.* then view the structure as supporting repeated internal reflections as in a Fabry–Perot etalon. Thinking in this way about the device function promotes intuitive understanding of recognizable changes in behavior brought about by applying a magnetic field along the *y*-direction, leading to curvature of charge motion along the *x*-direction.

The potential profile and carrier density induced by the top gate are important to analysis of the data. Beyond the width of the metal top gate, about 20 nm, the thickness and permittivity of the dielectric layer connecting the top gate to the graphene play a role in setting the gate voltage dependences. An approximate working formula used by Young and Kim for the induced carrier density as a function of *x* along the graphene sheet, with the Klein barrier at $x = 0$, is

$$n(x) = \{V_{\mathrm{TG}} C_{\mathrm{TG}} / [1 + (x/w)^{2.5}] + V_{\mathrm{BG}} C_{\mathrm{BG}}\}/e. \tag{8.23}$$

Here the exponent 2.5 was chosen to reasonably match the results of numerical simulations. Young and Kim give values $C_{\mathrm{TG}} = 1490$ aF/μm^2 and $C_{\mathrm{BG}} = 116$ aF/μm^2. Evaluated at $x = 0$, e.g., for $V_{\mathrm{TG}} = -10$ V and $V_{\mathrm{BG}} = 50$ V, the two terms in (8.23) give -9.3 and 3.6, respectively, in units of 10^{12}/cm^2. The choice of this induced carrier density profile is important in the simulations of the Fabry–Perot transmission because it sets the *L* parameter, the width between points with zero charge density, a value that varies substantially over the gate voltage ranges investigated.

Figure 8.16(b) shows the measured conductance in units of e^2/h (between source and drain in the above diagram) as a function of the top-gate voltage V_{TG} (which largely controls the carrier concentration in the central barrier region) and the back-gate voltage V_{BG}. These data are taken at zero magnetic field and 4.2 K. In upper inset, the global gray scale image is of the conductance in the ranges $-10 < V_{\mathrm{TG}} < 10$ V, $-80 < V_{\mathrm{BG}} < 80$ V. Data using barriers of both signs are symmetrically investigated. The bottom-gate voltage varies the carrier concentration in the whole

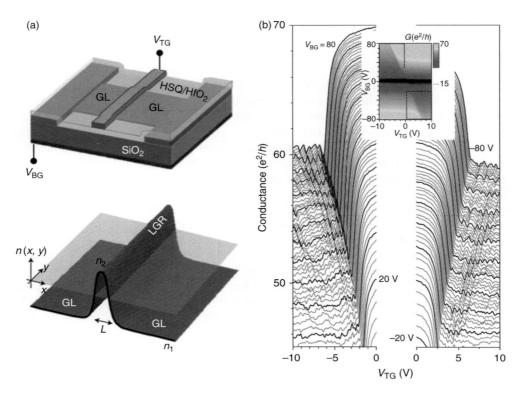

Fig. 8.16 (a) Layout of Kim and Young's Klein-barrier Fabry–Perot etalon device. A narrow, 20 nm, capacitively coupled top gate, crossing graphene layer GL, induces carrier concentration n_2, forming the locally gated region (LGR, or Fabry–Perot etalon). The dielectric for the top gate is poly-hydroxysilane and HfO_2. The measured quantity is conductance $G = dI/dV$ between source and drain that connect to the Klein barrier by graphene leads (GL). The gate-induced carrier densities n_1, n_2 are studied in the range between $\pm 5 \times 10^{12}$ cm^{-2}. The overall width W of the graphene sheet is around 10 μm, while the barrier width $L = D$ between charge neutrality points across the barrier is ≤ 100 nm. (b) Klein tunneling in summary of conductance data at 4.2 K over full range of electrode voltages, producing full range of carrier densities of both signs in the leads GL and in the barrier region (locally gated region, LGR). Upper inset shows gray scale image of conductance G over the whole investigated range. Marked regions in upper left and lower right quadrants are selected for detailed study, shown in sets of conductance curves vs. top-gate voltage V_{TG}, with back-gate voltage V_{BG} as parameter. Prominent are the lines of cutoff of diffuse sheet conductance as the Klein barrier forms and the transport becomes limited by Fabry–Perot interference/Klein tunneling across the narrow barrier structure. The weak conductance oscillations seen close to the barrier cutoff line, are singled out for enhancement and modeling, including their changes with applied magnetic field (not shown in this figure). (From Young and Kim, 2009, by permission from Macmillan Publishers Ltd., © 2009).

graphene sheet (that consists of the two graphene leads GL and the LGR locally gated-region) between approximately $\pm 5 \times 10^{12}$ cm^2, while the narrow top-gate, 20 nm in width, generates the barrier structure. This requires top-gate voltages rising from 2 V to about 5 V, as the bottom-gate voltage increases from about 20 V to about 80 V. The conductance is zero (horizontal black line in the inset gray scale image) for small V_{BG}, when the leads are devoid of carriers. The horizontal bands of moderate conductance above and below $V_{BG} = 0$, in the gray scale image, correspond to diffusive transport through the graphene sheet, before the barrier has formed. The diagonal lines in the selected portions of the upper left and lower right quadrants (upper inset) show the formation of the barrier. The main figure shows conductance traces vs. top-gate voltage, with back-gate voltage as a parameter. The sharp diagonal line with increasing gate voltages, marking a sharp falloff in device conductance from $\sim 70 \, e^2/h$ to $\sim 50 \, e^2/h$, reveals weak conductance oscillations, on the low conductance side of the boundary. These features, modified in important ways by magnetic field B, have been extracted and modeled by Young and Kim.

The sharp drop in conductance, as the barrier forms, may be puzzling if one accepts unit probability of Klein tunneling. The unit probability, however, applies only for *normally incident* carriers: the collimation effect then may be responsible for the sharp reduction in conductance.

Shytov *et al.* (2008) suggest that transmission through the Klein barrier structure is influenced by multiple internal reflections, as in a Fabry–Perot etalon. This idea, including the effect of an applied magnetic field B, is suggested in Fig. 8.17a. The conductance G of the Klein barrier, viewed as a Fabry–Perot etalon, is

$$G = 4e^2/h \sum\nolimits_{\mathrm{ky}} |[t_1 t_2 \exp(-L/2\lambda_{\mathrm{LGR}})]/[1 - |r_1| \, |r_2| \exp(i\theta) \exp(-L/\lambda_{\mathrm{LGR}})]|^2, \tag{8.24}$$

where the sum is over transverse wavevector, L is the width of the barrier (and depends parametrically on the bias voltages), λ_{LGR} is the mean free path in the locally gated region and $\theta = \Delta\theta$ is the phase change of the particle bouncing between the two junctions of transmission coefficients $t_{1,2}$ and reflection coefficients $r_{1,2}$. (Looking at Fig. 8.17(a), we see two trajectories, one transmitted through both barriers and the second reflected at the second barrier, making two further traversals of the length L to interfere with the initial trajectory. On this basis $\Delta\theta = 2\theta_{\mathrm{WKB}} + \Delta\theta_1 + \Delta\theta_2$, where $\theta_{\mathrm{WKB}} = \hbar^{-1} \int p_x (x') dx'$ and $\Delta\theta_{1(2)}$ are the back-reflection phases for the interfaces 1 and 2.) Young and Kim approach modeling the conductance data by extracting the oscillatory part of the conductance G. This is made easier by realizing that the probabilities $r_{1,2}$ in the denominator are small, allowing an expansion of eqn (8.24). The resulting G_{osc} is given as

$$G_{\mathrm{OSC}} = 8e^2/h \sum\nolimits_{\mathrm{ky}} |t_1|^2 \, |t_2|^2 \, |r_1| \, |r_2| \cos(\theta) \exp(-2L/\lambda_{\mathrm{LGR}}). \tag{8.25}$$

In the Fabry–Perot model, transmission has an oscillatory dependence on the phase difference, $\Delta\theta = 2\theta_{\mathrm{WKB}} + \Delta\theta_1 + \Delta\theta_2$ that electron waves accumulate in bouncing

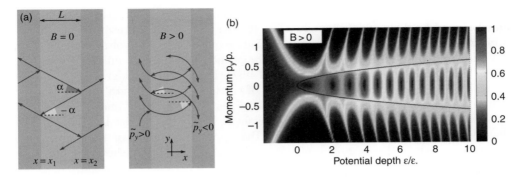

Fig. 8.17 (a) Schematic of electron transmission through PNP structure at (i) $B = 0$ and (ii) $B > 0$. Here angle of incidence is labeled α and $x_2 - x_1 = L$. The sign of the back-reflection amplitude reverses at $\alpha = 0$. At $B = 0$ the sign of the incidence angle α is opposite at interfaces 1 and 2 [see panel (i)], so that the phase changes upon back-reflection cancel. At a certain field B^* (see text) the incidence angles at interfaces 1 and 2 have the same sign that adds π to the electron phase accumulated between reflections. This results in a half-period shift of Fabry–Perot interference fringes. (b) Theoretical oscillatory ballistic Klein tunneling probability (gray scale) of transmitted transverse momenta p_y through narrow potential barrier (p-n-p structure in graphene), of depth $\varepsilon/\varepsilon_*$, with $\varepsilon_* = 14$ meV, in modest magnetic field $B = 0.3$ T. Probability of forward transmission is seen to be oscillatory with respect to barrier height, especially at barrier heights in the range 2 to 10. (The barrier height is controlled by a top gate voltage in the proposed structure.) The black parabola is locus of points where sign of reflection reverses, a characteristic of Klein tunneling. This effect leads to an oscillatory resistance across the device with varying top-gate voltage, with a characteristic half-fringe shift at modest magnetic field. (From Shytov *et al.*, 2008. © 2008 by the American Physical Society).

between the two interfaces. Here $\theta_{\text{WKB}} = \hbar^{-1} \int p_x(x')dx'$ (8.22) and $\Delta\theta_{1(2)}$ are the back-reflection phases for the interfaces 1 and 2, exhibiting a π-jump at zero incidence angle α. As shown in the Fig. 8.17(a) panel (ii), the contribution $\Delta\theta_1 + \Delta\theta_2$ to the net phase can be altered by a magnetic field. At zero B the incidence angles at interfaces 1 and 2 have opposite signs, so that the jumps in $\Delta\theta_{1(2)}$ cancel. However, for curved electron trajectories at a finite B the signs of the incidence angles can be made equal. Shytov *et al.* show that, for a given transverse momentum p_y, there is a magnetic field B^* (satisfying a condition $-B^*L/2 < p_y < B^*L/2$), such that the phase difference $\Delta\theta_1 + \Delta\theta_2$ equals π. This is shown to give a *half-period shift* in the Fabry–Perot fringes [this occurs along the parabola in Fig. 8.17(b)].

Shytov *et al.* did further modeling assuming for the Klein barrier a parabolic potential $U(x) = ax^2 - \varepsilon$, to create PN interfaces at $x = \pm x_\varepsilon$, where $x_\varepsilon = \sqrt{\varepsilon/a}$. This barrier differs slightly from the one used by Young and Kim, giving the $n(x)$ dependence described above.

To return to the modeling of G_{osc}, Young and Kim use the following expressions, largely based on the work of Shytov *et al.* for the various factors that enter. Assuming a magnetic field B, the WKB phase is given as

$$\theta_{\text{WKB}} = Re \left\{ \int_{-L/2}^{L/2} [\pi|n(x)| - (k_y - eBx/\hbar)^2]^{1/2} \, dx \right\}. \tag{8.26}$$

(The integrand has units of $p/\hbar$ which are inverse length, because $n(x)$ is expressed in cm^{-2}.) Evaluation of this integral depends on the fitted values $n(x)$ derived from the applied voltages that allows also determination of the local value of L.

The reflection phases are given in terms of the Heaviside function: $H(x) = 1$ for $x > 0$ and H is zero otherwise:

$$\Delta\theta_1 = \pi[H(-k_y + eBL/2\hbar)]; \quad \Delta\theta_2 = -\pi[H(-k_y - eBL/2\hbar)]. \tag{8.27}$$

The transmission and reflection amplitudes $t_{1,2}$ and $r_{1,2}$, for junctions located at $\pm L/2$, are given as

$$\begin{aligned} t_{1,2} &= \exp[-\pi(\hbar v_{\text{F}}/2eE)(k_y \pm eBL/2\hbar)^2]; \\ r_{1,2} &= \exp[i\pi H(-k_y \mp eBL/2\hbar)][1 - |t_{1,2}|^2]^{1/2}. \end{aligned} \tag{8.28}$$

One can see from the first expression that transmission is perfect when

$$k_y = B = 0, \tag{8.29}$$

and also whenever

$$k_y = -eBL/2\hbar. \tag{8.30}$$

In the first expression in (8.28), E is the electric field at the junction that is difficult to determine (Zhang and Fogler, 2008).

Young and Kim give arguments that this can be approximated as

$$eE \approx 2.1 \, \hbar v_{\text{F}} \, (dn/dx)^{2/3}, \tag{8.31}$$

where dn/dx is the carrier density gradient at the junction. This quantity, as well as L, depends upon the fitted dependence of $n(x)$ on the electrode voltages discussed earlier. Using this approach, Young and Kim model the observed conductance oscillations, presented, for example, as dG/dn_2 vs. n_2, with calculated images derived from the function G_{OSC}. Here n_2, the induced carrier density at the center of the locally gated-region, is related to the electrode potentials through the relation

$$n(x) = \{V_{TG}C_{TG}/[1 + (x/w)^{2.5}] + V_{BG}C_{BG}\}/e, \tag{8.32}$$

and w is a parameter chosen to be in the range $w \sim 45$–47 nm. Gray scale images vs. carrier densities 1 and 2 are closely related to images vs. V_{TG} and V_{BG}.

The experimental image shown below, representing the dG/dn_2 vs. magnetic field B and carrier density n_2, shows oscillations with respect to magnetic field B, as well as versus carrier density n_2. The oscillations are indeed closely predictable from the theory outlined above, based on the Klein tunneling phenomenon, including the magnetic field effects. It appears that these data are among the first to show oscillations in any Klein tunneling experiment. The authors have been able to show that one of the key predictions of the model of Shytov *et al.* (2008), a fringe-shift in the Fabry–Perot oscillations over a magnetic field range, is present in their data. This is shown in Figs. 8.18 and 8.19.

The oscillations, which include contributions from mesoscopic fluctuations, as well as from the ballistic Klein tunneling, are found to weaken as the temperature is raised above 4.2 K. It appears that these data and analysis clearly confirm the Klein tunneling phenomena in graphene. An independent thorough analysis by Sonin (2009) also concludes that Young and Kim have clearly observed the Klein tunneling effects, but see also Rossi *et al.* (2010).

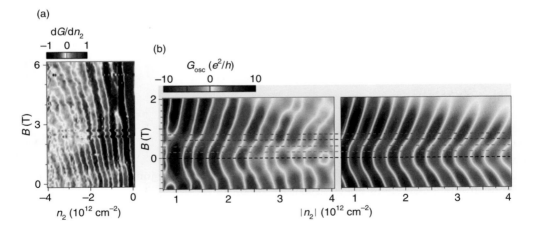

Fig. 8.18 (a) Klein tunneling oscillations. Magnetic field dependence of dG/dn_2 at $V_{BG} = 50$ V. See gray scale at top, the data are taken at 4.2 K. (b) Gray scale image of measured oscillating conductance G_{OSC} vs. carrier density n_2 (left) compared with model (right). Dashed lines indicate magnetic fields (bottom to top) of 0, 200, 400 and 800 mT, which span the Klein-scattering-specific half-fringe shift present in data and modeling. (From Young and Kim, 2009, by permission from Macmillan Publishers Ltd., © 2009).

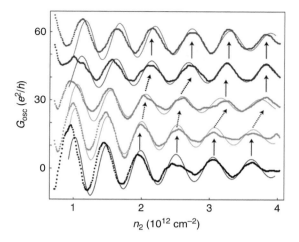

Fig. 8.19 Fringe shift in observed and modeled traces of G_{OSC} corresponding to magnetic fields (bottom to top) of 0, 200, 400, 600 and 800 mT. Strong oscillations are seen at zero-field (bottom trace), that weaken and then strengthen, at maximum field 800 mT, top traces. Four arrows per trace, proceeding from bottom to top of this figure, show the half-fringe phase shift. These curves, in range 0 to 800 mT, span the Klein-scattering-specific half-fringe shift present in data and modeling. (From Young and Kim, 2009, by permission from Macmillan Publishers Ltd., © 2009).

The extreme difficulty in observation of Klein tunneling effects in graphene, however, casts doubt that any device application of the effects will soon appear.

8.8 Superconducting proximity effect, graphene Josephson junction

It has been shown by Heersche *et al.* (2007) that graphene performs extremely well as the N-member of an SNS (superconductor-normal-superconductor) Josephson junction. This may seem anomalous because the effect appears symmetrically at positive and negative bias where the carriers switch between electrons and holes. However, it may not be well known that performing well in an SNS junction does not actually imply that the N material becomes superconducting in a traditional sense, but only that it transmit carriers without scattering.

For review, a superconducting Josephson junction passes a supercurrent J (with no voltage difference), according to

$$J = J_0 \sin \varphi, \quad \text{where} \quad \varphi = \theta_1 - \theta_2, \tag{8.33}$$

where θ_1 and θ_2 are the coherent phases of superconducting pairs on sides 1 and 2 of the junction. Similar to a Josephson junction is the SNS junction, a narrow normal metal layer, instead of a barrier, between two superconductors.

A relation similar to eqn (8.33), periodic in φ, applies to SNS structures. In essence, at low enough current density, the whole structure becomes a single superconductor, a macroscopic quantum state. This state is stabilized by a coupling energy between the two superconducting electrodes provided by the "weak link", the N-layer in the SNS junction or by electron-coupling across the barrier in the tunnel junction. The whole effect is built on the fact that a superconductor is a single quantum state, where macroscopic numbers of coupled electrons (superconducting pairs) are in the same state, having a specified phase θ. This is described by a pair wave function

$$\psi(\mathbf{r}, t) = \sqrt{n_{\mathrm{s}}} \, \exp(i\theta), \tag{8.34}$$

to describe the density of pairs of charge $e^* = 2e$ and can be used in a Schrödinger-like theory to derive flux quantization and the Josephson effects. In this description

$$n_{\mathrm{S}} = \psi^*(\mathbf{r}, t) \, \psi(\mathbf{r}, t) \tag{8.35}$$

is the *pair density*. In the simplest case, with no magnetic field, the phase θ has a common value over the whole superconducting region. One way to analyze the SNS Josephson junction is to learn how the superconducting pairs transfer, without loss of phase coherence, across the N region. To understand this, it is useful to learn more about the pairs.

Pairs in the traditional superconducting metals, like aluminum, used in graphene SNS junctions, are formed as two electrons bind together by slightly distorting the crystalline lattice, in a dynamic sense. (The effect is in principle like two bowling balls on a mattress, where the distortion favors having the two balls together in the same depression.) How electron pairing overcomes the electrostatic repulsion of the electrons is one of the puzzles explained in the successful theory of superconductivity, by Bardeen *et al.* (1957). To simplify, the pairing succeeds because the pair is large in dimension, reducing the direct electron repulsion. (Background on tunnel junctions and the SNS Josephson effect is contained in Wolf, 2012).

The mechanism of electron pairing in the conventional superconductors is "exchange of virtual phonons" (a phonon is a quantized wave of lattice distortion) shown in Fig. 8.20(a). Electron $\mathbf{k_1}$ (lower left) emits a phonon of wavevector $\mathbf{q}$, to change its momentum to $\mathbf{k_1} - \mathbf{q}$ (upper left). The phonon $\mathbf{q}$ is absorbed by the second electron $\mathbf{k_2}$, which deflects to $\mathbf{k_2} + \mathbf{q}$. It was shown by Bardeen *et al.* (1957) that this process lowers the energy of two electrons if they move oppositely with opposite spins and are spaced by a distance, ξ, called the superconducting coherence length. This optimum spacing is set by the idea that the first electron, of speed v_{F}, gives an impulse of momentum to one of the positive ions of the metal and, when that ion has returned to its original position, after about $^1/_2$ of the time period τ_L of the lattice motion, a second electron will see a local positive charge and will be attracted. This suggests a spacing

$$\xi \sim 2v_{\mathrm{F}}\tau_L = 2h \, v_{\mathrm{F}}/(h\nu). \tag{8.36}$$

(a)

(b)

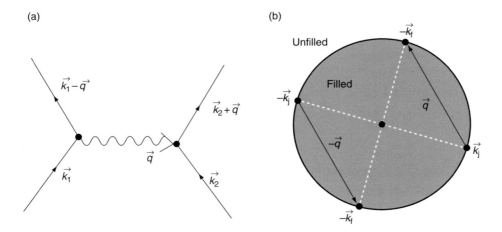

Fig. 8.20 (a) Sketch of electron–electron interaction by exchange of a virtual phonon, here labeled **q**. The process conserves energy and momentum and rotates pairs around the Fermi surface, as suggested in (b). (b) Suggestion of Fermi surface, a closed curve (here a circle) in 2D, with zero of momentum ($\hbar\mathbf{k} = 0$) at the center. States inside are filled, outside empty, at zero T. Dashed lines across the circle indicate pairs $(\mathbf{k}_i\uparrow, -\mathbf{k}_i\downarrow)$ and $(\mathbf{k}_f\uparrow, -\mathbf{k}_f\downarrow)$. The sketch indicates how the first pair rotates into the second position by absorbing a phonon **q**. The pairs all have net zero momentum and thus are Bosons all occupying the same quantum state.

With lattice oscillation frequency ν, typical phonon energy $h\nu = 20$ meV and typical Fermi velocity 10^6 m/s, this gives an electron spacing about 0.41×10^{-6} m. The actual formula for the coherence length in the superconductor is

$$\xi = 2\hbar\, v_\mathrm{F}/(\pi\Delta), \tag{8.37}$$

where 2Δ is the superconducting gap, approximately the pair binding energy and the value for Al is $\xi = 1.6$ μm. In the superconducting state all electrons are paired and all pairs have the same center-of-mass momentum—zero if no current is flowing. The Fermi surface (Fermi "circle" in 2D) is suggested in Fig. 8.20b.

The superconductor has n_s pairs all in the same state at $\mathbf{k} = 0$. A current occurs by slightly displacing the momentum of the whole set of pairs, from $\mathbf{k} = 0$, to a small value $\mathbf{k} = \delta\mathbf{k}$. The most common superconductors exhibit an energy gap 2Δ: the value for aluminum is about 0.87 meV. The pair density n_s is related to the energy gap parameter Δ, also known as the pair potential, by a relation that can be simplified as $\Delta = V n_s$, with V the pairing interaction. At the SN boundary, in the SNS junction, the pair density does not change abruptly but extends smoothly into the N region. This is true only if the interface is atomically ordered, but this is achievable. At the SN boundary we assume the pairing interaction V does jump abruptly to a smaller value or even zero, so that the pair potential will exhibit a downward jump at the interface. Thus in an NS junction, a piece of metal where the pairing interaction falls to zero across a boundary to the N side, expects that the pair wavefunction and pair density n_s to smoothly change, from its value in S, to zero at a point deep in the N region. Superconducting pairs are predicted to penetrate the N region on a length scale not greatly different from the coherence length: this penetration is the superconducting proximity effect. In the SNS junction, those pairs that penetrate through to the opposite SN boundary, provide the interaction energy between the two pair systems that gives the Josephson effect.

The SNS Josephson effect can operate even when the pairing interaction V is zero in the N layer. The transport across the N region is provided by "Andreev reflection." This is a process simply described as "an electron reflected as a hole." What happens is that an electron in the N region hits the S interface, where, under the influence of the pairing interaction V, it joins with a second electron from the normal region to form a pair, thus creating a hole in N at the NS interface. This does not require a pairing interaction V in N. This process creates an array of weakly bound quasiparticle states across the N region, with boundary conditions at the interfaces, $z = \pm d$ in Fig. 8.21. Since the pair has zero momentum, the created hole must have exactly the opposite momentum of the original electron. The combination of the electron motion and (oppositely directed) hole motion transfers 2 electron charges, as in a pair, across the N-region. Andreev reflection provides a mechanism for supercurrent flow in SNS structures. A model provided by Bardeen and Johnson (1972), is sketched in Figure 8.21. This specular Andreev model has been extended specifically for graphene by Beenakker (2006).

Beenakker finds that the electron-pair conversion efficiency at the NS interface is 100%, as assumed by Bardeen and Johnson, even though the electron and hole in graphene are in different valleys $\mathbf{K}$. The upper panel depicts the SNS structure,

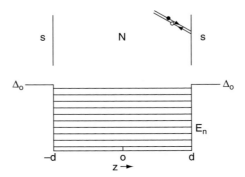

Fig. 8.21 Model for an SNS sandwich with abrupt pair-potential barriers at $z = -d$ and $+d$. There is an array of bound quasiparticle states in the normal region, each consisting of equal probabilities of particle and hole states. A particle is reflected (upper right) into a hole at a pair-potential boundary, with no change in current. (After Bardeen and Johnson, 1972).

indicating an Andreev retro-reflection event that transfers 2e (one pair) into the right hand S region at $z = d$. The lower panel sketches the energy levels in the N-region of width $2d$.

In this specular model, reflection at the pair-potential boundary changes a particle into a hole. Both particle and hole move to give a current in the same direction and the resultant quasiparticle current converts to supercurrent at the boundary. The quasiparticle states (shown in Fig. 8.27 and labeled E_n) are linear combinations of particle and hole states, selected by a matching wavevector condition

$$q = (n + \tfrac{1}{2})\pi/d. \tag{8.38}$$

This leads to allowed quasiparticle energies

$$E_n = \hbar^2 k_{zF}\left(n + \tfrac{1}{2}\right)\pi/(2m_e d^*), \tag{8.39}$$

where n is an integer. Here $d^* \approx d$ allows for slight penetration of the particle into the superconducting region and k_{zF} is the z-component of the Fermi wavevector.

The maximum value of the current density for $T = 0$ is found to be

$$J_{\max} = \hbar n_e \pi e / 4 m_e d^*, \tag{8.40}$$

estimated by Bardeen and Johnson as 10^6 A/cm^2 for $n_e = 10^{28}$ electrons/m^3 in N and $d^* = 10$ μm. (If this is converted to graphene by taking the thickness of the N region as 0.34 nm, the concentration becomes 4.3×10^{14} cm^{-2}.)

Finally, a temperature-dependent estimate is obtained for the Josephson current density carried by the set of quasiparticle states, based on Andreev scattering:

$$J_1 = (6\,\hbar n_e e / 2 m_e d^*)\exp[-4(d^*/\xi_o)(k_B T/\Delta_o)], \tag{8.41}$$

where $\xi_o < 2d^*$ has been assumed.

The current-phase relationship is shown by Bardeen and Johnson to be periodic in phase difference φ, gradually changing from a periodic ramp at $T = 0$ to the Josephson form, $J = J_1 \sin\varphi$, for higher temperatures. The authors comment that this current would "decrease much more slowly with increase in thickness of the normal layer than would be expected by the proximity effect. The effect is due to quantization of the quasiparticle states in the normal region."

Supercurrents up to to about 50 nA were observed by Heersche *et al.* (2007) (their Fig. 2a,b and they say in the text that supercurrents up to 800 nA were observed at high gate-voltage) in monolayer graphene SNS devices with S electrodes about 3 μm in length, with the 50 nA corresponding to a 3D supercurrent density around 4900 A/cm^2. This value is certainly compatible with the Bardeen–Johnson estimate, with adjustment of the carrier density downward by perhaps 100, but accompanied by the much smaller N region width d (in the range 0.1 μm to 0.5 μm) and a smaller effective mass in graphene, perhaps 0.1 m_e. The Josephson effect was fully identified

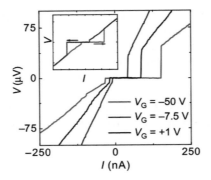

Fig. 8.22 Josephson voltage-current relation in monolayer graphene SNS device using aluminum S electrodes, measured near 30 mK in zero magnetic field. The horizontal portions of the V–I curves indicate supercurrents up to around 50 nA. The parameter is the voltage on the back-gate that varies the carrier concentration in the graphene layer. It is clear that the gate voltage affects the supercurrent, so the device can be called a superconducting graphene transistor. The authors find that a supercurrent is always present even if the graphene film is at the Dirac point that could be an effect of the electron-hole puddles expected because the un-annealed graphene is on an oxidized silicon surface. (From Heersche *et al.*, 2007, by permission from Macmillan Publishers Ltd., © 2007).

by Heersche *et al.* (2007), who observed the characteristic detailed Fraunhofer-like magnetic field interference pattern and RF-induced voltage steps known as Shapiro steps. The gate-voltage-dependent I–V characteristics of one of their devices is shown in Fig. 8.22. The typical hysteresis of such measurements is explained in the inset to the Figure.

The main requirement of the N-region is that it provide a density of delocalized states with long mean free path. It seems that the observation of Josephson's effect across a graphene layer is really not in the category of "anomalous electronic properties," unless one focuses on the observation of the effect near the Dirac point. From a basic point of view, independent of the doping level in the graphene, the Fermi level (uppermost filled electronic state) lies in a region of delocalized electronic states (a hole is equivalent to an oppositely moving electron) and the graphene provides this. A graphene-specific analysis of the SNS Josephson effect has been given by Titov and Beenakker (2006).

In the experiment of Heersche *et al.* (2007), the source of the superconductivity is Al, with a transition temperature about 1.3 K. The aluminum is deposited onto the graphene, with a thin Ti intervening layer to promote adhesion to the graphene. The authors say that the Al/Ti electrodes (below the 1.3 T_c of Al) induce the graphene layer into a state of proximity-induced superconductivity. They state that the observed Fraunhofer magnetic field induced pattern is proof of superconductivity in the graphene. This may be true, especially directly under the Al/Ti contacts, but for the intervening graphene N region it appears from the Bardeen–Johnson analysis that the only requirement of the graphene for observation of the supercurrent is ballistic transmission of electrons and holes from one S region to the other. Methods of "proximity electron tunneling spectroscopy" (PETS) as reviewed, e.g., in Wolf (2012), might be applied to further determine the state of graphene when in intimate contact with a superconductor like Al. An analogous tunneling spectroscopy experiment was reported by Burnell and Wolf (1982) and Burnell (1982), where a thin layer of the normal metal Mg was induced into the superconducting state by proximity and voltage-dependent tunneling spectra revealed the electron-phonon coupling V of Mg in great detail.

Confirming and extending the work of Heersche *et al.* (2007), larger supercurrents, over 50 μA, have recently been observed by Coskun *et al.* (2012) in wider (up to 214 μm) SNS junctions using Pb S-electrodes on monolayer graphene. These authors confirm the applicability of the specular Andreev scattering model, as outlined above, in general and extended to graphene SNS junctions (Beenakker, 2006).

Coskun *et al.* carefully study the distribution of switching currents in the devices as the current is ramped upward. The supercurrents persist up to about 1 K and their switching is analyzed using a model of Kurkijarvi (1972), which they extend in their paper.

The electric tunability, via the back-gate voltage, of graphene devices, has been exploited by Lee *et al.* (2011) in a demonstration of electrically tunable macroscopic quantum tunneling in a graphene-based Josephson junction. These authors suggest that such devices might be useful as qubits for quantum computing.

8.9   Quasi-Rydberg impurity states; Zitterbewegung

This section covers two areas of anomalous predicted behavior in graphene, both of which, it appears, still await clear experimental confirmation.

The Dirac 2D nature of the electrons that is unquestionably a feature of monolayer graphene, severely modifies the conventional idea of a donor or acceptor impurity that, when ionized, appears as a charged scattering center. An analysis of anticipated novel effects has been given by Shytov *et al.* (2007), but so far the effects do not appear in data.

The basic prediction of Shytov *et al.* (2007) is that massless particles in 2D cannot form bound states at a charge center, but that "an infinite family of quasibound (Rydberg) states appears abruptly when the Coulomb potential strength exceeds a critical value $\beta = 1/2$." Here β is defined as $Ze^2/\kappa\hbar v_F$, so that, taking dielectric constant $\kappa = 5$, the condition is met for $Z \geq 1$. Predictions for the conductivity and scattering cross section are shown in Fig. 3 of Shytov *et al.* 2007. In this Figure (not shown) is plotted the conductivity, calculated for graphene with 3×10^{11} cm^{-2} attractive impurities with β values (negative for attractive potential) in range -0.7 to -1.2, labeled on the curves. The conductivity curves show strong structures for doping levels in the range 0 to -3×10^{12}/cm^2 (hole doping). The inset to their figure shows pronounced resonances in the scattering cross-section. The authors state that an overall

$1/k$ dependence has been factored out and that the cutoff parameter r_0 has been set at 0.25 nm.

The predicted resonances in the conductivity have not been seen experimentally, for example see the set of curves in Tan *et al.* (2007). The authors predict that the resonant structures should appear in other transport coefficients, such as the thermopower. Shytov *et al.* (2007) propose an STS (scanning tunneling spectroscopy) experiment to measure the electronic density near an individual attractive impurity that sounds quite feasible. [Examples of STS studies are reviewed in Wolf (2012).] They suggest impurities with Z larger than 1, to be fully in the "supercritical regime." Examples of such impurities that have been introduced as intercalants in graphite, include Ca and Yb at $Z = 2$ and La and Gd at $Z = 3$. So it appears that the authors feel their predictions apply not only to substitutional impurities like Boron and Nitrogen but also to surface adsorbates.

Observation of the effects described by Shytov *et al.* (2007) has perhaps been hindered by the changing experimental situation from air exposed samples on "electrostatically noisy" silicon dioxide, to suspended samples with annealing to remove surface impurities, to samples isolated by electrostatically quiet boron nitride, to samples on boron nitride with electronic screening by metallic intervening layers. Only the last-mentioned samples have revealed the expected metal-insulator transition. There has been a tendency at each stage for an amenable theoretical literature to appear, only a portion likely applicable to intrinsic graphene as most recently revealed by Ponomarenko *et al.* (2011). A review up to 2010 of this complex situation has been included in the excellent review article of Das Sarma *et al.* (2011).

An assessment of this charge impurity scattering problem has been given also by Das Sarma *et al.* 2011, who agree that the resonant impurity scattering effects have not been identified experimentally. Very recently, however, Crommie and coworkers (Wang *et al.*, 2013) in a scanning tunneling microscope study entitled "Observing Atomic Collapse Resonances in Artificial Nuclei on Graphene" have seen some evidence of the effect.

Zitterbewegung (roughly, "jittery motion"), most clearly manifested in the propagation of a wavepacket (and also in noise properties) in systems with electron and accessible hole states, has been treated in general terms by Cserti and David (2006), David and Cserti (2010) and by Rusin and Zawadski (2007). The latter authors point out that Gaussian wavepackets properly contain admixtures of states at positive and negative energies in systems like graphene and bilayer graphene (indeed in many other cases, including narrow bandgap semiconductors and spin systems where positive and negative energy states appear symmetrically). In these cases the sub-packets of positive- and negative-energy states propagate in opposite directions, leading to abnormal decay of the packet. Rusin and Zawadski (2007) predict that the lifetimes of Gaussian wavepackets in graphene are femtoseconds and in carbon nanotubes are picoseconds. A later paper by the same authors (Rusin and Zawadski, 2009) offers theoretical analysis of a proposed femtosecond laser experiment that might confirm these suggested effects. A somewhat related hypothetical effect, "revivals" of electric current, predicted for graphene in a magnetic field (specifically for Gaussian wave-packets constructed only of positive energy states), has been analyzed by Romero and

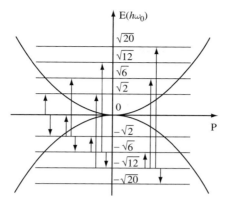

Fig. 8.23 Sketch of Landau levels in bilayer graphene at magnetic field in range 0.2 T – 2 T, showing transitions (arrows) that will occur out of states $n = -1, -2, -3$ and -4 (respectively, from left to right in Figure) under application of laser pulse. (Reprinted with permission from Wang *et al.*, 2010. © IOP Publishing 2010).

de los Santos (2009). The Zitterbewegung effect, proposed (without observation) early in the history of quantum mechanics, has also been touched upon and put into the context of graphene research by Katsnelson *et al.* (2007) and Castro Neto *et al.* (2009). Numerical simulations of oscillations in Zeeman transitions for bilayer graphene with a 1.6 fs laser pulse have been made by Wang *et al.* 2010.

Figure 8.23 shows the assumed Landau level structure in bilayer graphene (Wang *et al.*, 2010) with a perpendicular field in the range 0.2 T to 2 T. The level structure here has the electron-hole symmetry that is the basis of the Zitterbewegung effect. The laser pulse is assumed to represent a perturbing Hamiltonian term,

$$W(t) = -exE_0 \exp(-\alpha t^2/\tau^2) \cos(\omega_L t), \qquad (8.42)$$

where $\alpha = 2\sqrt{2}$; τ and E_0, respectively, are laser pulse duration and the electric field strength; and ω_L is laser frequency.

The simulations assume that the laser will drive transitions out of the Landau levels of indices -1 to -4, as indicated in Figure 8.23. Interferences between the positive and negative energy states are generated. The pulse duration is assumed to be 1.6 fs, but coherent oscillations are calculated to last considerably longer. The calculated quantities are the electric dipole moment and electric field generated by the carrier motions in the graphene bilayer, as suggested in Fig. 8.24.

In Fig. 8.24, the strength of the plotted induced electric field is in the order of μV/m, subject to scaling up to the order of 1 V/m by a degeneracy factor described by Wang *et al.*, while the laser field itself is stated as 0.1×10^7 V/m. The authors suggest that the effects are large enough, in bilayer graphene, to be measured at the state of the art. They also state that the corresponding effects in monolayer graphene are smaller.

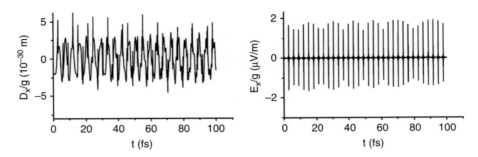

Fig. 8.24 Simulated oscillations in dipole moment and electric field of bilayer graphene with magnetic field $B = 0.2$ T after application of laser pulse. (Reprinted with permission from Wang *et al.*, 2010. © IOP Publishing 2010).

Some discussion is given of situations where oscillations are subject to periodic decay and revival, the decay being to a state of chaotic oscillation. No measurements appear to have been reported at present.

9
Applications of graphene

Graphene is available in a wide range of quality, from single crystals of micrometer and larger scale, to deposits of nanometer-scale chemically exfoliated flakes of variable thickness, from many layer to few layers, with a wide range of costs. The low-cost forms of graphene are finding application as additives, to improve electrical conductivity and to enhance flexibility and strength of composite materials. Graphene now supplements carbon fiber and carbon nanotubes as an additive, where a large market exists. A review of these applications is that of Segal (2009), who reports annual production in the scale of tons per year.

A recent excellent survey of the whole spectrum of graphene applications has been given by Novoselov *et al.*, 2012, who are able to project costs for the various types of graphene and to project where opportunities for market penetration will appear.

Our discussion is less broad, more limited to the higher-cost forms of graphene, more anecdotal and more focused on the operating mechanisms of the proposed devices. Work on devices based on the higher-cost, higher-quality monolayer and few-layer graphene is in largely in research phases. The most assured applications of high quality graphene seem to be in electronics, where specific devices (for instance, interconnects, flash memory and transistor-like devices) may soon be integrated into silicon device technology. Early reviews, emphasizing high-frequency transistors, were given by Chen *et al.* (2007) and Avouris *et al.* (2007). Graphene devices benefit from the uniquely high mobility and current density capabilities of graphene. The most revolutionary aspect, the small thickness, without incurring roughness or electrical discontinuity, may eventually allow scaling to smaller device sizes in FET field-effect transistor logic, but little experimental activity, specifically on switching devices, has been published so far. The unique property of adjustment of the work function by an electric field (modulating the Fermi level by a gate electrode) is a new possibility in electron devices (Yu *et al.*, 2009). A few prototype devices, exploiting this electric-field gating, are photoconductive detectors using quantum dots (see Section 9.2 below) and the Barristor barrier modulation transistor device that may become a possibility for large-scale, fast, on-chip logic. More devices exploiting the basic and unique electric field effect may appear in time. Larger display device applications seem promising, for example, as a touch-screen, while even larger-scale applications of transparent conductors (Nair *et al.*, 2008), notably as transparent electrodes in solar cells, are possible, but not assured in terms of market. It remains to be seen whether the 30-inch size, demonstrated for catalytically-grown-on-copper graphene, can be realized in roll-to-roll production at reasonable cost. Chemical vapor deposition is widely used in research devices, but the high temperatures so far needed for graphene, about 1000°C, appear problematic for large scale production.[1] Even so, the economy of minimal material usage (only one to four atomic layers), in conjunction with the superior electrical properties, mechanical resilience and the large abundance of carbon, might, in the long run, with likely improvements in fabrication techniques, turn out to be disruptive of earlier approaches. The possibility of graphene-based complementary logic devices, analogous to CMOS, will be discussed in Section 9.8.

9.1 Transistor-like devices

The primary semiconductor device (beyond the PN junction) is the field-effect transistor (FET) that has evolved with Moore's Law and is now produced in astonishingly large scale, reported (Pinto, 2007) as 10^{18} per year. These devices [see Fig. 9.1(a) for a graphene version] have a gate-electrode that draws carriers to a channel connecting the source and drain electrodes, to establish the On condition of the device. Graphene has superb conduction properties for the channel of such devices, with high mobility and high tolerated current density. FET devices have two main applications: as logic and memory devices that function as switches; and as radio-frequency (rf) amplifiers. However, a central fact is that the basic graphene FET device, composed of gapless graphene, essentially *cannot be turned OFF*, as required for a switch [see the review of Schwierz (2010)]. The ratio of On to Off current is desired in the range 10^4 to 10^7

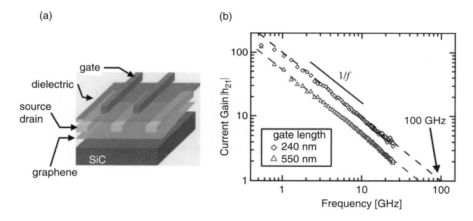

Fig. 9.1 (a) Sketch of graphene field-effect transistor FET fabricated on a 2-inch SiC wafer. Graphene grown on SiC is inherently conductive by electrons. The Source and Drain are shown and the top gate electrode. (b) Current gain for graphene FET as sketched in Fig. 9.1a, as function of frequency. (From Lin *et al.*, 2010, with permission from AAAS).

---

[1]The possibility of using benzene as the source gas, with pre-formed six-fold rings and allowing catalytic growth at 300°C (Z. Li *et al.*, 2011), may be a breakthrough in growing chemical vapor deposition (CVD) graphene at reasonable cost. It has even been suggested that coronene, containing seven benzene rings, might be a source vapor.

by logic circuit designers. [On the contrary, it is true that Wang *et al.* (2008a) have reported large On/Off ratios in <10 nm-wide graphene nanoribbon (GNR) field-effect transistors. Unfortunately, their method of GNR production is not scalable to the mass scale, nor did the devices have a top gate, considered essential by Schwierz (2010) for a practical FET switching device.] We defer to the end of this chapter further discussion of graphene FET devices intended for logic, as contenders to replace the present CMOS (complementary metal oxide semiconductor) logic family that has led the Moore's Law scaling progression.

9.1.1   High-frequency FET transistors

We now focus on graphene FET high-frequency devices that definitely are contenders in a post-silicon era for rf amplifiers. At the time of the Schwierz review, the fastest graphene FET had a cutoff frequency of 100 GHz with a gate length 240 nm (Lin *et al.*, 2010). A sketch of this device is shown in Fig. 9.1(a). The Dirac point of the film (grown in SiC) always required a highly negative gate voltage $V_G < -3.5$V, indicating a conductive n-type metal in the channel at zero bias voltage $n > 4 \times 10^{12}$cm^{-2}. The authors say that this helps to obtain a low series resistance in the device. The cutoff frequency is defined as the highest frequency where device current gain is unity or above, as shown in Fig. 9.1(b).

The graphene was grown on the Si-face of semi-insulating high-purity SiC, by thermal annealing at 1450°C and exhibited an electron carrier density $\sim 3 \times 10^{12}$cm^{-2} and a Hall-effect mobility in the range 0.1 to 0.15 m^2/Vs. The source and drain are long, closely spaced parallel electrodes [Fig. 9.1(a)] formed by sequential deposits of Ti (1 nm) and Pd (20 nm) followed by Au (40 nm), using e-beam lithography. To define the graphene channel (and to isolate individual devices), regions of graphene were etched in an oxygen plasma using poly(methyl methacrylate) (PMMA) as etch mask. To form the gate electrode, an interfacial polymer layer, made of a derivative of poly-hydroxysilane, was spin-coated onto the graphene before atomic layer deposition (ALD, see Chapter 6) of 10 nm thick HfO$_2$. These steps were chosen to least degrade the graphene mobility that afterwards was in the range 0.09 to 0.1520 m^2/Vs for devices across the 2 inch wafer. The measured cutoff frequency exceeds those of previously reported (to 2010) graphene FETs, as well as of Si metal-oxide-semiconductor FETs of the same gate length. The summary plot (shown in Fig. 9.2) suggests that 1 THz may be available at gate length 20 nm. In his review, Schwierz (2010) suggests that graphene FET devices may achieve superior performance by allowing a shorter channel.

Drawbacks to the design of Lin *et al.* include the high temperature needed (1450°C); the mobility degradation due to the graphene growth on SiC; the processing needed for the gate deposition; and the difficulty in precisely aligning the top gate to match the channel. One of these problems, gate-alignment, is solved by a self-aligned gate, in the work of Liao *et al.* (2010).

A field-effect transistor with narrow channel width, achieved by a self-aligned nanowire gate, has been demonstrated by Liao *et al.* (2010). This device is not a candidate for production, however, because it is based on a flake of micro-mechanically cleaved graphene transferred to an oxidized silicon wafer, as shown in Fig. 9.3. This

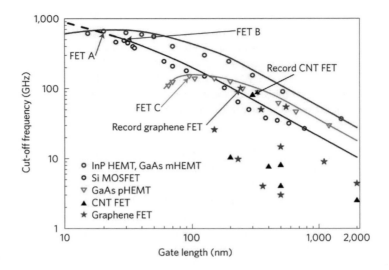

Fig. 9.2 Cutoff frequencies in GHz vs. Gate length in nm, for transistors. Here the device of Lin *et al.* is indicated as "Record graphene FET," indicated by star. Circles on upper line are InP HEMT and GaAs HEMT devices, triangles are GaAs HEMT devices. (HEMT indicates high electron mobility transistor.) The lower line represents carbon nanotube devices. In this figure CNT refers to carbon nanotube, and MOSFET refers to metal oxide semiconductor field-effect transistor. (From Schwierz, 2010, by permission from Macmillan Publishers Ltd., © 2010).

Fig. 9.3 Self-aligned nanowire gate design for high-speed graphene FET. Micro-mechanically cleaved graphene transferred to oxidized Si wafer, evaporated metal source and drain and Co_2Si-Al_2O_3 core-shell nanowire top gate, covered with self-aligning Pt gate electrode. Diameter of nanowire sets channel length with 5 nm Al_2O_3-shell functioning as gate dielectric. The self-aligned Pt thin-film pads extend the source and drain electrodes optimally to the location of the gate. (From Liao *et al.*, 2010, by permission from Macmillan Publishers Ltd., © 2010).

preserves the high mobility of the graphene and a portion of that high mobility may also be achieved with CVD-based production scale processes. The devices of Liao *et al.* (2010) achieve a higher cutoff frequency, about 300 GHz and higher current density, 3.32 mA/μm, compared to ~0.8 mA/μm for the devices of Lin *et al.*

Extremely high current densities, on the order of several mA/μm, are reported in these measurements. Assuming the thickness of the graphene is 0.34 nm, the current density is several times $3 \times 10^8 \mathrm{A/cm}^2$. A discussion of current density is given by Efetov and Kim (2010).

DaSilva *et al.* (2010) have addressed questions of carrier velocity and its saturation, relevant to these devices. They find that the primary scattering mechanism at high-current-density is emission of surface optical phonons of the substrate. It appears that one of the advantages of the diamond-like-carbon substrate (DLC) is that its phonons are at a higher energy, 165 meV and thus less likely to be excited. High electric field effects on carriers in graphene are also discussed by Tani *et al.* (2012).

Graphene transistors using ferro-electric gates have been analyzed by Zheng *et al.* (2010), including a review of different possible dielectrics.

Bilayer graphene FET devices have been fabricated and modeled by Szafranek *et al.* (2012). In comparison to similar devices with monolayer graphene, they find a voltage gain of 35, a factor 6 higher than they can attain in monolayer graphene FET devices.

Growth of graphene on SiC is appealing, because wafers of SiC are available and large arrays of graphene devices can be fabricated. The drawbacks are the high temperature needed and the lower mobility in graphene grown on SiC. Liao *et al.* have demonstrated FETs of better characteristics by going back to the micro-mechanically cleaved graphene, not a practical method. A different method for obtaining higher quality graphene on a large substrate has been explored by Wu *et al.* (2011). This is based on CVD growth of graphene on Cu, with subsequent transfer of large areas (cm scale) of graphene onto DLC substrates. These two innovations led to cutoff frequencies as high as 155 GHz at room temperature with gate length of 40 nm. The device of Wu *et al.* is imaged in Fig. 9.4(a), with summary of several devices in Fig. 9.4(b).

The procedure of Wu *et al.* (2011) follows the method of Li *et al.* (2009) for the graphene growth.[2]

The CVD graphene, after transfer to a DLC coated surface, was characterized by Raman spectroscopy, before device fabrication. A film of DLC was grown on an 8-inch Si substrate using cyclohexane (C_6H_{12}) with a vapor pressure of 1.8 psi in a CVD chamber. The flow rate was typically 25–40 cc STP per min at 100 milliTorr pressure. The DLC growth rate is 32 Å/s at 60°C and this was followed by anneal at 400°C for four hours. The source and drain electrodes were 20 nm of Pd followed by

---

[2]Namely, after evacuation of the chemical vapor deposition (CVD) chamber, a Cu foil was heated to 875°C in forming gas (H_2/Ar) for 30 min. After this reduction, the Cu foil was exposed to ethylene at 975°C for 10 min. and then cooled. PMMA was spin-coated on top of the graphene layer that had formed on one side of the Cu foil. The Cu foil was then dissolved in 1 M iron chloride solution. The remaining graphene/PMMA layer was washed and transferred to the desired substrate. Subsequently the PMMA was dissolved by treatment with hot acetone for one hour.

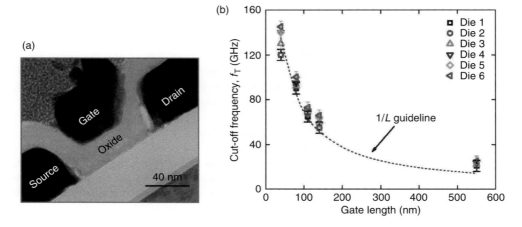

Fig. 9.4 (a) TEM cross-section of graphene FET, with gate width 40 nm. The oxide thickness is about 20 nm. The graphene layer is invisible in the TEM image and the diamond-like carbon layer (the base in the schematic diagram, appears to be about 50 nm thick. The TEM image suggests that this design cannot be made much smaller. (b) Gate-length-dependence of cutoff frequency. Summary of cut-off frequencies for family of graphene FET devices with varying gate lengths. (From Wu *et al.*, 2011, by permission from Macmillan Publishers Ltd., © 2011).

30 nm Au using e-beam evaporation. The gate was an oxidized Al layer evaporated by electron-beam deposition, followed by 15 nm atomic layer deposition (ALD, see Chapter 6) of Al_2O_3. The authors state that the DLC support for the CVD graphene is superior to the traditional oxidized Si support. They attribute the superiority to the higher phonon energy in diamond (165 meV) and a lower surface trap density than SiO_2, since DLC is non-polar and chemically inert, in their estimation.

High-frequency voltage amplifiers were fabricated by Han *et al.* (2011) using CVD-grown graphene transferred to a Si wafer. The devices, with 500 nm gate length, showed 5 dB low-frequency gain and 3 dB bandwidth greater than 6 GHz.

Wafer-scale graphene integrated circuits have recently been reported by Lin *et al.* (2011), again based on SiC growth of graphene. These circuits employ graphene FETs and include inductors, to act as broadband radio-frequency mixers at frequencies up to 10 GHz.

9.1.1.1 Properties of graphene grown on SiC.

In connection with FET devices, attention must be paid to the peculiar properties of epitaxial graphene grown on 4H-SiC(0001) (i.e., the Si polar surface). These properties have been studied by Sonde *et al.* (2009). in a scanning probe microscope study, comparing grown graphene with single layer graphene deposited after mechanical exfoliation. The adherence of the graphene, in both cases, is similar, dominated by van der Waals attraction. However, observation of the local I(V) characteristics in the scanning probe method enabled

Sonde *et al.* to conclude that a Schottky barrier is formed when mechanically-cleaved graphene is deposited on this SiC surface. The local I(V) obtained by the probe was strongly rectifying and the forward-bias exponential characteristic allowed an estimate of the barrier height. This method is explained by Giannazzo *et al.* (2006). These probed features were characteristically different in the three cases: on pure SiC, on SiC covered with micro-mechanically cleaved monolayer graphene and on epitaxial graphene grown *in situ*.

The barrier height, evident on the SiC surface, was reduced substantially in the epitaxially grown cases. The situation is suggested by Fig. 9.5. Graphene has an inherent work function 4.5 eV, while the electron affinity of 4H SiC is reported as 3.7 eV (Na *et al.*, 2004). This would imply a Schottky barrier height of 0.8 eV, compared with measured value 0.85 eV (Sonde *et al.*, 2009), in agreement. These measurements are consistent with theoretical work of Giovanetti *et al.* (2008), who conclude that in metal–graphene contacts of Al, Ag and Cu, the graphene becomes n-type, while contacts of Au and Pt make graphene p-type. These workers find that the graphene band structure is preserved by weak adsorption of Al, Cu, Ag, Au and Pt, but is significantly distorted by chemisorption of Co, Ni and Pd. A thorough study of Fe on epitaxial graphene is reported by Binz *et al.* (2012). In all of these cases the graphene is doped by the metal it contacts, as in the case of the epitaxial SiC.

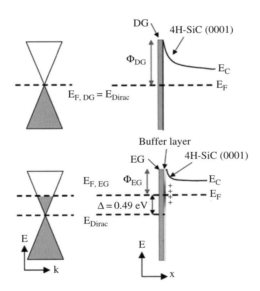

Fig. 9.5 Graphene doping by interaction with SiC. (top panel) Band diagram for micro-mechanically cleaved graphene (DG) deposited on SiC, showing Schottky barrier of height 0.85 eV. (bottom panel) Epitaxial graphene has Schottky barrier of height 0.36 eV, lower by 0.49 eV, due to Fermi level pinning to donor states in carbon-rich buffer layer on SiC surface. (From Sonde *et al.*, 2009. © 2009 by the American Physical Society).

As shown in Fig. 9.5 (top panel), no doping occurs with "deposited graphene" (DG) flakes placed (with no heat treatment) on SiC. The growth of epitaxial graphene (EG) on the Si face of SiC proceeds [see Section 5.2.1 and Seyller (2012); Bostwick *et al.* (2009)] by surface reconstruction: first, a Si-rich ($\sqrt{3} \times \sqrt{3}$) R30° phase, followed by the C-rich "buffer" ($6\sqrt{3} \times 6\sqrt{3}$) R30° reconstruction phase when the temperature increases. The latter is not yet a real graphene layer, rather an intermediate (buried) buffer layer, C-rich, with a large percentage of sp^2 hybridization. (Emtsev *et al.*, 2008). It appears that the Fermi level is raised by donor centers in this buffer layer, indicated in Fig. 9.5 (bottom panel).

These peculiarities of epitaxial graphene grown on SiC have been analyzed in the context of FET devices, by Kopylov *et al.* (2010).

The model of Kopylov *et al.* (2010) is indicated in Fig. 9.6. Work functions A and A′ induce transfer of electrons from surface states of density γ and/or bulk donor SiC states of density ρ. The dielectric constant of SiC, κ, is taken near 10, work functions A, A′ near 1 eV and surface state densities taken near $\gamma \sim 1 \times 10^{13}$ cm^{-2}ev^{-1}. Kopylov *et al.* estimate that the saturation density of n-type doping of monolayer graphene is around 10^{13}/cm^2 (corresponding to $E_F \sim 0.4$ev) for $d = 0.3$ nm, in case bulk donor density is 10^{19} cm^{-3} in the SiC buffer layer, or if the surface state density

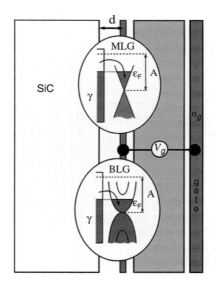

Fig. 9.6 Sketch of a SiC/G-based field-effect transistor. Insets illustrate relevant part of the MLG (monolayer graphene) and BLG (bilayer graphene) bandstructure. Surface charge in gate is denoted $n_G = CV_G/e$, d is the spacing between the SiC surface and the graphene layer. A is the work function, γ the surface state density, l the depletion layer width on the SiC and ρ the bulk donor density in cm^{-3} in the SiC buffer layer. (Reprinted with permission from Kopylov *et al.*, 2010. © 2010 AIP).

is $\gamma \sim 1 \times 10^{13} \text{cm}^{-2}\text{ev}^{-1}$. In device work, the SiC is semi-insulating and the charge density referred to exists only in the buffer layer just under the graphene, a consequence of the annealing process to generate graphene ("graphitization").

Kopylov *et al.* (2010) offer coupled equations that allow the carrier density in the graphene to be solved numerically:

$$\gamma[A - e^2 d(n + n_G)/\varepsilon_0 - E_F(n)] + \rho l = n + n_G \tag{9.1}$$

$$A' = E_F(n) + U + e^2 d(n + n_G)/\varepsilon_0. \tag{9.2}$$

Here $E_F(n)$ is the Fermi level induced in the graphene with carrier concentration n, $n_G = CV_G/e$, A and A' are the barrier heights in eV and $U = e^2 \rho l^2/\kappa \varepsilon_0$ is the height of the Schottky barrier at the surface of the SiC, whose permittivity is $\kappa \sim 10$, with l the related depletion layer width. Barrier height A relates to the graphene and A' to the donor levels in the SiC surface layer. The authors provide plots of the carrier concentration vs. assumed doping density ρ and surface state density γ. It is suggested that variations in the growth parameters can reduce, to some extent, the level of n-type doping remaining in the epitaxial graphene.

9.1.2   Vertically configured graphene tunneling FET devices

The conventionally configured FET devices described above, using a graphene channel, suffer from the metallic nature of the graphene, so that a high resistance off state cannot be achieved. This makes such graphene FET devices unsuitable for logic applications. (Strategies, e.g., to introduce bandgaps, to allow conventional FET operation in graphene devices are discussed more fully in Section 9.8.)

A different approach to graphene FETs, with vertical tunnel current flow between parallel graphene layers, acting as source and drain, has been recently presented by the Manchester group (Britnell *et al.*, 2012). Several devices, with accurate modeling, are constructed with single-crystal layers of graphene and hexagonal boron nitride, with additional devices using tunnel barrier MoS_2.

The source and drain are separated by a tunnel barrier, the whole sandwich placed horizontally on the oxidized Si wafer that serves as gate electrode. This FET device does not have a conventional "channel," but the forward device current flows vertically, by tunneling, from the source to the drain. The gate voltage modifies the carrier concentrations in the graphene layers and also, to some degree, modifies the tunneling probability across the single crystal tunnel barrier. It has been possible, remarkably, to create nearly ideal tunnel barriers and tunneling I(V) curves between two graphene monolayer electrodes obtained by micromechanical cleavage. The tunnel barrier is also a few-layer, mechanically exfoliated, film of BN or MoS_2.

The devices of Britnell *et al.* can be described as operating by electrostatic doping of the source and drain electrodes.

Because of this relevance, we turn to explore electrostatic doping in more detail, before fully describing the Britnell *et al.* devices.

9.1.2.1 Some details of electrostatic doping of graphene. The interactions leading to electrostatic doping in a similar structure, comprising [metal electrode (Cu)/h-BN/graphene], have been studied, in density functional theory, by Bokdam *et al.* (2011). The simplified Cu/BN/graphene structure analyzed by Bokdam *et al.* is shown in Fig. 9.7.

The Cu and graphene layers have different work functions, $W_M = 5.25$ eV and $W_G = 4.48$ eV, respectively, a difference of 0.77 eV. In the simplest analysis, when the Cu and graphene are brought together, the Fermi levels would become equal. That would require moving electrons to the Cu from the graphene, making the graphene p-type. This analysis, however, is refuted by density functional theory modeling of the Cu/h-BN/graphene sandwich that indicates that the h-BN layer plays an important role. The modeled result is indicated in Fig. 9.7, where sharp drops in the electrostatic potential, going from the left to the right, occur at the Cu-(h-BN) and (h-BN)-graphene interfaces. These drops in potential are the result of electric dipole layers. It appears that electrons are pushed into the Cu by proximity to the BN, presumably by Pauli principle repulsion between electrons in the two materials, with the Cu accommodating a negative charge on this basis. The potential jump at this interface is large, $\Delta_{Cu/h-BN} = 1.12$ eV, with a smaller similar effect at the h-BN/graphene interface, $\Delta_{h-BN/graphene} = 0.14$ eV. When these potential drops are included, a revised net potential difference becomes

$$V_0 = W_M - W_G - \Delta_{Cu/h-BN} - \Delta_{h-BN/graphene}$$
$$= 5.25 - 4.48 - 1.12 - 0.14 = -0.49 \text{ eV}.$$
(9.3)

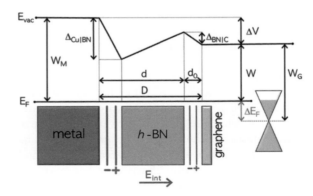

Fig. 9.7 Electrostatic doping of graphene. Electrostatic potential, as modeled, across Cu/h-BN/graphene sandwich, showing induced shift of graphene Fermi energy. E_{int} represents the total electric field across the h-BN slab of thickness d. The work functions of Cu (W_M) and graphene (W_G) are labeled on the left and right, respectively. (Reprinted with permission from Bokdam *et al.*, 2011. © 2011 American Chemical Society).

This non-trivial consideration of dipole layers has inverted the sign of the expected charge transfer. This number, -0.49 eV, is the shift between the Fermi levels of the Cu and graphene before charge transfer occurs, neglecting an internal electric field in the h-BN. To achieve equilibrium requires transferring electrons between Cu and graphene, to set up an electrostatic potential that compensates for V_0. To find what change in the Fermi level occurs in the graphene, Bokdam *et al.* (2011) calculate an effective work function W (see Fig. 9.7) for the Cu-h-BN stack, namely

$$W = W_M - \Delta_{Cu/h\text{-}BN} - \Delta_{h\text{-}BN/graphene} + e\,E_{int}d. \tag{9.4}$$

The charge density on the graphene, σ, is brought in, by the relation

$$\sigma = \varepsilon_0(E_{ext} - \kappa E_{int}), \tag{9.5}$$

including an external electric field E_{ext} driven by gate voltage $V_G = -e\,E_{ext}d/\kappa$. Here, κ is the permittivity of the dielectric BN of thickness d. Following the discussion at the beginning of Section 8.2, the charge density is related to the density of states by the relation:

$$\sigma = e \int g(E)dE. \tag{9.6}$$

Here $g(E) = g_0\,|E|\,/A'$, at energy E, where $A' = 5.18$ Ångström^2 is the area of the graphene unit cell and $g_0 = 0.09/(\text{eV}^2\text{unit cell})$. (Giovannetti *et al.*, 2008).

Combining these several relations, Bokdam *et al.* (2011) find for the Fermi-level shift in the graphene:

$$\Delta E_F = \pm \left\{ [(1 + 2\alpha g_0 d\,|V_G - V_0|\,/\kappa)^{1/2} - 1]/(2\alpha g_0 d/\kappa) \right\}, \tag{9.7}$$

where V_0 and V_G are given above and the sign of ΔE_F is the sign of $V_0 - V_G$. Here $\alpha = e^2/\varepsilon_0 A' = 34.93$ eV/Å, with A$'$ the cell area as above and the formula assumes dielectric width $d > 0$. A plot of ΔE_F from this formula was given previously as Fig. 8.7. The formula predicts a larger shift for smaller BN thickness, but requires at least two layers of BN.

These details do not apply directly to the vertical tunneling transistor of Britnell *et al.* but may illuminate some aspects of its operation.

The Britnell *et al.* device is sketched in Fig. 9.8(a), where the graphene layers are horizontal with vertical current flow.

The diagrams in Fig. 9.8 show the source/drain graphene layers separated by a single-crystal h-BN few-layer tunnel barrier. The isolation or encapsulation of the active graphene layers by h-BN (Dean *et al.*, 2010) largely avoids the spatial electrostatic fluctuations from SiO$_2$ (that would lead to charge "puddles") and allows nearly ideal tunneling characteristics to be observed. This is an unusual accomplishment indeed, but at considerable cost in complexity of manufacture.

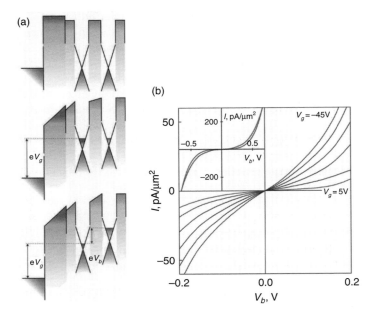

Fig. 9.8 (a) Sketches of the Britnell *et al.* (2012) vertical field-effect tunneling transistor. Layers, from left to right, are: doped Si, SiO₂, BN, Graphene$_{(Bottom)}$, h-BN (4 monolayers tunnel barrier), Graphene$_{(top)}$, h-BN encapsulant. (Top panel) Device with no bias applied. (Middle panel) Positive gate bias to silicon substrate, attracts carriers into both graphene layers (source and drain). (Lower panel) Source to drain bias added. The device will operate if Graphene$_{(top)}$ is replaced by a metal film. The central insulator, here hexagonal BN with variable number of monolayers, can be replaced with single crystal layers of MoS₂. (b) Tunneling characteristics at 240 K for a graphene-(h-BN)-graphene transistor device with 6 ± 1 layers of h-BN as the tunnel barrier. I–V curves for different gate voltages V_G in 10 V steps. Because of finite doping, the minimum tunneling conductivity is achieved with $V_G = 3V$. The inset compares the experimental I − V at $V_G = 5V$ (the lowest curve in main figure, upper curve in inset) with theory. (From Britnell *et al.*, 2012, with permission from AAAS).

Britnell *et al.* (2012) model the electrostatic doping in a similar fashion to that of Bokdam *et al.* Using F for electric field and subscripts b and t to designate bottom and top graphene layers, with subscript g for gate and μ as chemical potential (Fermi energy) Britnell *et al.* give

$$F_b - F_g = n_b e/\kappa\varepsilon_0 \tag{9.8}$$

and

$$-F_b = n_t e/\kappa\varepsilon_0. \tag{9.9}$$

With voltage V_b between the two graphene layers, one has

$$eV_b = eF_b d - \mu(n_t) + \mu(n_b). \tag{9.10}$$

Combining the relations gives the electrostatic doping condition

$$n_t e^2 d / \kappa \varepsilon_0 + \mu(n_t) + \mu(n_t + \kappa \varepsilon_0 F_g / e) + eV_b = 0. \tag{9.11}$$

This equation determines the carrier density in the top graphene layer induced by the field-effect from the gate voltage V_G.

For a conventional 2D electron gas the Fermi energy is proportional to the carrier density n and the first term in the previous equation, describing the classical capacitance of the tunnel barrier, is dominant for any realistic spacing d, larger than interatomic distances. In graphene, on the other hand, with its low density of states and Dirac-like spectrum, one finds $\mu(n) \propto \sqrt{n}$, leading to a qualitatively different behavior that can be described in terms of a quantum capacitance (Ponomarenko *et al.*, 2010). The new situation is also described in the work of Bokdam *et al.* (2011) discussed above.

Using the carrier densities as obtained from the analysis just described, the forward (vertical) tunnel current in the device is modeled using the conventional assumption for the electron tunneling conductance σ^T (at gate voltage V_G):

$$\sigma^T \propto g_{\mathrm{bottom}}(V_G) g_{\mathrm{top}}(V_G) T(V_G). \tag{9.12}$$

Here the factors are the densities of states g (see above) and the probability T of tunneling between bottom and top graphene layers at the appropriate bias conditions (Simmons 1963; Wolf, 1985). The forward current is obtained by integrating the conductance. The results are close to observation, as shown in inset to Fig. 9.9, an accomplishment in tunnel device construction.

The tunneling theory applied in Fig. 9.8(b) takes into account the linear (conical) density of states in the two graphene layers and assumes predominant hole tunneling with mass 0.5 and barrier height $\Delta = 1.5\,\mathrm{eV}$. Tunnel barrier thickness d is determined by atomic force microscopy. On/off current ratio near 50 is little changed at liquid helium temperature, consistent with tunneling mechanism. Current is normalized from device area, in the range 10 to 100 $\mu\mathrm{m}^2$.

The authors point out that this design, that has precedent in works of Luryi (1988), Heiblum *et al.* (1990), Simmons *et al.* (1998), Zaslavsky *et al.* (2003) and Sciambi *et al.* (2011), circumvents the limitation to on/off ratio imposed on the conventional graphene FET, by the lack of a bandgap and also the unimpeded electron transport through potential barriers caused by the Klein tunneling effect. (The Klein tunneling effect was discussed in Chapter 8, where it was found that its restriction to near normal incidence reduces its practical importance.) The authors have achieved on/off ratios close to 10 000 in similar devices where the h-BN insulator is replaced by MoS_2, with a smaller bandgap Δ. An additional tunneling device has been proposed by S. Banerjee *et al.* (2009) with no experimental results to date. Experiments of an elegant nature

(a)   (b)

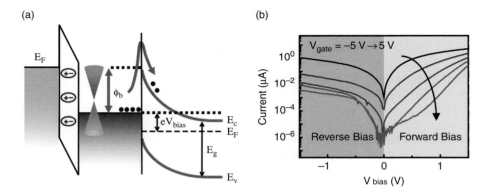

Fig. 9.9 (a) Band diagram of Graphene Barristor device, normally fabricated as horizontal layers on a silicon wafer, with gate electrode on top. Left to right, top-gate electrode, dielectric layer, graphene monolayer and hydrogenated N-type Silicon wafer. Bias between source (graphene) and drain (silicon) is in the reverse sense for a silicon Schottky device. The Barristor employs the top-gate (here shown negatively biased) to push the Fermi level down into the graphene valence band, thereby increasing the work function barrier φ_B. The current is thermionic, governed by the usual diode equation, with a strong temperature dependence. (b) Switching behavior of p-type Barristor device, note log scale for current. The current is plotted vs. bias at different gate voltages in 2 V increments. (From Yang *et al.*, 2012, with permission from AAAS).

on similar structures are described by Kim *et al.* (2012). In this work parallel graphene monolayers are used to allow accurate direct measurement of the Fermi energy. These workers conclude that the broadening of the zero energy Landau level is larger than that of the upper or lower Landau levels.

The work of Britnell *et al.* is striking in the high quality of the tunnel junctions, the apparent lack of defects such as pinholes that are common problems in working with tunnel barriers. While this work was carried out using micro-mechanically cleaved layers of graphene and h-BN, it may be possible that the devices could also be made using graphene and h-BN grown on Cu, as mentioned in Section 5.2.5 and transferred onto the device substrate. That said, a simpler (proposed and modeled, but not fabricated) graphene-based tunneling FET device (perhaps a contender for going beyond the Moore's Law era in CMOS devices) is described in Section 9.8, see Fig. 9.22. More generally, as we will see in Section 9.8, tunneling FET devices (T-FET) have a potential advantage over Silicon FET devices, in lower power consumption. Aspects of the electron–electron interaction between the graphene layers in the geometry of Fig. 9.8(a) have been examined by Song and Levitov (2012).

With regard to tunneling barriers on graphene, a different line of research, related to spin-injection and devices such as spin-valves and tunnel-valves, has led to innovation in fabrication methods. Han *et al.* (2010), have fabricated reliable electron tunneling barriers to observe tunneling spin injection into single-layer graphene. Their method involves a deposited MgO barrier, but with a prior deposition onto the graphene of a monolayer of TiO$_2$. Han *et al.* (2010) measure a "non-local magnetoresistance" $\Delta R_{\mathrm{NL}} = 130\Omega$ (alluding to their measurement of the Johnson–Silsbee voltage, to be discussed in Section 9.4). This is achieved by tunneling spin injection into graphene from Co through the MgO/TiO$_2$ barrier and is the largest such resistance value ever obtained. This topic is covered in more detail in Section 9.4.

Hybrid graphene/ferroelectric devices, for low-voltage electronics, have been described by Zheng *et al.* (2011). Graphene sheets, grown by chemical vapor deposition, have been transferred to the ferroelectric Pb Zirconate (PZT) to demonstrate field-effect transistors with on/off ratios of ten. These are deemed by the authors as candidates for ultra-fast non-volatile electronics.

9.1.3 The graphene Barristor, a solid state triode device

The tunneling FET device described by Britnell *et al.* has potential for gain and a wide operating range, both difficult to achieve in a conventional graphene FET device because it lacks an energy gap. A second approach, to using graphene to make a device with wide dynamic range, has been recently described by Yang *et al.* (2012). The Graphene Barristor is a hybrid silicon-graphene thermionic device that mimics the operation of the vacuum triode, with its control grid, as shown in Fig. 9.9. This device exploits the property of graphene that its Fermi level can be adjusted with a gate electrode, as the electric field induces carriers. This is a way to control the work function or barrier height φ_{b} in a graphene electrode. In the Barristor, a current-carrying graphene layer (the source) forms a Schottky barrier on the atomically clean surface of a hydrogenated Si wafer, acting as the drain. The Schottky barrier height φ_{b}, between the graphene and the silicon, is modulated by the top-gate electrode, controlling the thermionic current density J through the diode, described by the equation

$$J = A^* T^2 \exp(-e\varphi_{\mathrm{b}}/k_{\mathrm{B}}T)[\exp(-eV_{\mathrm{bias}}/\eta k_{\mathrm{B}}T) - 1].$$

Here A^* is the effective Richardson constant, k_{B} is Boltzmann's constant, η is an ideality factor near 1.0 and V_{bias} is the source-drain voltage, applied between the graphene layer and the silicon drain. A sketch of the bands in the device is shown in Fig. 9.9(a).

Figure 9.9(b) shows a family of curves for a p-type device with top-gate voltage as parameter. It is seen that positive increments (see arrow) in gate voltages, from -5V to $+5$V, under forward bias from source to drain, switch the device current from around 1 μA down to about 10^{-4}μA. The arrow shows the direction toward more positive gate voltage, in 2 V increments.

An extensive up-to-date review of graphene in nanoelectronics has been provided by Murali (2012).

9.2 Phototransistors, optical detectors and modulators

A recent development, a high-gain hybrid quantum dot graphene phototransistor, can be viewed as an example of electrostatic control of carrier density. In the recent devices of Konstantatos *et al.* (2012) a layer of PbS quantum dots of thickness 80 nm absorbs strongly in a wavelength region, between 0.9 μm and 1.4 μm, set by its bandgap and the quantum dot size. (The device geometry is similar to that of Fig. 9.4, absent the upper Pt coating and Co$_2$Si nanowire, now replaced by the layer of quantum dots, illuminated from above.) The result of light absorption is a negative charge accumulation in the affected quantum dot that electrostatically dopes the underlying graphene. A photoconductive multiplication effect occurs, at a level as high as 10^8 electrons per photon. This seems to be an important development, a device that is simple, scalable and adaptable to silicon technology. Inherent absorption of graphene is weak because it is so thin, making the material useful as a transparent conductor. Absorption in this device is provided by quantum dots, nanocrystals of semiconductor, often CdSe or PbS, where the fundamental absorption wavelength can be tailored by the nanocrystal dimension L. The mechanism of control of the absorption wavelength, by quantum dot size, is worth mention.

While quantum dot nanocrystals of different shapes occur in practice, the idea involved can be presented assuming a cube of side L. The energy of a photon to create an electron–hole pair (exciton) by breaking a valence bond, now exceeds the bandgap energy E_G and is increased by the containment energies of the resulting electron and hole. The lowest-energy electron state contained in the cube has one electron half-wavelength along each cube direction, requiring a confinement energy

$$\Delta E(L) = \tfrac{1}{2}\,\hbar^2 3(\pi/L)^2/m_e^*. \tag{9.13}$$

We can call this the (1, 1, 1) state of the confined (size-quantized) electron, whose effective mass (in kg) is m_e^*. A similar energy is required to confine the hole, so that the total energy to create an exciton is

$$E' = E_G + \Delta E_e(L) + \Delta E_h(L). \tag{9.13a}$$

The wavelength of the light absorbed is thus "blue-shifted" from $\lambda = hc/E_G$ to $\lambda' = hc/E'$. In the devices shown in Fig. 9.14 the exciton breaks up by donating the hole to the graphene, leaving the quantum dot nanocrystal negatively charged. The electric field from this negative charge, with a rather long lifetime, even 0.1 s, enhances the conductivity of the graphene, leading to large photoconductive gain.

The large enhancement of the responsivity R (in units of Amperes/W) is modeled in terms of a photoconductive gain

$$G_{\text{photo}} = [\tau_{\text{Lifetime}}(QE)\mu V_{\text{SD}}]/A, \tag{9.13b}$$

where τ_{Lifetime} is the lifetime of the negative remanent charge on the quantum dot (QE) the efficiency (about 0.25) between incident photons and trapped negative charges effective in electrostatically doping the graphene, μ the mobility in the graphene, on the order of 0.1 m^2/Vs, V_{SD} the source to drain voltage and A the area of the graphene flake. This gain G_{photo} is estimated to be on the order of 2×10^8.

In this work the graphene was micro-mechanically cleaved and placed on an oxidized Si wafer such as shown in Fig. 9.3. The PbS nanocrystals were deposited from solution in sequential steps to reach a thickness 80 nm. Methods of making these quantum dot nanocrystals are given by Hines *et al.* (2003).

Konstantatos *et al.* (2012) state that chemical processing of the deposited PbS nanocrystals shortened the ligands coating them, to make the PbS crystal layer more compact and more electrically conductive. These methods are described by McDonald *et al.* (2005).

The doped silicon wafer supporting the device (See Fig. 9.4) acts as a gate electrode and the back-gate voltage, V_{BG}, adjusts the responsivity of the phototransistor. The gate voltage acts to move the Dirac point of the graphene, to tailor the density of states and carrier density. The responsivity of the device is shown in Fig. 9.10(b), showing control by back-gate voltage.

The advance that has been made in this work can be appreciated by comparing with an earlier study of photoresponse in a similar device without the quantum dots (Xia *et al.*, 2008). In testing their earlier FET device with a scanned light spot it was found the optical responsivity was localized near the contacts. The peak responsivity found was 0.001 A/W, compared to up to 10^8 A/W in the devices of Konstantatos

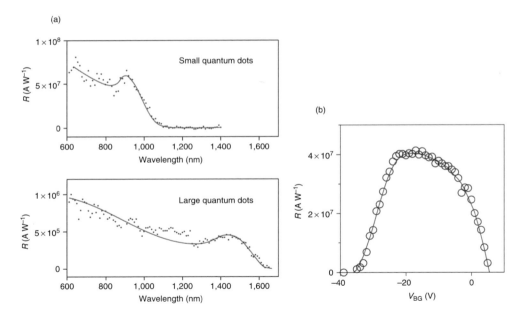

Fig. 9.10 (a) Data showing high photoconductive gain with hybrid quantum dot graphene devices with small quantum dots (upper) and large quantum dots (lower). Quantum dots are PbS deposited from solution to thickness about 80 nm. (b) Responsivity of quantum dot graphene phototransistor as a function of back-gate voltage V_{BG}. The gain of the photo-detector is directly controlled by the applied back-gate voltage. (From Konstantatos *et al.*, 2012, by permission from Macmillan Publishers Ltd., © 2012).

et al. (2012). Of course the frequency response is greatly reduced, to the order of $1/\tau$, where τ is the lifetime of the photoproduced charge on the nanocrystal.

We now turn to an ultra fast optical detector, realized in the work of Xia *et al.* (2009a). Their device is a single- or few-layer graphene sheet connected with source and drain electrodes, placed above oxidized silicon acting as a gate. The configuration is basically the graphene FET and the action of the light is to create electron–hole pairs in the graphene. The authors point out that electron–hole pairs created in the majority of the device, away from the source and drain, recombine on a time scale of tens of picoseconds, so that a useful photodetection does not result. However, near the contacts, where large electric fields exist in the Debye layer or Schottky barrier between graphene and metal source and drain contacts, the photo-electron and photo-hole can be separated. The device in this article depends upon such effects. Within the regions of the device that are photodetecting, the quantum efficiency is 6 to 16% efficient and no external bias voltage is needed. The authors find that the photoresponse does not degrade for optical intensity modulations up to 40 GHz and their analysis suggests that the intrinsic bandwidth may exceed 500 GHz. The advantages the authors see for the device are very high bandwidth optical detection, very wide detection wavelength range, zero dark current, good internal quantum efficiency and ease of fabrication.

Photodetection at a *monolayer–bilayer junction* in graphene was demonstrated by Xu *et al.* (2010). The geometry is, again, basically a graphene FET on an oxidized silicon wafer, the additional feature being a monolayer-bilayer junction midway between the source and drain electrodes. In this study, the photo-thermoelectric effect at the graphene interface junction was scanned with photocurrent microscopy, to indicate that the built-in electric field at the junction is central to the observed effect. The authors analyze their experiment using the band diagram in Fig. 9.11.

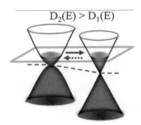

Fig. 9.11 Sketch of bands at *bilayer–monolayer junction*, assuming that constant carrier density is mandated by the back gate electric field. This means the Fermi level shifts as shown by the dashed line, representing an electric field at the junction. The upper arrow indicates the direction of an excited electron if driven by the built-in electric field, while the dashed lower arrow indicates the motion of an excited electron if dominated by the larger entropy available on the left that directly results from the larger density of states $D_2(E) > D_1(E)$. The data suggest that the latter entropy-driven process, termed the photo-thermoelectric effect, is dominant. (Reprinted with permission from Xu *et al.*, 2010. © 2011 American Chemical Society).

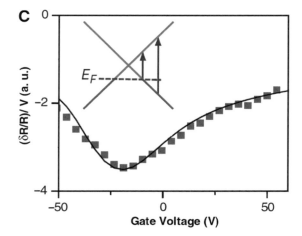

Fig. 9.12 Observed and modeled changes in reflectivity of 0.35 eV light (3.5 µm wavelength) from single-layer graphene, under change of back-gate voltage that shifts Fermi level. Inset band diagram shows optical transitions and their change with gate voltage. (From Wang *et al.*, 2008 with permission from AAAS).

Control of the Fermi level position in graphene by a gate-electrode allows control of optical absorption and reflectivity. This provides opportunities for devices for infrared optics and optoelectronics. A basic study of the gate-variability of optical transitions in graphene has been reported by Wang *et al.* (2008).

The data shown in Fig. 9.12 (squares) represent a cut through a gray scale image (not shown) of the measured quantity $\partial(\delta R/R)/\partial V$ vs. gate voltage and probing photon energy in the range 330 meV to 400 meV. The smooth curve in Fig. 9.12 is also a cut through a theoretically generated gray scale image of the same $\partial(\delta R/R)/\partial V$ differential reflectivity. Experimentally, tunable infrared (IR) photons in the range 330 meV to 400 meV were generated by an optical parametric amplifier pumped by a femtosecond Ti:sapphire amplifier system. The femtosecond pulses were attenuated to about 1 nJ and were then spectrally narrowed through a monochromator to about 10 cm^{-1} before they were focused to an area, about 10 µm on side, of a monolayer (or bilayer) graphene sample about 20 µm on a side. The back-reflected light was collected and detected by a mercury–cadmium telluride optical detector. The maximum absorption of the monolayer (bilayer) samples were about 2% (6%), respectively, of the normally incident infrared radiation. The reflectivity is small, but its relative changes are large and clearly detectable.

As suggested in the inset to Fig. 9.12, the most important absorptions were band-to-band transitions originating near the Fermi level in the graphene. The observations were consistent with the previously presented relation between the Fermi energy and the carrier concentration, n induced by the field electrode as

$$n = \alpha(V + V_0) \qquad (9.14)$$

that is a signed quantity, positive for hole doping. (Here the parameter $\alpha = 7 \times 10^{10}$ cm^{-2}V^{-1}, determined from a capacitor model.) The relation is

$$E_{\mathrm{F}} = \mathrm{sgn}(n)\hbar\nu_{\mathrm{F}}[\pi\,|n|]^{\frac{1}{2}}. \qquad (9.15)$$

where again n is the carrier concentration. The most important optical transitions satisfy the relation

$$\hbar\omega = 2\,|E_{\mathrm{F}}| = 2\hbar\nu_{\mathrm{F}}[\pi\alpha\,|V + V_0|]^{\frac{1}{2}}. \qquad (9.15a)$$

as sketched in the inset to Fig. 9.12. The authors made a plot of $(\hbar\omega)^2$ vs. V that gave a value $\nu_{\mathrm{F}} = 0.8 \times 10^6$ m/s. This is consistent with other values, but here is dependent on the accuracy of the scale factor α. These features were incorporated into the model for the gray scale image of the differential reflectivity vs. gate voltage and phonon energy, from which the solid line in Fig. 9.12 was extracted. To match the gray scale data image, the authors found it necessary to introduce an uncertainty $\delta V_0 = \pm 16$V, to account for apparent inhomogeneity in carrier density in the exfoliated flakes mounted on amorphous silica. This may be independent evidence of the electron–hole puddles that were gradually recognized as distorting the intrinsic nature of the conductivity near the Dirac point. The authors have carried out the same analysis on bilayer graphene (not shown), with distinctly and appropriately different results, but again quite consistent with bilayer graphene as a semimetal with parabolic bands. The possible bandgap of bilayer graphene was not seen.

With regard to the small values of absorption and small values of reflectivity in the context of device design (the maximum absorption at normal incidence on monolayer graphene in the IR is only 2%) the authors suggest that multiple passes of the light, or use of a waveguide to carry the light past the graphene surface, could alleviate this weakness. The authors estimate that a waveguide of length 100 μm (placed above undoped graphene) would allow complete absorption of the infrared light. The authors suggest that tunable IR detectors, modulators and emitters could be based on the properties that they have investigated.

An *optical modulator* based on graphene has been demonstrated by Liu *et al.* (2011), see also Lin *et al.* (2011).

The waveguide-integrated electroabsorption device of Liu *et al.* comprises a rectangular, single-mode waveguide (of doped silicon) placed above a SiO$_2$ surface. The 250 nm thick Si bus-waveguide is optimized for the communication range 1.35–1.6 μm and allows modulation at frequencies over 1 GHz. The modulator is basically a monolayer of graphene overlaid upon the waveguide (above a thin Al$_2$O$_3$ insulator), with electrical connection to the graphene by a gold pad on one side. Electrical connection to the silicon waveguide is brought out, on the opposite side, to a similar gold pad, by a 50 nm thick layer of the same doped silicon as forms the optical bus-waveguide. Bias voltage, between the graphene overlay and the silicon bus, adjusts the Fermi level

(a)

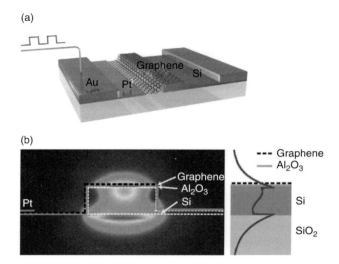

(b)

Fig. 9.13 (a) Geometry of broadband optical modulator. Si waveguide (upper center), crossing back to front (overlain with graphene) is 250 nm wide and optimized for communication range 1.35–1.6 µm. The waveguide assembly is mounted on thick SiO_2. Contact to the right Au electrode is by 50 nm thick Si, of the same doped composition used for the waveguide. The graphene crossing overlay is separated from the silicon waveguide by a thin layer of Al_2O_3. Voltage pulses to the graphene serve to induce charge and shift the Fermi level of the graphene, changing its absorption of the single mode photons, whose energy is around 0.55 eV. (b) Left panel: Geometry and single-mode 0.55 eV light pattern in and around the rectangular single-mode Si waveguide (center), mounted on wafer above SiO_2. Connection to left terminal (Au pad to Pt overlay to graphene) is indicated as dark dashed line. Contact to the right Au pad is by 50 nm thick extension of the Si forming the waveguide. Right panel: Electric field strength in the structure, to induce Fermi-level-shift in the graphene (top, dashed line). (From Liu *et al.*, 2011, with permission of The Royal Society of Chemistry).

within the graphene and sets up conditions for electroabsorption of the photons in the single optical mode. The overall geometry is shown in Fig. 9.13(a).

The optical mode fields and the static electric field imposed by the pulses to the graphene electrode are depicted in Fig. 9.13(b).

Operation of the device over a range of voltages applied to the graphene is demonstrated in Fig. 9.14, with insets to show the nature of the transitions induced in the graphene by the optical photons. In the insets to Fig. 9.14, the graphene is depicted by the usual conical band structure. The overall modulation is 0.08 dB/µm in this figure, so for a 40 µm device the modulation would be 3.2 dB.

Analysis and assessment of graphene-based terahertz modulators has been offered by Sensale-Rodriguez *et al.* (2011).

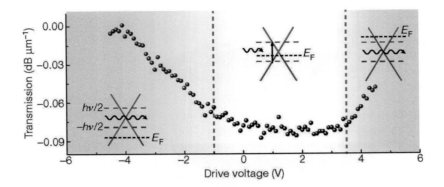

Fig. 9.14 Static electro-optical response of the device at different drive voltages. Data points show the modulation depth at different drive voltages. In the central region near zero voltage the Fermi level in the graphene is near the Dirac point, such that an interband transition can be driven by the 0.55 eV photons in the waveguide, leading to absorption (see inset). At strong negative and positive drive voltages the Fermi level is situated such that an interband transition is not possible. (From Liu *et al.*, 2011, with permission of The Royal Society of Chemistry).

Blue-light emitting graphene-based materials and their use in generating white light, have been described by Subrahmanyam *et al.* (2010). These materials are derived by chemical methods, for example, few-layer graphene was prepared by thermal exfoliation of graphite oxide at high temperature, following a method of Subrahmanyam *et al.* (2008). Using prolonged sonication and reduction, dispersions of nanoparticles of graphite oxide were obtained that give fairly intense blue emission around 440 nm.

Addition of ZnO nanoparticles to the suspension resulted in a second emission peak around 550 nm, such that the total emission is nearly white in appearance. These articles give many references to chemical preparation methods. The light emissions are from solutions containing particles that we can view as quantum dots. The photon energies may be understandable as a bandgap energy plus confinement energy of electrons and holes as described at the beginning of Section 9.2. A recent report of giant optical non-linearity of graphene in strong magnetic field has been given by Yao and Belyanin (2012).

9.3 Wide area conductors, interconnects, solar cells, Li-ion batteries and hydrogen storage

Methods for making large area graphene sheets were described in Section 5.2.3. [A review of graphene electrodes is given recently by Jo *et al.* (2012).] Large-area copper foil was demonstrated as a catalytic substrate for CVD growth of graphene monolayers up to 30 inches in diagonal measure by Bae *et al.* (2010), see Section 5.2.3. These authors reported $30\Omega/\square$ measured on a four-layer stack of CVD-grown graphene monolayers, chemically doped and sequentially transferred. Conductive transparent touch screens, for devices such as cell phones and tablet computers, are applications for this type of low-resistance transparent graphene. Thirty inches is large on the scale of touch screens and interconnects (as well as for transistors) but is less clearly adequate for large-scale solar cells, a potentially huge application also requiring low sheet resistance. (It is not clear that dispersed graphene flakes, as are available from liquid-phase or chemical exfoliations, give, in a planar deposit, a sufficiently conductive layer for solar cell application. The refractory nature of graphene probably makes impractical an annealing/recrystallizing step, such as is used in nano-ink-deposited CuInGaSe solar cell manufacture, to strongly increase the conductivity of the thin film deposit.)

Large areas are needed for solar cells, because the incoming energy is 1000 W/m^2 at best and around 200 W/m^2 averaging over time. A typical silicon cell, sold for rooftop mounting, is, e.g., 14 inch by 9.3 inch in size, deposited continuously at 14 inch width on a foil moving at 2.3inches/minute, see A. Banerjee *et al.* (2009).

The market for thin-film solar cells is large and growing: in 2010 the capacity of thin-film cells produced was about 2000 MW. The corresponding market value, by rule of thumb, would be around \$2 billion. Crudely, assuming a nominal efficiency, the sun-facing area of this capacity would be ~ 2.8 miles $\times$ 2.8 miles. Solar cells thus set a different size scale than the previously mentioned electronic applications. It is not clear that present methods to make low-resistance graphene can compete on the size scale of the main solar cell market.

Another potentially large-scale application for graphene is in improving the properties of lithium-ion batteries. In this application graphite and $LiCoO_2$ are typical materials for the anode, where Li ions are reversibly stored in intercalated locations. The charge capacity of the battery is related to the accessible surface area and the power and recharging rates are related to the rate of ion motion in and out of the storage locations. Relating to graphene, the graphite intercalation compound C_6Li is a benchmark: the charge capacity at this stoichiometry corresponds to 372 mAh/g. Higher-performance carbon-based anode materials, with a higher Li-ion accommodation number than in C_6Li, have been studied by Yang *et al.* (2010), while Yoo *et al.* (2008) investigated "graphene nano-sheet [(GNS)] families" for use in rechargeable lithium ion batteries. The paper of Yoo *et al.* presents experimental results comparing reversible charge–discharge capacities of several anodes. These are: crystalline graphite (measured as 320 mAh/g), GNS (6 to 15 layers thick, measured as 540 mAh/g), "GNS + carbon nanotubes" (measured as 730 mAh/g) and "GNS + C_{60}" (measured as 784 mAh/g). It appears that inserting carbon nanotubes or C_{60} molecules into the nano-sheet structures opens them to wider interplanar spacing, making more room for the Li ions, whose radius is estimated as 0.06 nm. The authors suggest that the limiting capacity might correspond to the LiC_2 stoichiometry, that would evidently correspond to 1116 mAh/g. The nanosheet structures are chemically exfoliated and it is found that enhanced storage capacity correlates with an increased average interlayer spacing, from 0.34 nm in graphite, to nearly 0.4 nm in the "GNS + C_{60}" material.

Chemical methods to produce low sheet resistance graphene films are actively being pursued, as we will see. These methods however, presently produce films of inferior electrical conduction and are actually quite limited in the lateral dimension available.

For example, chemically derived graphene as a transparent conductor for light emitting diodes has been assessed by Wu *et al.* (2010), in an article containing many useful references. These authors demonstrate organic light-emitting diodes built on chemically processed graphene that are suggested to perform comparably to conventional devices using indium tin oxide (ITO).

Wu *et al.* (2010) describe a chemical process starting from graphite and ending with spin-coating a water dispersion of graphene oxide particles onto a substrate, followed by reducing the film by heating to 1100°C in vacuum for 3 h. (An alternative reduction, using hydrazine, followed by a 400°C anneal in Ar, was also reported, with somewhat higher sheet resistances.) In a 7 nm thick film, a sheet resistance $R_{\mathrm{sh}} = 800\Omega/\square$ was measured with an optical transmission at 550 nm of 82%. The measured sheet resistance is more than 100 times that of an ideal graphene conductor that the authors estimate as $R_{\mathrm{sh}} = 62.4\Omega/N$, where N is the number of layers. They attribute the 100 times excess in sheet resistance to "multiple grain boundaries, lattice defects and oxidative traps that limit carrier mobility." It appears to the present author that steps of spin-coating and heating to 1100°C in vacuum are difficult steps for production, especially for sizes more than centimeters on a side. The severe anneal seems incompatible with any organic substrate as might be used for flexible electronics. Spin-coating becomes inconvenient for surfaces more than several centimeters in dimension and alternative means, such as a doctor-blade approach, likely produce thicker and less-even deposits on a large substrate. If a large surface area, without electrical connectivity, is sufficient, for example, for adsorbing hydrogen in a storage cannister, then a commercial chemical product like Grafoil® could be a starting point.

An extreme attempt toward a large surface area (non-conductive) material, an integrated graphene-nanotube structure, has been reported by Zhu *et al.* (2012). Graphene monolayers are held apart by arrays of perpendicular single-wall carbon nanotubes. The product, demonstrated on a small scale using CVD deposition onto copper foil, is characterized by a surface area $>2000\,\mathrm{m}^2/\mathrm{g}$ and allows ohmic contacts internally between the vertically aligned single walled carbon nanotubes and the graphene. This value is close to the theoretical limit for fully expanding graphite into graphene, mentioned at the end of Section 5.1.2, as 2630 m^2/g. The structure of Zhu *et al.* would not collapse, to block access of the surface to gases, as typically happens when flakes of graphite are spun onto a surface. This structure is suitable for hydrogen storage, but is likely not available, using the cited methods, in the large scale that a practical application would need.

Primitive organic photovoltaic cells, based on graphene deposited on quartz, were described by Wu *et al.* (2008). Their chemical preparation was similar to that used by Wu *et al.* (2010) (see above), but the stated resistance per square, for <20 nm thick graphene, was quoted in the range 5kΩ to 1MΩ, much higher than the 800Ω/$\square$ given by Wu *et al.* (2010). The performance of these cells was poor that was partly attributed to the high sheet-resistance, a feature evidently improved upon in later work by the same authors. The early report of Wu *et al.* (2008) may be more of an indication of learning curves in processing chemically-derived graphene and also in its incorporation in organic solar cells, than that graphene films, even those chemically derived, cannot be envisioned in this application. Rather similar results for dye-sensitized solar cells on chemically-derived graphene are given by Wang *et al.* (2008) and by Eda *et al.* (2008). Sheet resistances around 1kΩ per square were obtained by Wang *et al.* (2008) after spin-coating graphite oxide particles and reducing the deposit by heating to 1100°C. A chemical method for making boron-doped graphene flakes is described by Lin *et al.* (2011) who also show that their flakes can be used to make improved back-contacts to CdTe solar cells.

Superior solar cell performance has recently been reported by Miao *et al.* (2012) in Schottky barrier devices of graphene on n-type Silicon. An efficiency of 8.6% was observed in devices where the graphene, grown by CVD on copper foils and transferred to the n-silicon surface, was subsequently doped with TFSA. The latter, bis (trifluoromethanesulfonyl-amide ($(CF_3SO_2)_2NH$), transfers charge to the single graphene layer. In this device, the work function difference between the graphene and the Si results in electrons being transferred from the Si to graphene, leaving a Schottky barrier in the Si. The resulting built-in potential separates the holes and electrons in the functioning of the solar cell. Quite ideal solar cell I(V) curves are shown in the paper of Miao *et al.* The authors suggest that the same direct transfer of graphene, to make a Schottky barrier, with results enhanced by TFSA doping, might be applicable to Schottky barrier solar cells on other semiconductors, including CdS and CdSe. The paper gives experimental details and also references to earlier work.

Doping of graphene by organic polymers has also been reported by Sojoudi *et al.* (2012), who also confirm the importance, in practice, of p-type doping by typical atmospheric exposure (Schedin *et al.*, 2007). Sojoudi *et al.* use amine-rich 3-aminopropyltriethoxysilane (APTES) for electron doping and fluorine-containing perfluorooctyltriethoxysilane (PFES) for hole doping. They actually transferred CVD-grown graphene from a copper surface onto a Si/SiO_2 substrate, further prepared with adjoining self-assembled monolayers (SAM) of APTES and PFES, in the hope of creating a p–n junction at the boundary. They found that the intended dopings of the graphene were observed only after the resulting structure was annealed in nitrogen at 200°C. The need for the anneal is suggested as to remove the typical p-type doping from the atmospheric exposure, but possibly also improving atomic contact between the graphene and the underlying dopant layers, to release or activate the intended doping. While it appeared that the adjoining layers, after the processing, were indeed doped, the I(V) of the p-n structure did not show rectifying behavior.

Millimeter-sized solar cells are used in watches and calculators but large scale applications for roofs and solar farms call for single cells on the order of 6 inches on a side, as mentioned above. Solar influx is a diffuse energy resource, on the order to 200 W/m^2 (on a time-averaged basis), so that cell areas of square miles would be needed for a serious impact on the energy economy. A recent comment on solar panels to supplement grid power is given by Crane and Kennedy (2012).

CVD growth on Cu or Ni foil, as described in Section 5.2.3, may be amenable to roll-to-roll growth and processing of graphene on a metal substrate but the resulting graphene/(copper-substrate) combination seems not directly useful in solar cell manufacture. (For a description of roll-to-roll production of silicon thin film solar cells, see A. Banerjee *et al.* (2009). For a tutorial discussion and review of solar cell types and their operation, see Wolf (2012a).

To extract the catalytic-on-copper graphene and deposit it, e.g., onto a PN junction structure as a transparent upper electrode, is a more difficult task. The default method at present is to dissolve the copper foil and lift the graphene, using a polymer layer such as PMMA (later to be dissolved) onto the new location, as was done by Miao *et al.* (2012) mentioned above. Variations on the basic CVD growth, to reduce the cost, might avoid the copper foil by evaporating (Tao *et al.*, 2012, see Section 5.2.4) the copper in a much thinner layer, not needing to be removed and to grow the CVD layer at reduced temperature, perhaps 300°C, using benzene, already containing six-fold rings, as source gas (see Z. Li *et al.*, 2011).

9.3.1   Interconnects

Metal wires or interconnects are an essential aspect of silicon chip design and manufacture and graphene is a conductor much superior to any metal. Silicon computer chips commonly have six layers of metallization. Much publicity was attached to the change from Al to Cu interconnects in chip manufacture several years ago (Rosenberg *et al.*, 2000). A comparative analysis of graphene interconnects was offered by Xu *et al.* (2009a). The more recent report of Bae *et al.* (2010) of the low sheet resistance of 30Ω/□, measured on a four-layer p-doped graphene sheet, may be encouraging from the point of view of making on-chip interconnects. The growth of the semiconductor industry is paced by ITRS, the International Technology Roadmap for Semiconductors that predicts that by 2020 the basic width of wiring will be 22 nm and will require current density for interconnects of 5.8×10^9 A/cm^2. This current density is regarded as impossible in Cu wiring, suggesting a definite opportunity for graphene interconnects.

A practical study of the breakdown-current density in multilayer CVD graphene grown on an evaporated Ni film on an oxidized silicon surface was reported by Lee *et al.* (2011). These authors used a 500 nm thick Ni film on a Si/SiO$_2$ substrate to grow CVD graphene at 1000°C in 10 min. using 5–30 sccm (standard cc per minute) of methane and 1300 sccm of hydrogen at atmospheric pressure. It was found that the thicknesses of the graphene layers, measured by atomic force microscopy (AFM) increased with methane concentration. The resulting films were transferred to a clean oxidized Si substrate and were patterned into wires of 1 and 10μm widths and lengths from 2 to 1000 μm, provided with Cr/Au or Ti/Au contacts. The contacts were determined to have low resistance and measurements on the wires established the resistivity of the graphene. The study emphasized 10–20 nm thick graphene films and found resistances per square in the range 500–1000 Ω. Breakdown current densities up to 4×10^7 A/cm^2 were measured, exceeding, by at least an order of magnitude, the current capacity of Cu. The breakdown mechanism was identified as heating.

Growth directly on silicon would be desirable from a production point of view, but using present methods will limit the performance of the graphene. [As mentioned earlier, Hackley *et al.* (2009) have had some success growing graphitic carbon directly on Si 111.] Higher values of conductivity have recently been reported by Jain *et al.* (2012) for graphene on hexagonal BN substrates. The graphene is grown by CVD on Cu foil and then transferred to micro-mechanically cleaved flakes of hexagonal BN that, in turn, had been transferred onto an oxidized Si wafer. The electrical characteristics

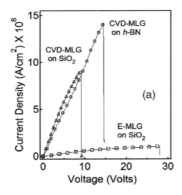

Fig. 9.15 Current density vs. Voltage for monolayer graphene on (top to bottom) h-BN, CVD graphene on SiO_2 and mechanically exfoliated graphene on SiO_2. (Reproduced with permission from Jain *et al.*, 2012. © 2012 IEEE).

of Cu-grown CVD graphene on this surface are excellent, including $\mu = 1.5 \mathrm{m^2/Vs}$ at carrier density $1 \times 10^{12} \, \mathrm{cm^{-2}}$. The breakdown current density is recorded as $1.4 \times 10^9 \, \mathrm{A/cm^2}$ as shown in Fig. 9.15, in comparison to Cu-CVD grown graphene on SiO_2 and to mechanically exfoliated graphene on SiO_2.

An extensive report demonstrating the use of graphene interconnects for arrays of inorganic microscale light- emitting diodes on stretchable rubber substrates has been given by Kim *et al.* (2011). The authors find that linear distortions up to 100% can be accommodated in a properly designed system.

9.3.2 Hydrogen storage, supercapacitors

Large surface area is desirable for storage by adsorption or chemisorption of hydrogen and also in enhancing the capacitance of a supercapacitor. Monolayer graphene offers the largest surface area per gram of any possible substrate. The potential of few-layer graphene for hydrogen storage has been described recently (Subrahmanyam *et al.*, 2011).

These authors note that hydrogenated graphene containing ∼5 wt% hydrogen is stable and can be stored over long periods. Spectroscopic studies of such samples reveal the presence of sp^3 C–H bonds in the hydrogenated graphenes. The hydrogenated graphenes, suitably, decompose readily on heating to 500°C, releasing all of the hydrogen, thereby demonstrating the use of few-layer graphene for chemical storage of hydrogen. It was found that release of hydrogen starts at 200°C and is complete by 500°C. The applications of such hydrogen storage are potentially large. In simple terms, a conventional automobile (such as a BMW sedan) with minor modification can be operated with hydrogen replacing petrol. The "gas tank" of such an auto could be a sorption cell for hydrogen to be released (according to the accelerator foot-pedal operation, related to the temperature of the storage region) to run the engine.

The conventional engine can burn the hydrogen as if it were petrol but in the more advanced scenario one uses a fuel cell and electric motor to replace the four-cycle combustion engine (of low efficiency). The fuel cell then powers a high efficiency electric motor drive. The fuel cell/electric motor drive is superior in efficiency and avoids undesired emissions. The choice of an electric power train also offers opportunity to reclaim kinetic energy in the braking process. This opportunity is available in present hybrid autos such as the Toyota Prius.

The problem of intermittency in the solar cell production of energy requires that energy be stored; that might be locally accomplished by using the energy to decompose water, resulting in hydrogen. Hydrogen is itself a valuable commodity that can be sold on an open market. Its primary commercial uses are in production of ammonia and in the production of gasoline from crude oil. Hydrogen is a portable storage medium for energy and the chance to store this energy until it is needed is offered by graphene, as explained by Subrahmanyam *et al.* (2011). The conventional approaches to transportation of hydrogen (as an energy medium) are via pipelines or in containers of high pressure gas or of cryogenic liquid. The promising added approach to transporting hydrogen is in sorption cells as described.

The authors used a Birch reduction process to add hydrogen to the pristine graphene. The actual process involved lithium in liquid ammonia and thus is not practical for reversible storage as in the context of an auto. It was not shown how such a process could be adapted to conveniently add hydrogen to graphene, as would be needed in a potential automotive application. Once added, the hydrogen release was shown to be convenient: simply by heating the hydrogenated graphene.

It has been recently shown (Sigal *et al.*, 2011) that adsorption sites for H_2 on graphene or carbon nanotubes can be blocked by even small amounts of O_2. These authors say that to accomplish hydrogen adsorption, in practice, where some oxygen will always be present, access to the surface by oxygen would have to be blocked, perhaps by a polymeric coating or by a nanoporous structure that would not allow molecular oxygen to enter. This appears to be a significant drawback, although Palladium–silver alloys are permeable to hydrogen but not other gases and could in principle block entry of oxygen to a cask containing the activated surfaces.

Blake *et al.* (2008) have discussed the use of exfoliated graphene flakes in a liquid crystal device.

Supercapacitors (or ultracapacitors) based on graphene have been described by Stoller *et al.* (2008) and by Liu *et al.* (2010). The latter authors describe high energy density: 85.6 Wh/kg at room temperature and 136 Wh/kg at 80°C, at a current density of 1 Ampere/gram. These high energy densities are comparable to Ni-metal-hydride batteries, but the charging/discharging times are reduced to seconds or minutes, according to the authors. Related work is reported by Hiraoka *et al.* (2010) and Zhu *et al.* (2011).

Hiraoka *et al.* (2010) describe supercapacitor electrodes made from carbon nanotubes, not easily available in large quantity, but Zhu *et al.* (2011) describe a commercially scalable process starting from graphite oxide that is exfoliated in a microwave heating step, resulting, finally, in a porous material stated as having a surface area of 3100 m^2/g.

9.4   Spintronic applications of graphene

The superior conductivity of graphene is accompanied by a large spin mean free path, a small tendency for the spin direction of the electron to be changed in transit. The long mean free path for the spin, enhanced by the small spin-orbit interaction in the carbon atom, makes graphene a good medium for "spin-tronic" devices. Electron devices that exploit the spin direction of the electron (spintronic devices) include, notably, the magnetic tunnel junction (MTJ) magnetic disk reader, used in most computers. This state-of-the-art device actually followed and improved upon a class of "giant magnetoresistance" (GMR) devices, for which the Nobel Prize in Physics was awarded in 2007 to Albert Fert and Peter Grunberg. In these earlier "spin valve" devices, ferromagnetic electrodes, such as Co and Ni, are used as sources of spin-polarized electrons that transit the device preserving spin direction, finding smaller resistance when the magnetizations of the two electrodes are parallel. A key element in the superior and commercially important Tunnel Valve (or Tunnel Magnetoresistive, TMR) device, is a tunnel barrier that will not disrupt the spin-alignment of tunneling electrons. MgO is an excellent tunnel barrier for this purpose. A tutorial description of many of these devices is given by Wolf (2009). An expert review of real and potential spin-based logic devices is given by Dery *et al.* (2007). Materials for use in such devices are characterized by a long spin lifetime, corresponding to a long spin- diffusion length. Micrometer spin-diffusion lengths have been reported in graphene. The role of small spin-orbit coupling in graphene is detailed by Kane and Mele (2005) and Huertas-Hernando *et al.* (2006). The potential of graphene nanoribbons for spintronics was described by Son *et al.* (2006).

A graphene spintronic device, a spin-valve, has been demonstrated by Tombros *et al.* (2007), whose work is extended by Han *et al.* (2010). Tombros *et al.* (2007) have nicely measured essential spin injection and diffusion effects in graphene, using ferromagnetic Co as electrodes, that we now describe.

The geometry in Fig. 9.16(a) and the measurement of Fig. 9.16(b) (Tombros *et al.*, 2007) are based on the Johnson–Silsbee effect (Johnson and Silsbee, 1988; see also Jedema *et al.*, 2002). That is, a non-equilibrium magnetization in a paramagnetic metal, such as graphene, can be detected as an open-circuit voltage across an interface between the paramagnet (graphene) and a ferromagnet (Co). In Fig. 9.16(a) electrodes 1 and 2 are ferromagnets and the (Johnson–Silsbee) voltage difference $V_{2,1}$ is proportional to the difference in spin polarization in the graphene at locations 2 vs. 1, assuming both electrodes have positive magnetizations. As the spin polarizations are of either sign, the detected voltage is a signed quantity, nevertheless recently treated as representing a "non-local resistance" when divided by the driving current. The sign of the detected voltage is reversible, by reversing the magnetization direction of the ferromagnetic electrode. In Fig. 9.16(a) contacts 3, 4 inject electrons into the graphene, with polarization depending upon the magnetization direction of the individual Co ferromagnetic electrode. The electrode polarizations are controlled by the experimenters, by programming an applied magnetic field B.

Figure 9.16(b) shows measurements on the non-local spin-valve device conceptually sketched in Fig. 9.16(a). These measurements, at 4.2 K, reveal at least 4 different

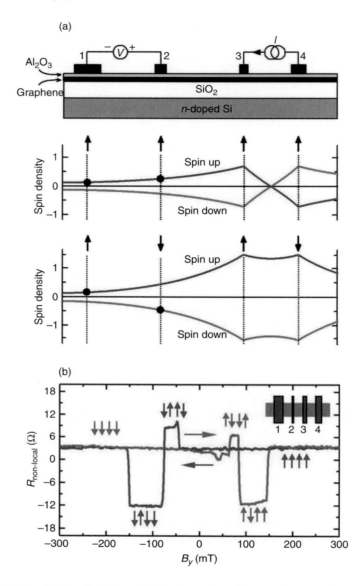

Fig. 9.16 (a) (Upper panel) Spin-valve device in graphene defined by ferromagnetic electrodes 1–4, separated from graphene by aluminum oxide barriers to allow spin injection from the Co into the graphene. Length of the device (left to right) is about 8 μm, while depth (into the page) is about 2 μm. Electron current is driven through the graphene from terminal 3 to terminal 4, while voltage corresponding to non-equilibrium magnetization is measured between terminals 2 and 1. Electrodes are long and thin, so that only a single magnetic domain is contained. The different small widths that vary the critical field for switching the magnetization directions that point into and out of the page. At relatively large magnetic

"resistance" levels[3] in the switching of the electrode magnetizations, as indicated by the horizontal arrows. The dominant voltages are obtained when the current electrodes 3, 4 have the same magnetization direction, as suggested in the lowest panel of Fig. 9.16(a). The distinct magnetization reversals of the Co electrodes, with magnetic field change, are a consequence of their distinct coercivities, primarily determined by the electrode widths, stated to be in the range 80 to 800 nm.

In Fig. 9.16(b) the most negative region in B_y shows zero voltage, corresponding to the situation shown in middle panel of Fig. 9.16(a). The largest change in Johnson–Silsbee voltage, corresponding to the largest change in non-equilibrium magnetization between electrodes 1, 2, occurs when the magnetizations of the current electrodes 3, 4 are changed from parallel to anti-parallel. In the latter case, with current flow, injection is of spin-up electrons, while extraction is of spin-down electrons, enhancing

---

Fig. 9.16 (*continued*) field into the page, all electrode magnetizations are parallel into the page. The narrowest electrode, #3, requires the largest reversed magnetic field to switch its magnetization to out of the page. (Lower panels) Conceptual sketches of the device operation. The upper and lower spin density curves are mirrored around spin density zero, as increase of spin-up density implies loss of spin-down density. The electrode magnetization directions, indicated as ↑ and ↓ (corresponding to into and out of the page) in and under the four Co electrodes, are controlled by sequencing a magnetic field B_y into and out of the page. Arrows in bottom panels show direction of spins at locations in the device. The definition of the non-local "resistance" is the voltage detected non-locally by the left pair of electrodes, divided by current between the right hand pair of electrodes. The measured quantity has units of resistance but it is an inherently signed quantity. Current is limited to the region between electrodes 3 and 4. The indicated decay of graphene magnetization in region between electrodes 1 and 3 is due only to electron spin diffusion in a region free of electric field. The spin diffusion length is suggested by the sketched decay to be on the order of 2 μm. In the middle panel, the falling spin-up density, going from electrode 3 to electrode 4, is the effect of the imposed current, extracting spin-up electrons (injected at electrode 3) at electrode 4. The large increase (bottom panel) in the measured voltage when the injection electrodes have opposite magnetization, is due to the fact that electrode 3, in that case injects spin-up electrons while electrode 4 extracts spin-down electrons, to enhance the spin imbalance between electrodes 3 and 4. Some of this imbalance then diffuses to the measuring electrodes. (b) Graphene Spin-valve device: Measurements at 4.2K of the voltage V_{21} divided by drive current I_{43} as polarizing magnetic field B_y [into the page, as in (a) or upward, in inset] is changed from −300 mT (all Co electrodes magnetized ↓↓↓↓) to + 300 mT (electrodes magnetized ↑↑↑↑). Hysteresis is indicated by horizontal arrows. Spacing between electrodes 2 and 3 is 0.33 μm. (From Tombros *et al.*, 2007, by permission from Macmillan Publishers Ltd., © 2007).

---

[3]We use quotes, because resistance is not normally a signed quantity; these are measurements of the Johnson–Silsbee voltage.

the non-equilibrium in the region between electrodes 3, 4. This situation is suggested in the lowest panel of Fig. 9.16(a).

The Hanle effect, relevant to the device in question, is based on the fact that the magnetization, $M = N\mu_B$, the result of individual magnetic spins, in the presence of an applied magnetic field B, precesses around a static field at the Larmor frequency,

$$\omega_L = g\mu_B B/\hbar. \tag{9.16}$$

This variation, the Hanle effect, is observed as a reversal of the Johnson–Silsbee voltage.

Tombros *et al.* (2007) also confirmed a Hanle precession effect in their graphene devices that agreed with the analysis earlier provided by Jedema *et al.* (2002). This precession effect occurs when a magnetic field perpendicular to the plane of the graphene device is applied [vertical in the representation of Fig. 9.16(a) top panel] such that the electron spins precess at the Larmor frequency, eqn (9.16), around the field direction. Here $g = 2$ for the electron and μ_B is the Bohr magneton. They found that $B = 100$ mT reversed the sign of the Johnson–Silsbee voltage. The interpretation is that when spin-up electrons are injected at contact 3 [see Figs. 9.16(a), (b)], as they diffuse toward contact 2 they precess around the vertical magnetic field and have accumulated an angle 180°, a spin reversal (at 100 mT) as they reach contact 2. The authors used this information to deduce the spin diffusion length as between 1.5 and 2.0 μm.

The magnitude of the achieved spin non-equilibrium is predicted to be enhanced when the contacts are not too transparent and it has been suggested by Rashba (2000) (see also Fert and Jaffres, 2001) that a tunnel barrier is optimum for contact between the paramagnet (here graphene) and a ferromagnet. Han *et al.* (2010) have improved on the magnitude of the achieved polarization, by fabricating reliable electron tunneling barriers and observing tunneling spin injection into single layer graphene. Their method involves a deposited MgO barrier, as was analyzed by Jiang *et al.* (2005) but with a prior deposition onto the graphene of a monolayer of TiO_2. Han *et al.* (2010) have observed a non-local magnetoresistance $\Delta R_{NL} = 130\,\Omega$, by tunneling spin injection from Co through the MgO/TiO_2 barrier. This is the largest non-local resistance value ever obtained. These authors provide an analysis of the effects of changing the tunnel barrier contact to a transparent contact. Recent work on spin injection has been reported by Han *et al.* (2012) and by McCreary *et al.* (2012). The effect of localized states on spin transport in epitaxial graphene on SiC has been studied experimentally by Maassen *et al.* (2013).

9.5   Sensors of single molecules, "electronic nose"

Sensing of single molecules on graphene was reported by Schedin *et al.* (2007) and by Zhou *et al.* (2008), as mentioned in the text following eqn (7.32). The electric field effect is the basis for exceptional sensitivity of detection, as the local Fermi energy is moved by an adsorbed molecule, modulating the carrier density and resistance. This is a corollary of the observation that cleaning of graphene by heating to remove adsorbed species strongly changes the electrical properties, as pointed out by

Bolotin *et al.* (2008). Such detectors can be configured similarly as FET devices, as demonstrated by Ang *et al.* (2008). Their device is described as a SGFET (solution gated field-effect transistor) and is demonstrated to measure the solution quantity pH with a sensitivity of 99 meV/pH. The dependence of the performance of such devices on the type of electrode cleaning used was discussed by Dan *et al.* (2009). This is an active area of interest to chemists and has been reviewed specifically by Shao *et al.* (2010) and, as part of a large and useful review of graphene from a chemical point of view, by Allen *et al.* (2010). Use of graphene in a tunable single electron transistor was described by Stampler *et al.* (2008).

Gas-phase sensors based on graphene, termed "electronic noses," have been described by Lu *et al.* (2010) and by Park *et al.* (2012).

9.6 Metrology, resistance standard

The possibility of a resistance standard based on graphene was noted early by Novoselov *et al.* (2007), in connection with their observation of the quantum Hall effect at room temperature. This topic has been treated more recently by Giesbers *et al.* (2008) and by Tzalenchuk *et al.* (2010). The latter authors demonstrate measurement of the von Klitzing constant $R_{\mathrm{K}} = h/e^2 \approx 25.813$ kΩ to an accuracy of three parts per billion at 300 mK using a graphene sample. Nanometrology based on Raman effect measurements of graphene has been reported by Calizo *et al.* (2007) and Ferralis (2010). These topics are also covered in the excellent edited volume of Raza (2012).

9.7 Memory elements

A flash memory cell is essentially a storage capacitor of minimum lateral dimension, currently in the vicinity of 45 nm, with read and write capabilities and storage time for charge measured in years. High performance laptop computers use flash memory in place of magnetic disk memory and USB flash-memory drives are ubiquitous. The current state-of-the-art in flash memory is the polysilicon floating-gate device on a p-type Si wafer. (A review is offered by Lu *et al.*, 2009).

A promising graphene flash memory device has been described by a consortium of workers at IBM, UCLA, Samsung and Aerospace Corp. (Hong *et al.*, 2011). (This paper also gives reference to earlier approaches to non-volatile memory based on graphene.) These authors have patterned capacitor-like storage devices on large-area CVD-grown graphene transferred onto an oxidized p-type Si substrate.

Individual capacitor devices of form: graphene/(5nm SiO_2 tunnel oxide)/p-type Si, have been shown by the authors to retain charge for ten years at eight percent loss of charge [see Fig. 9.17(a)]. The height of the graphene/quartz barrier is about 3 eV. The storage capacitor is charged from an upper gate electrode, separated by a gate insulator. The upper electrode (graphene) of the storage capacitor is covered by 35 nm of sapphire (Al_2O_3), with a write/read gate-electrode on the top. The devices are reported to have a window of voltage, on the read/write electrode, of ± 6 V for secure charge storage, with program/erase voltages at ± 7 V. The authors have determined that the stored charge resides on the graphene layer. The fabrication of the devices requires extreme care with respect to the properties of the tunnel barrier of SiO_2

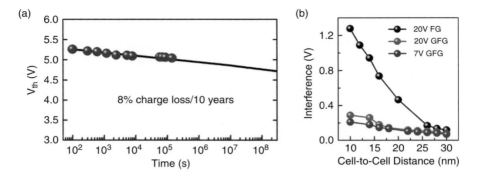

Fig. 9.17 (a) Demonstration of storage time of graphene flash memory device. (b) Superiority of graphene flash memory: Limits of flash memory (FM) cell packing from cross-talk, comparing simulations of traditional floating gate (FG) polysilicon flash memory (upper curve) with graphene flash memory (GFM) of tri-layer graphene in lower two curves. Interference voltage is defined as shift in threshold voltage of unprogrammed cell by its two nearest-neighbor cells. The superiority of GFM in this regard, below cell spacing of about 25 nm, comes from smaller inter-cell capacitance consequent on the lower cell height. This is an advantage in miniaturization of graphene flash memory vs. traditional polysilicon flash memory. (Reprinted with permission from Hong *et al.*, 2011. © 2011 American Chemical Society).

and the gate electrode barrier of Al_2O_3. The care is needed to preserve the desired retention time of charge and to achieve reproducibility in the writing and erasing voltages, applied respectively, across the tunneling- and gate-barriers.

In more detail, the p-type silicon wafer was completely cleaned of native oxide and then exposed to flowing oxygen for seven seconds at 1000°C. The graphene layer (or multilayer), grown on Cu foil or deposited Ni, was then transferred onto the carefully oxidized surface. The gate oxide on the graphene was prepared by initial deposit of 1.1 nm of Al, oxidized in air for two days, followed by 300 cycles (monolayers) of Al_2O_3 by atomic layer deposition (ALD). The gate-electrode on this oxide barrier is Ti/Al/Au (10 nm/500 nm/10 nm) using photolithography and electron-beam evaporation. The gate-electrodes were varied in area (2.5×10^{-5} cm^2 to 7.4×10^{-4}cm^2). These electrodes were used as a mask to etch the individual gate stacks. The Al_2O_3 and graphene, outside the device area, are removed with 30 s reactive-ion-etching using Cl_2, followed by 3 minutes of O_2 plasma.

The packing density of flash memory cells is limited by interference voltage between neighboring cells, termed crosstalk. The crosstalk interference voltage is defined as shift in threshold voltage, of an unprogrammed cell, by its two nearest neighbor cells. The essential feature leading to smaller crosstalk is the vertical height of the cell, that controls the inter-cell capacitance. The covalent bonding of the graphene in these devices allows a stable, diffusion-free and electrically continuous conductive thin layer, as small as 0.34 nm in thickness, but here assumed to be 1 nm (graphene trilayer) in Fig. 9.17(b). This is an important reduction in height, reducing crosstalk, compared to an evaporated metal that must be much thicker, lest it break into islands and not conduct well.

The authors conclude that the graphene flash memory (GFM) has significant advantages in miniaturization below the 25 nm "node" in semiconductor scaling, based on the lower interference and also in lower operating voltage to achieve the minimum window of stable storage voltage, 1.5 V accepted in current practice. The energy per bit is predicted to be reduced by 75% in the GFM device. The advantages of graphene in this application are the high work function, the high density of states (relative to polysilicon) and the low dimensionality.

An all-graphene electromechanical switch has been fabricated, using chemical vapor deposition, by Milaninia *et al.* (2009).

9.8 Prospects for graphene in new digital electronics beyond CMOS

It is clear from the examples above that graphene devices will play an increasing role in electronics. Several examples have been given of graphene elements that can be integrated into silicon chips e.g., as interconnects and flash memory elements. A review of graphene's supplemental role in silicon electronics is given by Kim *et al.* (2011).

A larger question is whether graphene will play any role in the Moore's Law scaling of semiconductor logic. The scaling progression is near to its fundamental limits, as reduced size changes the functionality of conventional silicon devices. The logic device family at the heart of silicon electronics is CMOS: complementary metal oxide semiconductor field-effect transistors (FETs). At present (2013) a high level chip contains up to 2 billion FETs, with annual FET production reported as 10^{18} (Pinto, 2007) much of it carried out in facilities that cost more than one billion U.S. dollars. It seems inevitable that any successful innovation, to be widely adopted, would have to be largely compatible with existing production methods and plants. The silicon FET devices are presently fabricated with basic line widths at or below 45 nm with plans to go to 22 nm.

9.8.1 Optimizing silicon FET switches

The band diagram of a field-effect transistor device is shown in Fig. 2.3, in the N-FET form. The basic form of the FET and aspects of its essential switching performance, are shown in Fig. 9.18.

The FET device operation depends on control of the channel conductivity and thus of the drain current, by the gate voltage, applied between gate and source. The steepness of the $(\log I)/V_{\mathrm{g}}$ plot is related to the switching (digital logic) capability of the device. In the nomenclature of logic devices, a small value of the "inverse sub-threshold slope $S = (d \log I_{\mathrm{D}}/dV_{\mathrm{gs}})^{-1}$" (also referred to as SS), is a figure of merit [see Fig. 9.18(b)]. The ideal value for S for a conventional MOSFET is $k_{\mathrm{B}}T \ln(10) = 59.6$ mV/decade at room temperature. As we will see, smaller values can be obtained,

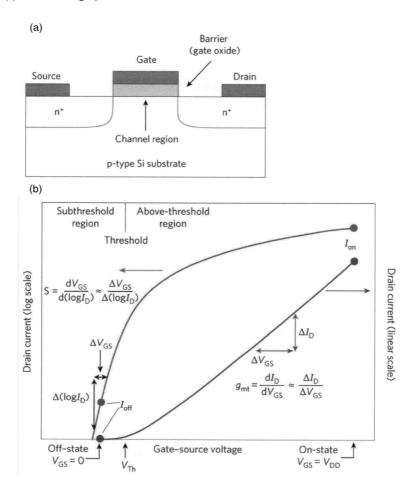

Fig. 9.18 (a) Sketch of N-FET device, based on p-type Si crystal. Positive gate voltage draws electrons to the upper surface, thus inverted to n-type, forming a conductive channel connecting source and drain. The vertical dimension is influenced by the diffused n^+ contacts, much thicker than graphene, at least 10–15 nm, constituting one of the limits to miniaturization of the conventional FET device. (b) Electrical switching of the field-effect transistor FET. Schematic of drain current (log scale on left, linear scale on right) vs. gate voltage. In linear scale, drain current appears at threshold gate voltage, V_{th} and rises to full On-state value at maximum value of gate voltage, V_{DD}. A figure of merit is the linear slope above threshold, $g_t = dI_d/dV_g$. A second figure of merit, defined on the logarithmic scale, is the "inverse sub-threshold slope" $S = (d \log I_D/dV_{gs})^{-1}$. A small value of S, ideally 60 mV/decade at room temperature, leading to a steep curve on the logarithmic scale, is considered important. Smaller values are available with tunneling devices. (From Schwierz, 2010, by permission from Macmillan Publishers Ltd., © 2010).

if the transport mechanism is changed, from thermal injection, to tunneling. For high-speed applications, the device should respond quickly to change in V_{gs}, requiring short gates and fast carriers in the channel. Devices with short channels (the working length will soon be 22 nm) suffer from degraded electrostatics and other problems, collectively referred to as "short–channel-effects." The scaling theory (Frank *et al.*, 1998; Ferain *et al.*, 2011) predicts that an FET with a thin gate oxide barrier and a thin gate-controlled region (measured in the vertical direction in Fig. 9.18a) will avoid the short-channel-effects down to very short gate lengths (measured horizontally in Fig. 9.18a), perhaps as short as 12 nm. Therefore, according to Schwierz (2010), the possibility of having a channel just one atom thick is one of the most attractive features of graphene, in possible application to FET transistors. If channels are made thin in conventional silicon technology, surface roughness is usually introduced, to reduce mobility, but such an effect is absent using graphene that is both thin and smooth. In CMOS logic, a large number of devices are always in the "off" state, yet drawing some power. Low current in this state, corresponding to a large on/off current ratio, is desired and the expectation is that a successful device family will have a ratio at least 10^4. This leads to the estimate, for a conventional semiconductor FET that the bandgap should be at least 0.4 eV. Further, since the devices are used in complementary P-FET/ N-FET pairs, it is desired that the P and N devices shall have symmetrical properties. This condition is well met by graphene, inherently symmetric above and below the Dirac point, while the substantial difference in electron and hole effective-masses in conventional semiconductors has required adjustments in design to restore operational symmetry.

Silicon research FET devices have been demonstrated at gate lengths as small as 5 nm (Wakabayashi *et al.*, 2006). But clearly there is a limit and soon the scaling progression, at least in Si logic devices, will stop. A set of likely final steps in the Moore's Law scaling, based on multi-gated Si-on-insulator (SOI) devices, is given by Balestra (2010) in an article entitled "Silicon-Based Devices and Materials for Nanoscale CMOS *and Beyond-CMOS*" (italics added). A similar position is arrived at by Ferain *et al.* (2011), who predict that multi-gated silicon devices will reach the 3 nm node and that this will take about 20 years. (These authors are certain that devices in 20 years' time will in fact be silicon multi-gated FETs, sometimes called FinFETs, since the Si is photolithographically patterned and etched into fins to allow more electrode area.)

A second aspect, however, is the power dissipation that continues to rise in the projected remaining steps of the Moore's Law progression. It has been established (Ionescu and Riel, 2011 and work cited therein) that *tunneling* FET devices (T-FET) can lower the power dissipation by a factor of 100 from the Si CMOS family of devices.

For suitably small off-state current I_{off}, the energy usage is largely comprised of energy per switching-operation. Switching-energy scales with the supply voltage, V_{dd} as

$$p \propto I_{off} V_{dd}^3. \qquad (9.17)$$

According to Ionescu and Riel (2011), the tunnel FET could allow the supply voltage V_{dd} to fall from 1 V to 0.2 V, giving a reduction in power of 125. It appears that T-FET devices can be made in several different materials, including Si, III-V compounds and carbon (nanotube or graphene) and it may well be that the conventional materials would be easier to incorporate into the silicon technology. We will return in Section 9.8.3 to more detail about T-FET operation, as a preface to describing the proposed graphene T-FET devices.

A rough assessment might be that the silicon technology is successful beyond any expectation; computing power is a commodity and indeed the final state of the art has probably been achieved, or nearly so. This evolution may resemble those of automobiles and jet aircraft, having reached nearly static but very favorable levels.[4] In cars and airplanes, apart from incremental improvements, as in reducing pollution and improving efficiency, there is little suggestion of a disruptive change, as long as oil is available at low cost.

Cloud computing, in very large installations, may still change toward superconducting logic to save energy (Bunyk *et al.*, 2001), but the mass chip market is likely to reach a static state, nearly as it is presently. Even so, the CMOS logic market is vast and there remains a chance that a portion of it may be encroached upon by a logic family of lower energy cost, such as the tunnel FET that might be based on graphene. Before considering possible graphene tunneling logic, we give an indication of where the Si CMOS devices stand, as suggested by the work of Wakabayashi *et al.* (2006). (A separate overview of the present status of CMOS is given by Khakifirooz and Antoniadis, 2008.)

Using several fabrication innovations within the silicon technology, Wakabayashi *et al.* (2006) produced a family of MOSFET research devices with channel lengths no more than 5 nm and showed that they operate in a useful fashion. Drain current vs. gate voltage curves for N- and P- type devices, with drain voltage as parameter, are shown in Fig. 9.19(a). The drive currents reached are on the order of 0.1 mA/μm. On the other hand, the drain currents do not saturate at high drain voltage, but are linear to 0.4 V (data not shown). These are non-ideal effects, but may not remove the utility of the devices.

Drain voltage is limited to 0.4 V and the authors state "the saturation characteristics are poor; these might be due to the parasitic resistance and poor electrostatic." Good cutoff characteristics were observed, however, in both N- and P- devices at 0.4 V. These look like useful devices. The gate electrodes in these devices are polysilicon, reduced in thickness by etching steps. Proof of the structure is shown in the TEM image in Fig. 9.19(b).

In this figure source and drain contacts are $CoSi_2$ that is metallic in its electrical conduction. Different silicide stoichiometries $CoSi$ and Co_2Si form nanowires and the latter, a Co_2Si nanowire, was shown, in Fig. 9.3, as the self-aligning gate of a graphene FET device.

The scaling rules of Frank *et al.* (1998) and Ferain *et al.* (2011) favor thin structures of minimum vertical extent (perpendicular to the channel). The image in Fig. 9.19(b)

---

[4]A discussion of the development of such technologies that bears on the future of growth in electronics, is given in *The Economist*, January 12, 2013, pp. 21–24, in an article entitled "Has the ideas machine broken down?"

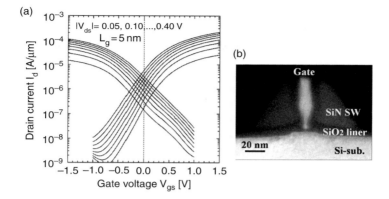

Fig. 9.19 (a) I_{DS}–V_G curves of N-FET and P-FET devices with gate length of 5 nm. Note logarithmic current scale. (b) Transmission electron microscope image of gate electrodes with 5 nm gate length on silicon substrate. The gate electrode is polysilicon, reduced in width at its lower end and enclosed by a sidewall layer of silicon nitride and by a silicon dioxide liner. This device also involves CoSi$_2$ silicide source and drain contacts, augmented with "shallow source and drain extensions" to contact the extremely short channel. (Reproduced with permission from Wakabayashi *et al.*, 2006. © 2006 IEEE).

does not appear to be thin; in fact the gate assembly extends at least 50 nm vertically and the depths of the conductive regions under the source and drain, although not visible, are also large compared to the gate length, 5 nm, suggesting that short channel effects may be important. These devices show promise, but deviate substantially from expectation for an ideal MOSFET, primarily in the failure to reach saturation at elevated drain voltage and in the lower value of saturation current. The authors suggest a parasitic resistance in the drain system. It is not clear that these shortcomings are fundamental or the result of an imperfect step in fabrication. In any event, the device appears far from an ideal of a narrow horizontal channel and offers no chance of further miniaturization.

A review of many possible approaches to replace CMOS has been given by Bernstein *et al.* (2010), in addition to the reviews of Schwierz (2010) and Ionescu and Riel (2011).

To return to graphene FET devices, we have seen earlier (Section 9.1) that the most actively pursued have been micrometer-scale devices optimized for high-frequency performance. Because graphene is metallic, these devices do not turn off, but this may be tolerable in the rf amplifier capacity. For logic applications, as we have seen, it is essential that an FET device have a large ratio of resistance between the off and on state. In the context of the traditional FET device, this requires a large bandgap. While graphene inherently has no gap, a bandgap does appear, from size quantization, in a narrow ribbon. Thus, the main approach discussed to graphene-based logic FET has been via nanoribbon devices. The difficulty is that to achieve a suitably large bandgap, at least 0.4 eV, the ribbon has to be extremely narrow, less than 10 nm wide. Such ribbons have been selected from suspensions resulting from sonication of graphite flakes, in the work of Hongjie Dai and collaborators (Li *et al.*, 2008) yielding FET prototype devices of with on/off ratios on the order of 10^6 (see Fig. 5.3). From the point of view of logic FET devices, these ribbons are better than carbon nanotubes, in that they do reliably have energy gaps, but they suffer from the same problem as do carbon nanotubes. Namely, there is no production method of getting the right size GNR or nanotube to the right location on the chip that in current practice may contain a billion FET transistors in a centimeter square. This forces a conclusion that a realistic device will have to be deposited and patterned on the planar chip surface, there being no prospect of importing and locating accurately vast numbers of selected nanoribbons. On the other hand, present and future lithography (with a caveat to be mentioned) does not permit patterning 10 nm wide nanoribbons that, further would need atomically accurate edges. A possible exception to this negative assessment may be offered by the chemical route to nanoribbons described above in the work of Cai *et al.* (2010) and Koch *et al.* (2012). These methods do clearly offer atomically precise edges for narrow nanoribbons.

These points, with a second possible exception that we will shortly mention, *exclude* GNR devices, within a logical point of view, based on mass manufacturability needed in the context of extending Moore's Law.

The second possible exception, to the non-manufacturable assessment of graphene nanoribbon devices, is recently suggested in the works of Sprinkle *et al.* (2010) and Hicks *et al.* (2012). Their innovation is to specify, on an atomic size scale, the widths of nanoribbons (and also to specify the edge configuration as armchair or zigzag) by growing the graphene on etched groove geometries pre-patterned on SiC. The latter paper is entitled "A wide-bandgap metal-semiconductor-metal nanostructure made entirely from graphene". The Hicks *et al.* (2012) nanostructure takes a form similar to Fig. 5.7, where arrays of 12 nm or 18 nm deep trenches are patterned into SiC, onto which about 400 parallel monolayer graphene ribbons of length 50 μm and widths 15 nm and 36 nm are then grown. The ribbons are centered on the trench sidewalls, using the advanced sublimation methods suggested in Section 5.2.1 and described by Seyller (2012) and in the cited works of Sprinkle and Hicks. The sidewalls of the Hicks *et al.* (2012) trenches are (2207) and (1103) facets of SiC with angles from the vertical [normal to the (0001) surface] about 30°. As shown by Sprinkle *et al.* (2010), an initially vertical sidewall will reconstruct into a sloping (110n) facet if the SiC is heated to 1250°C and, further, a monolayer of graphene will grow on that sidewall if the crystal is further briefly heated to 1450°C. Sprinkle *et al.* 2010 made an array of 40 000 transistor devices per cm² on SiC (0001) using this process. Hicks *et al.* give evidence that the sharp curvature at the join of the top (0001) plane and the sloping trench wall makes a 1.4 nm-wide semiconducting graphene nanoribbon. This ribbon is epitaxially connected to an n-type metallic nanoribbon on the (0001) top surface and a p-type conducting nanoribbon on the sloping facet wall. The result is a narrow nanoribbon whose armchair edges are aligned along the corner between the (0001) top and sidewall surfaces. Strong experimental evidence of the bandgap, restricted to the 1.4 nm-wide strip at the sidewall intersection, was provided by ARPES measurements (see Chapter 6). These epitaxial structures are of atomically controlled dimensions and atomically specified edge geometry and are single-crystalline (via epitaxy) over lengths of at least 50 μm. This class of nanoribbons, extending to nanostructures with semiconductor–metal junctions, seem candidates for a dense FET device technology based on graphene nanoribbons (Berger *et al.*, 2004). It is very early in such a possible development, but de Heer *et al.* (2006) obtained a U.S. Patent entitled "Patterned thin film graphitic devices and method for making same."

Beyond the SiC option just mentioned, the further remaining options for graphene dense logic devices therefore must lie with wider-area graphene devices that do not require a narrow width to obtain a gap. Three different types of such wide-area graphene devices have been proposed. These are: (1) based on an electrically-induced gap in bilayer graphene, in a traditional FET configuration; (2) a bilayer graphene tunnel FET device; and (3) a purely two-dimensional graphene-based tunnel transistor. The second two devices are based on interband tunneling and we provide some review of that topic after mentioning the first graphene device, of the conventional FET form.

9.8.2   Potentially manufacturable graphene FET devices

We are here concerned with FET devices that will work as logic switches, requiring a large ratio of current in the 'on' condition compared to that in the 'off' condition. A wide-area conventional graphene FET device, see for example Lemme *et al.* (2007) and Ouyang *et al.* (2008), does not qualify for switch applications, because no high resistance, zero current, state is available.

Nanoribbon devices, achieving a bandgap by size-quantization, have been reported (Li *et al.*, 2008), but we now reject these devices as non-manufacturable in large scale [with a possible caveat regarding the process of Koch *et al.* (2012)]. If one rejects graphene FET devices that depend on unattainable patterning and limit the devices to those that are potentially available with present and future photolithography, the available literature is quite limited.

The first graphene FET design that we describe is modeled by Fiori and Iannaccone (2009a). This device uses bilayer graphene, with a vertical electric field applied to induce a gap (McCann, 2006; Castro *et al.*, 2007, see Fig. 4.5; Zhang *et al.*, 2009a). The device, called the Tunable-gap Bilayer Graphene FET, is modeled numerically. As can be seen in Fig. 9.20, this proposed device is thin, about 5 nm in vertical extent.

The overall height of the proposed structure in Fig. 9.20 is much less than the channel length, quite unlike the silicon device shown in Fig. 9.19(b). This is an advantage, in general terms, to avoid undesirable short-channel effects. The assumption is that such a structure could be achieved by transferring bilayer graphene, grown on copper foil, for example, onto an insulating substrate followed by lithographic patterning. The width of the device (out of the page) is not a crucial parameter in this and following designs and should not conflict with present capabilities in lithographic patterning. It is predicted that this device operates but, with the bandgap available by electric field in bilayer graphene, does not have a large enough on/off ratio to be suitable for logic applications. This device is not promising for fabrication.

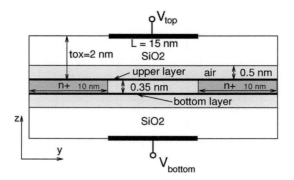

Fig. 9.20 Tunable-gap Bilayer Graphene FET as modeled by Fiori and Iannaccone (2009a). The vertical height of the device as shown is 4.35 nm, excluding the width of top and bottom metal (or graphene) gate-electrodes. The overall height is small compared to the channel length, shown as 15 nm. In this figure, bilayer graphene is represented as two single layers of graphene separated by 0.35 nm. It is assumed that source and drain are bilayer graphene, heavily doped to achieve strong n-type conductivity. (Reproduced with permission from Fiori and Iannaccone, 2009a. © 2009 IEEE).

9.8.3 Tunneling FET devices

We now turn to tunneling field-effect transistor devices that operate by interband tunneling. The acronym for such devices is T-FET. For background on interband tunneling, the Zener diode is a commercial silicon device, a PN junction that, when put into strong reverse bias (increasing the built-in shift of the bands on the opposite sides of the junction) abruptly becomes highly conductive as carriers tunnel from the valence band on one side across the depletion region to empty states in the conduction band on the opposite side. The physics of this working device was initially explained by Zener (1934). A more recent analysis of the interband Zener tunneling current is given by Kane (1959). This current is independent of temperature and given by Kane's formula for n, the rate of electron transfer at electric field E, with $F = eE$, for a semiconductor of bandgap E_G, as

$$n = [F^2 \, m_r^{1/2}/(18\pi\hbar^2 \, E_G^{1/2})] \exp[-\pi m_r E_G^{3/2}/2\hbar F]. \qquad (9.18)$$

Here the reduced effective mass m_r is derived from the electron- and hole-mass values, $m_r = m_e m_h/(m_e + m_h)$. The carrier mass changes in character from electron to hole as it crosses the depletion region. This eqn (9.18) strongly favors semiconductors of small bandgap and requires a high electric field E.

A related device is the Esaki (1958) diode, where highly degenerate n- and p-layers, with small depletion widths, are needed to make the tunneling probability usefully large at zero bias. This device was associated with the Nobel Prize in Physics in 1973.

The initial suggestion and analysis of T-FET devices based on interband tunneling was given by Quinn *et al.* (1978) who considered a p-i-n structure. In the Quinn structure, p and n are heavily doped regions on a weakly p-type silicon surface, with an upper field electrode shifting the bands in the connecting i region (weakly p-type Si) sufficient to allow Zener tunneling into the empty conduction band. The general shape of the energy bands in such a device is shown in Fig. 9.21.

Early reports of operational tunneling T-FET devices on silicon were given by Reddick and Amaratunga in (1995) and by Koga and Toriumi (1997).

Wang *et al.* (2004) first discussed the low power dissipation possibility of the T-FET and also the faster switching, with a smaller S factor, not limited by $k_B T/e$. Their p-i-n devices are built on a Si surface by conventional diffusion of dopants and a gate covers the intrinsic i region. The devices are operated in reverse bias of p to n and are described as a "MOS-gated reverse biased p-i-n diode. The band-to-band tunneling can be controlled by the gate voltage and the leakage current is minimized due to the reverse biased p-i-n structure." The authors find that complementary logic is possible using these devices: "complementary NTFET and PTFET similar to NMOS and PMOS are fabricated on the same silicon substrate." These aspects were further discussed and demonstrated in carbon nanotube T-FET devices, by Appenzeller *et al.* (2004).

Knoch *et al.* (2007) provided more analysis toward practical T-FET devices and found values of the parameter S smaller than 60 mV/decade, with a tradeoff between

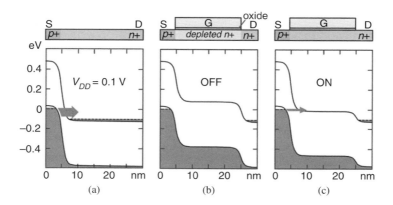

Fig. 9.21 Energy band diagram and layer structure for a Tunneling Field-effect Transistor T-FET consisting of a p$^+$ source, an n$^+$ drain and a gate (G). (a) The channel is shown without a gate and with sufficient source-drain bias $V_{DD} = 0.1$ V to drive Zener tunneling into the conduction band of the drain from the valence band of the source. (b) The gate fully depletes the channel creating the OFF condition of the gated T-FET. (c) A positive gate potential turns the channel ON, with current set by overlap of valence band electrons with unfilled conduction band states. (Reproduced with permission from Seabaugh and Zhang, 2010. © 2010 IEEE).

small S and high on-current for the device. An extensive review of the subject was given by Seabaugh and Zhang (2010).

In these structures (see Fig. 9.21) the n and p regions are designated n^+ and p^+, meaning that they are so heavily doped with donor and acceptor impurities, respectively that they are metallic, in the sense that the dopant atoms are fully ionized, freeing all of the carriers, even at $T = 0$. The doping level exceeds that of the Mott metal-insulator transition, mentioned in Section 5.5. The high doping means that the depletion region widths, strongly influencing the interband tunneling, are very small, as is also the case with the Esaki (1958) diode.

9.8.4 Manufacturable graphene tunneling FET devices

A tunneling field-effect transistor T-FET in graphene is described by Fiori and Iannaccone (2009b). This device [Fig. 9.22(a)] is geometrically similar to the graphene device in Fig. 9.20, but the mode of operation is changed. Now, forward current is by interband tunneling, between n^+ source and p^+ drain in bilayer graphene. Forward current is enabled at voltages, on the two gates that line up the valence band and conduction band energies in source and drain electrodes, as well as influencing a band gap in the bilayer graphene.

The principle illustrated in Fig. 9.21(a) and Fig. 9.22(a) was demonstrated by Appenzeller *et al.* (2004), in a device based on a carbon nanotube, with an aluminum top gate electrode. The conditions in the carbon nanotube are similar to those in the graphene device described by Fiori and Iannaccone (2009b).

Fiori and Iannaccone (2009b) find that, with a drain to source voltage of 0.1 V, an on/off ratio larger than 1000 can be reached, even within the limitations on achievable bandgap in the bilayer graphene (in the range of a few hundred millielectron Volts). The switching in this tunneling device is superior, with an S factor [see Fig. 9.18(b)] reported as 20 meV/decade, vs. 60 meV/decade for the best thermionic device. In Fig. 9.22(a), the large vertical field, established by the top- and bottom-gate electrodes, maintains the bilayer band gap in the length between those gates, but the bandgap of the bilayer graphene goes to zero in heavily doped regions (beyond the extent of the gates). The use of SiO_2 dielectric, as shown in total height of 6 nm, permits an electric field up to 14 MV/cm before breakdown. Because the device is based on bilayer graphene, with massive electrons and holes in parabolic bands, the tunneling probability is not impacted by the Klein tunneling phenomena mentioned in Section 8.7. Manufacturing such a device could be done, in principle, by transferring bilayer graphene, grown on Cu foil, onto the device substrate and then using masking to deposit n- and p-type dopants specifically onto the source and drain sections of the device, under typical CVD conditions. The width of the device (out of the page) is not a crucial parameter, so the suggested device is compatible with present silicon technology. The authors assess this device as promising for fabrication and circuit integration, based on their simulation results.

Finally, a monolayer-graphene double-gated tunneling transistor is described. This second conceptual approach (modeled but not experimentally demonstrated) to graphene tunneling T-FET transistors with large on/off ratio [Fig. 9.22(b)] is a single

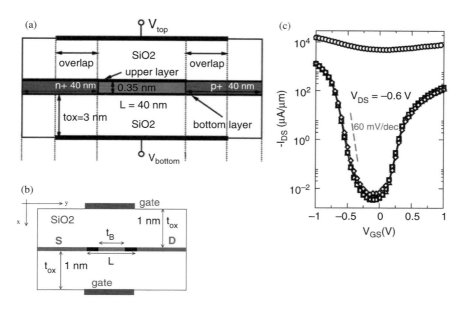

Fig. 9.22 (a) Bilayer graphene tunnel T- FET device. Bilayer graphene is shown as two graphene layers separated by 0.35 nm. In the central region, of width $L = 40$ nm, the bilayer is undoped and its bandgap is enhanced by electric field, established by the two gates. The main contribution to the shift of band energies between p- and n- electrodes results from the large fractional chemical dopings, on the order of 2.5×10^{-3} that are assumed in the bilayer graphene. (Reproduced with permission from Fiori and Iannaccone, 2009a. © 2009 IEEE). (b) Monolayer-graphene double-gated tunneling transistor with high I_{ON}/I_{OFF} ratio. This proposed device (here P-insulator-P or PFET), is monolithic single layer graphene, modified in the center by chemical change from pure C to BCN (in the same hexagonal lattice), introducing a conventional tunnel barrier. (From Fiori *et al.*, 2011). (c) Simulated drain current I_D vs. gate voltage, for graphene-h-BC$_2$N tunnel FET device (lower data points), compared with simple graphene FET transistor (upper points). The tunnel device exhibits I_{ON}/I_{OFF} ratio exceeding 10^4, with source-drain voltage 0.6 V at room temperature. The different symbols in the lower curve represent slightly different assumptions on the exact location of the middle of the barrier. (Insensitivity to exact location may be an indication that the design is manufacturable.) The length of the barrier t_B [see (b)] is 5 nm and the doping fraction in the p+ leads is 10^{-2}. (From Fiori *et al.*, 2011).

graphene sheet (wide ribbon, including source and drain) with an inserted barrier (Fiori *et al.*, 2011; Fiori *et al.*, 2012). The ends of the wide ribbon are heavily doped P-type and the narrow barrier is hexagonal BCN. This barrier is envisioned as an inserted BCN film (epitaxed to the graphene channel) with the same hexagonal structure and lattice constant as graphene, see Fig. 5.10(a), (b). (The modeled device is PFET but there is no basic problem to make the complementary NFET in this form.)

Methods that may be adaptable to inserting the tunnel barrier are given by Levendorf *et al.* (2012). It is pointed out that insertion of the BCN barrier adds flexibility to design of graphene transistors. The BCN barrier region can be tailored, by choice of composition, to have a barrier ranging from 1 eV to 5 eV (with some caveat to control large domain formation, vs. homogeneous films) and, further can be designed to pass one kind of carrier while blocking the other. The structure suggested in Fig. 9.22(b) is a hybrid h-BCN-graphene device, with a chemically-formed and gate-controlled barrier inserted between source and drain. The barrier comprises an epitaxial length tB of h-BCN. Hexagonal BCN, discussed in Section 5.2.5, can potentially have variable fractional C content from 0.1 approaching to 1.0, with energy gap adjustable in the range 1 eV to 5 eV. The h-BCN can be grown in two to three atomic layer films on copper foil by CVD, using methane and ammonia-borane (NH_3–BH_3) as source gases. It may be possible to introduce the h-BCN barrier region starting with pure graphene and masking, so that the desired barrier region is exposed briefly to NH_3-BH_3 under typical CVD conditions. In any case, the graphene in this device is of large area and fabrication, in principle, does not suffer from difficulty in reducing its width that would require (in a GNR device) atomically defined edges, unattainable in present or future lithography. The modeled device was of the PFET form, with source and drain both of heavily doped p-type graphene. The graphene–graphene interband tunneling probability involved here has presumably been estimated within the rules established in Section 8.7 where the barrier region is likely to have parabolic, rather than conical, bands, separated by the mentioned energy gap. The simulated performance of the device is shown in Fig. 9.22(c), with actual modeled barrier composition BC_2N.

The performance of the simulated device is acceptable from the point of view of CMOS logic. The dashed line marked "60 mV/dec" in Fig. 9.22(c) is a measure of the steepness of the $(\log I)/V_g$ plot, related to the switching (digital logic) capability of the device (see Fig. 9.18b). The S value, from the points in Fig. 9.22(c), is about 80 mV/decade, acceptable and clearly superior to that of the simple graphene FET [upper points in Fig. 9.22(c)].

Beyond the design for the logic T-FET device, Fiori *et al.* (2011) investigated resonant tunneling devices, capable of negative resistance behavior. The simulated resonant tunneling devices were fully planar and are similar to the device shown in Fig. 9.22(b) except that the barrier region is now split into two BN barriers (width t_B) with an intervening narrow graphene region (width w). Localized states are now possible on the interior graphene region of width w: when such states align in energy with the Fermi level of the source, a peak appears in the I_D–V_{GS} characteristic. The authors argue that this proposed family of resonant tunneling transistors achievable in graphene may be superior to those obtained in III-V materials, because the graphene device offers better electrostatic control of operation.

The design of the TFET graphene logic device in Fig. 9.21 and Fig. 9.22 is superior to any possible silicon device, from the point of view of having a conducting channel of thickness much smaller than its length, because the depletion depth in any silicon device is large on the scale of the graphene thickness, 0.34 nm. The height of the modeled device in Fig. 9.22(b) is about 2.5 nm, excluding the top and bottom gates. If the top and bottom gates are graphene (graphene as interconnects have been discussed above) the total height of the device would be less than 4 nm. The off-current in a tunneling device can be exponentially small and the forward current capability of graphene, as we know, is large. Polycrystalline silicon (polysilicon), employed to make gate electrodes in present device technology, is conveniently provided by chemical vapor deposition (CVD). The same technology is adaptable to provide graphene, or BCN or insulating BN, the materials that might replace silicon. It seems clear that such graphene tunneling devices are capable of extending the scaling past the limit of silicon devices.

10
Summary and assessment

The work of Geim and Novoselov (Novoselov *et al.*, 2004, 2005) demonstrates the existence and practical importance of a new, two-dimensional (2D) phase of matter. Graphene is the prime example, but many other examples, including hexagonal boron nitride, molybdenum disulfide, TaS_2 and $NbSe_2$ are now recognized. These are two-dimensional systems embedded in 3D space, or "2D-3". These 2D-3 systems, unlike 2D systems, can flex into the perpendicular direction and their resistance to such flexural distortion is small, a classical result of small thickness. As we saw in Chapter 1, a square of graphene of edge L can be characterized by a spring constant $K \approx Yt^3/L^2$, with t the thickness and Y Young's modulus, such that a transverse deflection y requires a force $F = Ky$. With t near a single atom size, one sees that graphene, in spite a large value of Young's modulus, $Y \sim 1$ TPa, is the softest possible material against transverse deflection. Even so, for small samples, with L approaching molecular dimensions, K can take values on the order of hundreds of N/m. Embedded in graphene is, in particular, a purely 2D electron system, since the electrons are contained within the nominal 0.34 nm thickness of the graphene layer that is planar on a lateral scale much larger than its thickness. The purely 2D electron system in graphene has novel features, as we have described, mostly due to the two distinct triangular sublattices in the honeycomb crystal. The electron system in graphene, with no free motion perpendicular to the plane, is analogous to electrons on the surface of liquid helium that showed the highest electron mobility ever measured, 2090 m^2/Vs at 0.4 K. That system also transforms to an insulating Wigner electron crystal below a critical temperature 0.45 K at an electron density $4.6 \times 10^8/cm^2$. At the Dirac neutrality point in pure and undisturbed graphene, the electron density should fall well below such a value. One can ask if Wigner electron crystallization might also be the origin of the observed insulator transition in carefully prepared graphene (Ponomarenko *et al.*, 2011) as we have described in Fig. 1.6 and Section 7.5. An alternative explanation of the insulating state at low temperature for graphene at the neutrality point is that of Abrahams *et al.* (1979). In any case, it is now known that the non-zero reported minimum conductivity is (as shown, for example, in Fig. 1.5) an artifact and not an intrinsic aspect. It arises by the sensitivity of graphene to develop local carrier concentrations in the presence of stray electric fields. The conductivity at the neutral point at low temperature is zero. Graphene is less mysterious than once thought, on this behalf. A second area of simplification relates to the question of small, short-wavelength undulations or waves or ripples that now appear always to have plausible extrinsic origin in mounting strain and/or adsorbates that slightly distort the local bonding. Pure graphene is *not* unstable to developing ripples, as was shown by Thompson-Flagg (2009), see Fig. 7.26.

In practical terms, graphene samples do not show any evidence of predicted limitations of 2D matter. Even though graphene is in category 2D-3, it should still suffer the predicted limitation on size of a purely 2D perfect crystal (as argued in the HLMPW theorem, Section 2.4.2) and summarized, following Landau and Lifshitz (1986) in Section 2.4.1.1. The purely 2D limitation is a $T \ln(L/a)$ divergence of the fluctuation of atom positions away from their lattice positions, a feature not present in the *relative* locations of pairs of atoms, as indicated in eqn (2.47). These considerations of formal crystallinity do not affect the existence or stability of 2D films, but indicate abnormal in-plane thermal fluctuations of atomic positions, such that an infinite system is not crystalline in its atomic positions. All of the predicted effects disappear at $T = 0$. The fact that none of the effects have been seen in graphene may relate to the high Debye temperature, 2500 K, for in-plane atomic motion in graphene, as well as the modest size, micrometers, of carefully observed samples.

A second 2D-3 effect that should apply to graphene is crumpling, out-of-plane motions on the order of the sample size that are favored by the TS term in the free energy, where S is the entropy. Fluctuations at finite temperature increase the entropy and the treatment (Peliti and Leibler, 1985) of sample size needed for such effects to occur, given in eqn (2.50), is a refinement of an earlier estimate [eqn (2.52)] originating from de Gennes and Taupin (1982). The eqn (2.50) gives an astronomically large available size, free from crumpling at 300 K, but estimates sample sizes of 48 and 4 μm, respectively at 3900 K and 4900 K. The former temperature is the experimental sublimation temperature of graphite and the latter is the temperature estimate for the disintegration of graphene, provided by the simulation of Zakharchenko *et al.* (2011). The fact that the size of an uncrumpled sample, at even 4900 K, is predicted to be much larger than the simulated sample sizes, on the scale of nanometers, suggests that graphene does not crumple before it disintegrates. Rather, it disintegrates into vacuum starting from local defects such as five-fold and seven-fold rings that first appear in locations that are "locally crumpled." It seems, however that the de Gennes-based estimate [eqn (2.50)] has not been tested for graphene by experiment or any sort of simulation. The fact that the local graphene lattice is predicted to remain intact up to 4900 K on the basis of several simulations does suggest that the classical treatments may be useful as up to nearly that temperature.[1]

In the end it appears that the 2D theorems have no practical impact on observed properties nor on likely applications of graphene. What does have a large obvious impact is the extreme softness, lack of rigidity against transverse deflection, of the material, even on micrometer sized samples. This is evident in the formula for the classical spring constant for a square of side L, $K = 32 \, Yt^3/L^2$ as discussed in Chapter 1 and in Section 7.4 in connection with the work of Bunch *et al.* (2007). Any graphene device of lateral dimension of 1 μm or more will have to be supported, but this is not unusual in electronics. "Silicon on Insulator" (SOI) wafers are common in advanced electronics

---

[1]The classically predicted waves and vibrations are closely checked for graphene by the work e.g., of Bunch *et al.* (2007). There is no experimental comparison, to the author's knowledge, of graphene behavior to classical crumpling theory, e.g. de Gennes and Taupin (1982) or Peliti and Leibler (1985).

and might be supplanted by graphene on SiC, graphene on silicon, or graphene on SiC on Si, as possible examples.[2]

The most important advantage of graphene in electronics is its continuity and indeed record breaking conductivity, mobility and tolerated current density at one atom thickness. In miniaturization, the thickness of conducting elements is the prime variable and graphene is the thinnest useful conductor, by a factor of ten or more. The covalent bonding of the material avoids a breakup into discontinuous islands that universally occurs otherwise. Applications where this ultimate minimum thickness could have great impact are in switching-logic transistors, to avoid the "short channel effects" (Sections 9.8.1–9.8.4) in flash memory, to reduce cross-talk between adjacent cells (described in Section 9.7) and in on-chip interconnects (discussed in Section 9.3.1). The advantage in the former cases is the reduced capacitance between adjacent elements when their thickness is reduced and it appears that a factor of ten reduction in height is very significant. In the case of interconnects, the advantage is in the tolerated current density, much greater than that of any metal.

Applications of graphene are severely hampered by its poor availability in adequate quality for all but the least demanding cases, for example, as a conductive and/or strong/flexible additive to a bulk material. For these applications, chemically exfoliated material of various types is being shipped in ton quantities, according to Segal (2009). Electronic applications require an "epitaxial" form of graphene, e.g., as grown on copper foil by chemical vapor deposition (CVD), since the micro-mechanical cleavage method is impractical except for research. The two basic epitaxial forms of graphene are grown on SiC by heating to 1450°C. causing Si sublimation, leaving one or two graphene layers; and growth by chemical vapor deposition at about 1000°C on Cu foil. These temperatures are high, irreducibly so in the case of SiC, but it may well be possible to reduce the CVD deposition temperature by making a plasma in the gas that transports the carbon into the reaction zone or by using a feed gas such as benzene or even coronene with several benzene rings in the molecule (Chen *et al.*, 2012). The high temperatures hinder adoption on the basis of energy cost and also in being incompatible with typical materials often used in chip manufacture. It is important to find ways to lower the CVD deposition temperature and to broaden the type of substrate that can be used. As it is, the typical Cu substrate has to be removed by chemical dissolution followed by transferring the graphene to the target substrate. Comparing to a silicon boule that is slowly grown and then sliced into many wafers then later processed to provide a large number of chips, only one layer of graphene results from a careful growth process and there is no slicing to follow and multiply the useful output. The actual growth time for a layer of graphene in a CVD reactor is a matter of minutes but typical annealing of the catalytic substrate may be more

---

[2]A preliminary report of growth of graphitic carbon on Si (111) was given by Hackley *et al.* (2009). Their UHV method used carbon atoms from an e-beam source of highly oriented pyrolytic graphite. After cleaning the Si and heating to obtain the 7×7 reconstruction, an amorphous carbon buffer layer was deposited at 560°C. This was followed by growing graphitic carbon at 830°C. The result was not electrically characterized but was composed of small grains of graphitic carbon, approximately 2 nm in size, using a method described in their text. These workers indicated that a similar growth could be obtained on Si (100).

time-consuming. Unlike the silicon wafer, where (after the slicing) surface layers can be removed, to restore atomic cleanliness and crystalline order, the single graphene layer of carbon atoms cannot be similarly refreshed. The great advance would be to develop CVD deposition methods that could be carried out *in situ* at moderate temperatures, with the kinetic energy, needed for good graphene growth, coming from an assist to the chemical vapor deposition, perhaps in a gas discharge or by microwaves, rather than from a high temperature. Another approach was suggested in Chapter 5, in the work of Z. Li *et al.* (2011) who found CVD growth of graphene on Cu was possible at 300°C using benzene as the source gas.

Large-scale potential applications based on the high conductivity and current density may also be reached, to some extent, by deposits of disconnected flakes as are produced by various chemical and/or liquid exfoliation methods. The small size of the flakes, micrometers at most, makes it critical in such applications to establish reliable low resistance electrical connection between flakes. The refractory nature of graphene makes it difficult to sinter the individual flakes together, as can be done in nano-ink-deposited semiconductor films in solar cell manufacture. It might be possible to achieve good electrical connection between overlapping flakes by suffusing the deposited flake layer with a metal, possibly calcium or potassium, known as graphite intercalants. The result might be mechanically stronger and flexible as well as electrically conductive. Power cables and high tension power lines, in principle (as well as on-chip interconnects, on the small scale) could benefit from the increased conductivity and current density capability of graphene relative to aluminum and copper. Such applications might be reached by large-scale deposition of graphene on a catalytic Cu layer that might itself be evaporated and very thin. Another large-scale application of planar graphene-on-copper (C/Cu) would be as the back-electrode of a solar cell.

It was shown in Chapter 9 that there are several applications of graphene within the existing silicon electronics industry. These include radio-frequency transistors, current-carrying interconnects, flash memory cells and optical modulators. The most revolutionary prospect for graphene has been that it might replace silicon electronics. This would mean, first finding an alternative to the complementary metal oxide semiconductor (CMOS) family of N- and P-FET Si switching transistors. It is often stated that this is impossible because graphene is nearly a semimetal and cannot be formed into a field-effect transistor device that in the off state will draw no current. The prospect is indeed severely diminished by the semi-metallic nature of graphene, but as was explored in Sections 9.8.1–9.8.4, tunneling FET devices are possible and graphene might find its place in switching logic based on T-FET devices. Using graphene, specifically, T-FET devices can have much smaller vertical dimensions, to avoid the "short channel effects", than are available in silicon or other semiconductors. The T-FET devices can also reduce the power dissipation that remains a major problem in continuing the Moore's Law miniaturization of silicon devices.

The chance that the graphene tunnel T-FET will replace the multi-gated silicon FET is probably small. In defense of the small probability that graphene will be utilized, note that silicon technology did replace the conventional dielectric (SiO_2) with the high permittivity dielectric (HfO_2) to solve the industry-wide problem of tunneling leakage from gate to channel (a tutorial discussion of this remarkable adjustment is given by Wolf, 2010). It is more likely that the future of silicon technology (i.e., the final steps at the end of Moore's Law) will follow the lines suggested by Balestra (2010) and Ferain *et al.* (2011) that involve, for example, Si designs with multiple and multiply-electroded channels (fin-FETS) perhaps based on insulating substrates. Such changes have been under consideration for a period of years and are less radical than a possible change from complicated silicon devices to simpler (and potentially superior) graphene devices. The advantages of electrical continuity consistent with vanishing thickness are great but in practice will be balanced against available quality and cost of fabrication, the complexity of silicon chips, tradition and questions of compatibility with existing plants. As we have suggested, there do seem to be routes to graphene switching devices in which the plant and technology, so highly developed in silicon, could in large part be retained. If changes toward graphene switching technology do occur, it is likely to be in areas where the high power dissipation of silicon FET devices is a major problem.

The chance that graphene devices will enter the semiconductor technology in various supporting and specialty applications is high. The first incorporation of graphene will be into niche areas such as high-frequency amplifiers, optical modulators, photodetectors and flash memory, as suggested above. It is also likely that the intense effort on graphene will make it easier to exploit the other two-dimensional compounds, particularly BN and MoS_2. (Xiao *et al.*, 2012).

Finally, the graphene phenomenon has indeed fostered a degree of unification between particle physics and condensed matter physics. The Klein tunneling effect has convincingly been observed and the claims that electrons near the neutral point in graphene display properties heretofore only attributed to photons and to neutrinos have certainly been borne out by work on graphene.

References

Abanin, D., Morozov, S., Ponomarenko, L., Gorbachev, R., Mayorov, A., Katsnelson, M., Watanabe, K., Taniguchi, T., Novoselov, K., Levitov, L. and Geim, A. (2011). *Science*, **332**, 328.

Abedpour, N., Neek-Amal, M., Asgari, R., Shahbazi, F., Nafari, N. and Reza Rahimi Tabar, M. (2007). *Phys. Rev. B*, **76**, 195407.

Abergel, D., Russell, A. and Fal'ko, V. (2007). *Appl. Phys. Lett.*, **91**, 063125.

Abraham, F. and Kardar, M. (1991). *Science*, **252**, 419.

Abraham, F. and Nelson, D. (1990). *J. Physique (Paris)*, **51**, 2653.

Abrahams, E., Anderson, P., Licciardello, D. and Ramakrishnan, T. (1979). *Phys. Rev. Lett.*, **42**, 673.

Adam, S., Hwang, E., Galitski, V. and Das Sarma, S. (2007). *Proc. National Acad. Sciences USA*, **104**, 18392.

Affoune, A., Prasad, B., Sato, H., Enoki, T., Kaburagi, Y. and Hishiyama, Y. (2001). *Chem. Phys. Lett.*, **348**, 17.

Allen, M., Tung, V. and Kaner, R. (2010). *Chem. Reviews*, **110**, 137.

Amaratunga, G., Gaskell, P. and Saeed, A. (1991). *Diamond and Related Materials*, **1**, 51.

Amini, S., Garay, J., Liu, G., Balandin, A. and Abbaschian, R. (2010). *J. Appl. Phys.*, **108**, 094321.

Amorim, R., Fazzio, A., Antonelli, A., Novaes, F. and da Silva, A. (2007). *Nano Lett.*, **7**, 2459.

Anderson, P. W. (1958). *Phys. Rev.*, **109**, 1492.

Ando, T. (1974). *J. Phys. Soc. Jpn.*, **37**, 622.

Ando, T. and Nakanishi, T. (1998b). *J. Phys. Soc. Jpn.*, **67**, 1704.

Ando, T., Nakanishi, T. and Saito, R. (1998a). *J. Phys. Soc. Jpn.*, **67**, 2857.

Ang, P., Chen, W., Shen, A. and Loh, K. (2008). *J. Am. Chem. Soc.*, **130**, 14392.

Apalkov, V. and Chakraborty, T. (2006). *Phys. Rev. Lett.*, **97**, 126801.

Appenzeller, J., Lin, Y.-M., Knoch, J. and Avouris, P. (2004). *Phys. Rev. Lett.*, **93**, 196805.

Aronov, A. and Pikus, G. (1967). *Soviet Phys. JETP*, **24**, 188.

Aronovitz, J. and Lubensky, T. (1988). *Phys. Rev. Lett.*, **60**, 2634.

Avouris, P., Chen, Z. and Perebeinos, V. (2007). *Nature Nanotech.*, **2**, 605.

Bae, S., Kim, Y., Lee, Y., Xu, X., Park, J.-S., Zheng, Y., Balakrishnan, J., Lei, T., Kim, H., Song, Y., Kim, Y.-J., Kim, K., Ozyllmaz, B., Ahn, J.-H., Hong, B. and Iijima, S. (2010). *Nature Nanotech.*, **5**, 574.

Balandin, A., Ghosh, S., Bao, W., Calizo, I., Teweldebrhan, D., Miao, F. and Lau, C. (2008). *Nano Lett.*, **8**, 902.

Balestra, F. (2010). "Silicon-Based Devices and Materials for Nanoscale CMOS and Beyond-CMOS", Section 2.1, pp. 109–126, in *Future Trends in Microelectronics: From Nanophotonics to Sensors and Energy* (eds. S. Luryi, J. Xu and J. Zaslavsky). IEEE Press, John Wiley & Sons, Inc., Hoboken.

Banerjee, A., DeMaggio, G., Lord, K., Yan, B., Liu, F., Xu, X., Beernink, K., Pietka, G., Worrel, C., Cotter, B., Yang, J. and Guha. S. (United Solar Ovonics, LLC) (2009). "Advances in cell and module efficiency of a- Si:H based triple junction solar cells made using roll-to-roll deposition" IEEE Conference Proceeding 978-1-4244-2950-9/09 p. 000116.

Banerjee, S., Register, R., Tutuc, W., Reddy, D. and MacDonald, A. (2009). *IEEE Electron Device Lett.*, **30**, 158.

Bao, W., Miao, F., Chen, Z., Zhang, H., Jang, W., Dames, C. and Lau, C. (2009). *Nature Nanotech.*, **4**, 562.

Bardeen, J. and Johnson, J. (1972). *Phys. Rev. B*, **5**, 72.

Bardeen, J., Cooper, L. and Schrieffer, R. (1957). *Phys. Rev.*, **108**, 1175.

Barone, V., Hod, O. and Scuseria, G. (2006). *Nano Lett.*, **6**, 2748.

Baym, G. (1969). *Lectures on Quantum Mechanics*. Benjamin, New York, p. 496.

Bedanov, B., Gadiyak, G. and Lozovik, Y. (1985). *Phys. Lett. A*, **109**, 289.

Beenakker, C. (2006). *Phys. Rev. Lett.*, **97**, 067007.

Beenakker, C. (2008). *Rev. Mod. Phys.*, **80**, 1337.

Berger, C., Song, Z., Li, S., Wu, X., Brown, N., Naud, C., Mayou, D., Li, T., Hass, J., Marchenkov, A., Conrad, E., First, P. and de Heer, W. (2006). *Science*, **312**, 1191.

Berger, C., Song, Z., Li, T., Li, X., Oghazghi, A., Feng, R., Dai, Z., Marchenkov, A., Conrad, E., First, P. and de Heer, W. (2004). *J. Phys. Chem. B*, **108**, 19912.

Bernstein, K., Cavin, R., Porod, W., Seabaugh, A. and Welser, J. (2010). *Proc. IEEE*, **98**, 2169.

Binz, S., Jupalo, M., Liu, X., Wang, C., Lu, W.-C., Thiel, P., Ho, K., Conrad, E. and Tringides, M. (2012). *Phys. Rev. Lett.*, **109**, 026103.

Blake, P., Brimicombe, P., Nair, R., Booth, T., Jiang, D., Schedin, F., Ponomarenko, L., Morozov, S., Gleeson, H., Hill, E., Geim, A. and Novoselov, K. (2008). *Nano Lett.*, **8**, 1704.

Blake, P., Hill, E., Castro Neto, A., Novoselov, K., Jiang, D., Yang, R., Booth, T. and Geim, A. (2007). *Appl. Phys. Lett.*, **91**, 063124.

Blake, P., Yang, R., Morozov, S., Schedin, F., Ponomarenko, L., Zhukov, A., Nair, R., Grigorieva, I., Novoselov, K. and Geim, A. (2009). *Solid State Commun.*, **149**, 1068.

Blakslee, D., Proctor, D., Seldin, E., Spence, G. and Weng, T. (1970). *J. Appl. Phys.*, **41**, 3373.

Boehm, H., Clauss, A. and Hofmann, U. (1961). *J. Chim. Phys. Physicochim. Biol.*, **58**, 141.

Boehm, H., Clauss, A., Hofmann, U. and Fischer, G. (1962). *Zeitschrift Fur Naturfoschung B*, **17**, 150.

Bokdam, M., Khomyakov, P., Brocks, G., Zhong, Z. and Kelly, P. (2011). *Nano Lett.*, **11**, 4631.

Bolotin, K., Ghahari, F., Shulman, M., Stormer, H. and Kim, P. (2009). *Nature*, **462**, 196.

Bolotin, K., Sikes, K., Hone, J., Stormer, H. and Kim, P. (2008). *Phys. Rev. Lett.*, **101**, 096802.

Bolotin, K., Sikes, K., Jiang, Z., Fudenberg, G., Hone, J., Kim., P. and Stormer, H. (2008a). *Solid State Commun.*, **146**, 351.

Bonnani, A., Bobisch, C. and Moller, R. (2008). *Rev. Sci. Instrum.*, **79**, 83704.

Bostwick, A., McChesney, J., Ohta, T., Rotenberg, E., Seyller, T. and Horn, K. (2009). *Progress in Surface Science*, **84**, 380.

Bowick, M. and Travesset, A. (2001). *Phys. Rep.*, **344**, 255.

Bradbury, F., Takita, M., Gurreri, T., Wilkel, K., Eng, K., Carroll, M. and Lyon, S. (2011). *Phys. Rev. Lett.*, **107**, 266803.

Braga, S. F., Coluci, V. R., Legoas, S. B., Giro, R., Galvao, D. S. and Baughman, R. H. (2004). *Nano Lett.*, **4**, 881.

Braghin, F. and Hasselmann, N. (2010). *Phys. Rev. B*, **82**, 35407.

Bransden, B. and Joachain, C. (2003). *Physics of Atoms and Molecules* (2nd edn). Pearson Education, Harlow.

Brar, V., Decker, R., Solowan, H.-M., Wang, Y., Maserati, L., Chan, K., Lee, H., Girit, C., Zettl, A., Louie, S., Cohen, J. and Crommie, M. (2011). *Nature Phys.*, **7**, 43.

Brenner, D., Shenderova, O., Harrison, J., Stuart, S., Ni, B. and Sinnott, S. (2002). *J. Phys.: Condens. Matter*, **14**, 783.

Britnell, L., Gorbachev, R., Jalil, R., Belle, B., Schedin, F., Mishcenko, A., Georgiou, T., Katsnelson, M., Eaves, L., Morozov, S., Peres, N., Leist, J. Geim, A., Novoselov, K. and Ponomarenko, L. (2012). *Science*, **335**, 947.

Brodie, B. (1859). *Phil. Trans. Roy. Soc. London*, **149**, 249.

Bruzzone, S. and Fiori, G. (2011). *Appl. Phys. Lett.*, **99**, 222108.

Bunch, J., van der Zande, A., Verbridge, S., Frank, I., Tanenbaum, D., Parpia, J., Craighead, H. and McEuen, P. (2007). *Science*, **315**, 490.

Bunch, J., Verbridge, S., Alden, J., van der Zande, A., Parpia, J., Craighead, H. and McEuen, P. (2008). *Nano Lett.*, **8**, 2458.

Bunyk, P., Likharev, K. and Zinoviev, D. (2001). *Intern. J. High Speed Electron Syst.*, **11**, 257.

Burnell, D. and Wolf, E. (1982). *Phys. Lett. A*, **90**, 471.

Burnell, D. M. (1982). Ph.D. Thesis, Department of Physics, Iowa State University (unpublished).

Cai, J., Ruffieux, P., Jaafar, R., Bieri, M., Braun, T., Blankenburg, S., Muoth, M., Seitsonen, A., Saleh, M., Feng, X., Mullen, K. and Fasel, R. (2010). *Nature*, **466**, 470.

Cao, H., Yu, Q., Jauregui, L., Tian, J., Wu, W., Liu, Z., Jalilian, R., Benjamin, D., Jiang, Z., Bao, J., Pei, S. and Chen, Y. (2010). *Appl. Phys. Lett.*, **96**, 122106.

Calizo, I., Miao, F., Bao, W., Lau, C. and Balandin, A. (2007). *Appl. Phys. Lett.*, **91**, 071913.

Carey, F. and Guiliano, R. (2011). *Organic Chemistry* (8th edn). McGraw Hill, New York.

Carlsson, J. and Scheffler, M. (2006). *Phys. Rev. Lett.*, **96**, 046806.

Carr, D., Evoy, S., Sekaric, L, Craighead, H. and Parpia, J. (1999). *Appl. Phys. Lett.*, **75**, 920.

Castro Neto, A. (2011). *Science*, **332**, 315.

Castro Neto, A. H., Guinea, F., Peres, N. M. R., Novoselov, K. S. and Geim, A. K. (2009). *Revs. Mod. Phys.*, **81**, 109.

Castro, E., Novoselov, K., Morozov, S., Peres, N., Dos Santos, J., Nilsson, J., Guinea, F., Geim, A. and Neto, A. (2007). *Phys. Rev. Lett.*, **99**, 216802.

Castro, E., Ochoa, H., Katsnelson, M., Gorbachev, R., Elias, D., Novoselov, K., Geim, A. and Guinea, F. (2010). *Phys. Rev. Lett.*, **105**, 266601.

Celzard, A., Mareche, J., Furdin, G. and Puricelli, S. (2000). *J. Phys. D: Appl. Phys.*, **33**, 3094.

Cerda, E. and Mahadevan, L. (2003). *Phys. Rev. Lett.*, **90**, 074302.

Checkelsky, J., Li, L. and Ong, N. (2009). *Phys. Rev. B*, **79**, 115434.

Cheianov, V. and Fal'ko, V. (2006). *Phys. Rev. B*, **74**, 041403.

Cheianov, V., Fal'ko, V., Altshuler, B. and Aleiner, I. (2007). *Phys. Rev. Lett.*, **99**, 176801.

Chen, C., Rosenblatt, S., Bolotin, K., Kalb, W., Kim, P., Kymissis, I., Stormer, H., Heinz, T. and Hone, J. (2009). *Nature Nanotech.*, **4**, 861.

Chen, G., Weng, W., Wu, D., Wu, C., Lu, J., Wang, P. and Chen, X. (2004). *Carbon*, **42**, 753.

Chen, W., Chen, H., Lan, H., Schulze, T., Zhu, W. and Zhang, Z. (2012). *Phys. Rev. Lett.*, **109**, 265507.

Chen, Z., Lin, Y.-M., Rooks, M. and Avouris, P. (2007). *Physica E*, **40**, 228.

Chuang, A., Robertson, J., Boskovic, B. and Koziol, K. (2007). *Appl. Phys. Lett.*, **90**, 123107.

Chung, D. D. L. (1987). *J. Mater. Sci.*, **22**, 4190.

Ci, L., Song, L., Jin, C., Jariwala, D., Wu, D., Li, Y., Srivastava, A., Wang, Z., Storr, K., Balicas, L., Liu, F. and Ajayan, P. (2010). *Nature Mater.*, **9**, 430.

Cooper, D., Geratt, J. and Raimondi, M. (1986). *Nature*, **323**, 699.

Coskun, U., Brenner, M., Hymel, T., Vakaryuk, V., Levchenko, A. and Bezryadin, A. (2012). *Phys. Rev. Lett.*, **108**, 097003.

Crane, D. and Kennedy, R., Jr. (2012). "Solar Panels for Every Home", *New York Times*, Dec. 12, 2012.

Cserti, J. and David, G. (2006). *Phys. Rev. B*, **74**, 172305.

Cserti, J., Csordas, A. and David, G. (2007). *Phys. Rev. Lett.*, **99**, 066802.

Dan, Y., Lu, Y., Kybert, N., Luo, Z. and Johnson, A. (2009). *Nano Lett.*, **9**, 1472.

Das Sarma, S., Adam, S., Hwang, E. and Rossi, E. (2011). *Rev. Mod. Phys.*, **83**, 407.

DaSilva, A., Zou, K., Jain, J. and Zhu, J. (2010). *Phys. Rev. Lett.*, **104**, 236601.

Dato, A., Radmilovic, B., Lee, Z., Phillips, J. and Frenklach, M. (2008). *Nano Lett.*, **8**, 2012.

David, G. and Cserti, J. (2010). *Phys. Rev. B*, **81**, 121417.

de Gennes, P.-G. ((1979). "Scaling Concepts in Polymer Physics". Cornell University Press, Ithaca, NY).

de Gennes, P.-G. and Taupin, C. (1982). *J. Phys. Chem.*, **86**, 2294.

De Heer, W., Berger, C., Ruan, M., Sprinkle, M., Li, X., Hu, Y., Zhang, B., Hankinson, J. and Conrad, E. (2010). *Proc. National Academy of Sciences*, **108**, 16900.

De Heer, W., Berger, C. and First, P. (2006). U. S. Patent 7015142.

DeVries, R. (1987). *Ann. Rev. Mater. Sci.*, **17**, 161.

Dean, C., Young, A., Meric, I., Lee, C., Wang, L., Sorgenfrei, S., Watanabe, K., Taniguchi, T., Kim, P., Shepard, K. and Hone, J. (2010). *Nature Nanotech.*, **5**, 722.

Dery, H., Dalal, P., Cywinski, L. and Sham, L. (2007). *Nature*, **447**, 573.

Dethlefsen, A., Mariani, E., Tranitz, H.-B., Wegescheider, W. and Haug, R. (2006). *Phys. Rev. B*, **74**, 165325.

Dienwiebel, M., Verhoeven, G., Pradeep, N., Frenken, J., Heimbert, J. and Zandenbergen, H. (2004). *Phys. Rev. Lett.*, **92**, 126101.

Dikin, D., Stankovich, S., Zimney, E., Piner, R., Dommett, G., Evmenenko, G., Nguyen, S. and Ruoff, R. (2007). *Nature*, **448**, 457.

DiVincenzo, D., Mele, E. (1984). *Phys. Rev. B*, **29**, 1685.

Dresselhaus, M. and Dresselhaus, G. (1981). *Adv. Phys.*, **30**, 139.

Dresselhaus, M. and Dresselhaus, G. (2002). *Adv. Phys.*, **51**, 1.

Dresselhaus, M., Dresselhaus, G. and Eklund, P. (1996). *Science of Fullerenes and Carbon Nanotubes*. Elsevier Science USA, Orlando.

Droscher, S., Roulieau, P., Molitor, F., Studerus, P., Stampfer, C., Ensslin, K. and Ihn, T. (2010). *Appl. Phys. Lett.*, **96**, 152104.

Du, X., Skachko, I., Barker, A. and Andrei, E. (2008). *Nature Nanotech.*, **3**, 491.

Du, X., Skachko, I., Duerr, F., Luican, A. and Andrei, E. (2009). *Nature*, **462**, 192.

Duong, D., Han, G., Lee, S., Gunes, F., Kim, S. E., Kim, S. T., Kim, H., Ta, Q., So, K., Yoon, S., Chae, S., Jo, Y., Park, M., Chae, S., Lim, S., Choi, J. and Lee, Y. (2012). *Nature*, **490**, 235.

Eda, G., Lin, Y.-Y., Miller, S., Chen, C.-W., Su, W.-F. and Chhowalla, M. (2008). *Appl. Phys. Lett.*, **92**, 233305.

Efetov, D. and Kim, P. (2010). *Phys. Rev. Lett.*, **105**, 256805.

Eisenstein, J., Pfeiffer, L. and West, K. (1994). *Phys. Rev. B*, **50**, 1760.

Elias, D., Nair, R., Mohiuddin, M., Morozov, S., Blake, P., Hatsall, M., Ferrari, A., Boukhvalov, D., Katsnelson, M., Geim, A. and Novoselov, K. (2009). *Science*, **323**, 610.

Ekinci, K. and Roukes, M. (2005). *Rev. Sci. Instrum.*, **76**, 61101.

Elias, D., Gorbachev, R., Mayorov, A., Morozov, S., Zhukov, A., Blake, P., Ponomarenko, L., Grigorieva, I., Novoselov, K., Guinea, F. and Geim, A. (2011). *Nature Phys.*, **7**, 701.

Elser, V. and Haddon, R. (1994). *Phys. Rev. A*, **36**, 4579.

Emtsev, K. Bostwick, A., Horn, K., Jobst, J., Kellogg, G., Ley, L., McChesney, J., Ohta, T., Reshanov, S., Rohrl, J., Rotenbert, E., Schmid, A., Waldmann, D., Weber, H. and Seyller, T. (2009). *Nature Mater.*, **8**, 203.

Emtsev, K., Speck, F., Seyller, T., Ley, L. and Riley, J. (2008). *Phys. Rev. B*, **77**, 155303.

Esaki, L. (1958). *Phys. Rev.*, **109**, 503.

Evers, F. and Mirlin, A. (2008). *Revs. Mod. Phys.*, **80**, 1355.

Fasolino, A., Los, J. and Katsnelson, M. (2007). *Nature Mater.*, **6**, 858.

Feldman, B., Krauss, B., Smet, J. and Yacoby, A. (2012). *Science*, **337**, 1196.

Ferain, I., Colinge, C. and Colinge, J.-P. (2011). *Nature*, **479**, 310.

Ferralis, N. (2010). *J. Mater. Sci.*, **45**, 5135.

Ferralis, N., Maboudian, R. and Carraro, C. (2008). *Phys. Rev. Lett.*, **101**, 156802.

Ferrari, A., Meyer, J. U., Scardaci, V., Casiraghi, C., Lazzeri, M., Mauri, F., Pascanee, S., Jian, D., Novoselov, K., Roth, S. and Geim, A. (2006). *Phys. Rev. Lett.*, **97**, 187401.

Fert, A. and Jaffres, H. (2001). *Phys. Rev. B*, **64**, 184420.

Feynman, R., Leighton, R. and Sands, M. (1964). *The Feynman Lectures on Physics*. Addison-Wesley, Reading, MA. Vol. II, Section 34-4.

Fiori, G. and Iannaccone, G. (2009a). *IEEE Electron Device Lett.*, **30**, 261.

Fiori, G. and Iannaccone, G. (2009b). *IEEE Electron Device Lett.*, **30**, 1096.

Fiori, G., Betti, A., Bruzzone, S., D'Amico, P. and Iannaccone, G. (2011). *IEDM Tech. Digest*, pp. 11.4.1–4. Washington DC, USA, ISSN: 0163–1918.

Fiori, G., Betti, A., Bruzzone, S. and Iannaccone, G. (2012). *ACS Nano*, **6**, 2642.

Fiori, G., Lebegue, S., Betti, A., Michetti, P., Klintenberg, M., Eriksson, O. and Iannaccone, G. (2010). *Phys. Rev. B*, **82**, 153404.

Fisher, D., Halperin, B. and Platzman, P. (1979). *Phys. Rev. Lett.*, **42**, 798.

Fliegl, H., Sundholm, D., Taubert, S., Juselius, J. and Klopper, W. (2009). *J. Phys. Chem. A*, **113**, 8668.

Forbeaux, I., Themlin, J., Langlais, V., Yu, M., Belkhir, H. and Debever, J. (1998). *Surf. Rev. Lett.*, **5**, 193.

Forbeaux, I., Themlin, J. and Debever, J. (1998). *Phys. Rev. B*, **58**, 16396.

Fradkin, E., Kivelson, S. and Oganesyan, V. (2007). *Science*, **315**, 196.

Frank, D., Taur, F. and Wong, H.-S. (1998). *IEEE Electron Device Lett.*, **19**, 385.

Fukui, K., Yonezawa, T. and Shingu, H. (1952). *J. Chem. Phys.*, **20**, 722.

Galitski, V., Adam, S. and Das Sarma, S. (2007). *Phys. Rev. B*, **76**, 245405.

Gall, N., Mikhailov, S., Ruf'kov, E. and Tontegode, A. (1987). *Surf. Sci.*, **191**, 185.

Gall, N., Rutkov, E. and Tontegode, A. (1997). *Int. J. Mod. Phys. B*, **11**, 1865.

Gass, M., Bangert, U., Bleloch, A., Wang, P., Nair, R. and Geim, A. (2008). *Nature Nanotech.*, **3**, 676.

Geim, A. (2011a). *Angew. Chem. Int. Edn.*, **50**, 6966.

Geim, A. (2011). *Rev. Mod. Phys.*, **83**, 851.

George, S., Ott, A. and Klaus, J. (1996). *J. Phys. Chem.*, **100**, 13121.

Geringer, V., Liebmann, M., Echtermeyer, T., Runte, S., Schmidt, M., Ruckamp, M., Lemme, M. and Morgenstern, M. (2009). *Phys. Rev. Lett.*, **102**, 076102.

Ghahari, F., Zhao, Y., Cadden-Zimansky, P., Bolotin, K. and Kim, P. (2011). *Phys. Rev. Lett.*, **106**, 046801.

Ghiringhelli, L., Los, J., Fasolino, A. and Meijer, E. (2005a). *Phys. Rev. B*, **72**, 214103.

Ghiringhelli, L., Los, J., Meijer, E., Fasolino, A. and Frenkel, D. (2005). *Phys. Rev. Lett.*, **94**, 145701.

Giannazzo, F., Roccaforte, F., Raineri, V. and Liotta, S. (2006). *Europhysics Lett.*, **74**, 686.

Giannazzo, F., Sonde, S., Raineri, V. and Rimini, E. (2009). *Nano Lett.*, **9**, 23.

Giannopoulos, G. (2012). *Comp. Materials Science*, **53**, 388.

Gibertini, M., Tomadin, A., Polini, M., Fasolino, A. and Katsnelson, M. (2010). *Phys. Rev. B*, **81**, 125437.

Giesbers, A., Rietveld, G., Houtzager, E., Zeitler, U., Yang, R., Novoselov, K., Geim, A. and Maan, J. (2008). *Appl. Phys. Lett.*, **93**, 222109.

Giovannetti, G., Khomyakov, P., Brocks, G., Karpan, V., van den Brink, J. and Kelly, P. (2008). *Phys. Rev. Lett.*, **101**, 26803.

Goerbig, M. (2011). "Quantum Hall Effects", Chapter 6 in *Ultracold Gases and Quantum Information*. (Eds. C. Miniatura, L.-C. Kwak, M. Ducloy, B. Gremaud, B.-G. Englert, L. Cugliandolo, A. Ekert, and K. Phua.) Oxford University Press, Oxford.

Goerbig, M. (2011a). *Rev. Mod. Phys.*, **83**, 1193.

Golubovic, L., Moldovan, D. and Peredera, A. (1998). *Phys. Rev. Lett.*, **81**, 3387.

Gomez-Navarro, C., Weitz, R., Bittner, A., Scolari, M., Mews, A., Burghard, M. and Kern, K. (2007). *Nano Lett.*, **7**, 3499.

Gorbachev, R., Mayorov, A., Savchenko, A., Horsell, D. and Guinea, F. (2008). *Nano Lett.*, **8**, 1995.

Grant, J. and Haas, T. (1970). *Surf. Sci.*, **21**, 76.

Grimes, C. and Adams, G. (1979). *Phys. Rev. Lett.*, **42**, 795.

Gupta, S. and Batra, R. (2010). *J. Comput. Theor. Nanosci.*, **7**, 1.

Hackley, J., Ali, D., DiPasquaale, J., Demaree, J. and Richardson, C. (2009). *Appl. Phys. Lett.*, **95**, 133114.

Han, M., Ozyilmaz, B., Zhang, Y. and Kim, P. (2007). *Phys. Rev. Lett.*, **98**, 206805.

Han, S.-J., Jenkins, K., Garcie, A., Frandkon, A., Bol, A. and Haensch, W. (2011). *Nano Lett.*, **11**, 3690.

Han, W., McCreary, K., Pi, K., Wang, W., Li, Y., Wen, H., Chen, J. and Kawakami, R. (2012). *J. Magn. Magn. Mater.*, **324**, 369.

Han, W., Pi, K., McCreary, K., Li, Y., Wong, J., Swartz, A. and Kawakami, R. (2010). *Phys. Rev. Lett.*, **105**, 167202.

Harada, A., Shimojo, F. and Hoshino, K. (2005). *J. Non-Cryst. Solids*, **74**, 2017.

Harigaya, K. and Enoki, T. (2002). *Chem. Phys. Lett.*, **351**, 128.

Hashimoto, K., Sohrmann, C., Wiebe, J., Inaoka, T., Meier, F., Hirayama, Y., Romer, R., Wiesendanger, R. and Morgenstern, M. (2008). *Phys. Rev. Lett.*, **101**, 256802.

Haskell, B., Andersson, N., Jones, D. and Samuelsson, L. (2007). *Phys. Rev. Lett.*, **99**, 231101.

Hass, J., Varchon, F., Millan-Otoya, J., Sprinkle, M., Sharma, H., de Heer, W., Berger, C., First, P., Magaud, L. and Conrad, E. (2008). *Phys. Rev. Lett.*, **100**, 125504.

Heersche, H., Jarillo-Herrero, P., Oostinga, J., Vandersypen, L. and Morpurgo, A. (2007). *Nature*, **446**, 56.

Heiblum, M. and Fischetti, M. (1990). *IBM J. Res. Develop.*, **34**, 530.

Hernandez, Y., Nicolosi, V., Lotya, M., Blighe, R., Sun, Z., De, S., McGovern, I., Holland, B., Byrne, M., Gun'ko, Y., Boland, J., Niraj, P., Duesberg, G., Krishnamurthy, S., Goodhue, R., Hutchison, J., Scardacii, V., Ferrari, A. and Coleman, J. (2008). *Nature Nanotech.*, **3**, 563.

Hicks, J., Tejeda, A., Taleb-Ibrahim, A., Nevius, M., Wang, F., Shepperd, K., Palmer, J., Bertran, F., Le Fevre, P., Kunc, J., de Heer, W., Berger, C. and Conrad, E. (2012). *Nature Phys.*, [DOI:10.1038/NPHYS52487].

Hines, M. and Scholes, G. (2003). *Adv. Mater.*, **15**, 1844.

Hiraoka, T., Izadi-Najafabadi, A., Yamada, T., Futaba, D., Yasuda, S., Tanaike, O., Hatori, H., Yumura, M., Iijima, S. and Hata, K. (2010). *Adv. Func. Mater.*, **20**, 422.

Hohenberg, P. (1967). *Phys. Rev.*, **158**, 383.

Hohenberg, P. and Kohn, W. (1964). *Phys. Rev.*, **136**, B864.

Homoth, J., Wenderoth, M., Druga, T., Winking, L., Ulbrich, R., Bobisch, C., Weyers, B., Bonnani, A., Zubkov, E., Bernhart, A., Kaspers, M. and Moller, R. (2009). *Nano Lett.*, **9**, 1588.

Hong, A., Song, E., Yu, H., Allen, M., Kim, J., Fowler, J., Wassel, J., Park, Y., Wang, Y., Zou, J., Kaner, R., Weiller, B. and Wang, K. (2011). *ACS Nano*, **5**, 7812.

Hossain, M., Johns, J., Bevan, K., Karmel, H., Liang, Y., Yoshimoto, S., Mukal, K., Koitaya, T., Yoshinobu, J., Kawal, M., Lear, A., Kesmodel, L., Tait, S. and Hersam, M. (2012). *Nature Chem.*, **4**, 305.

Huang, J., Ding, F., Yakobson, B., Lu, P., Qi, L. and Li, J. (2009). *Proc. National Acad. Sciences*, **106**, 10103.

Huard, B., Sulpizio, A., Stander, N., Todd, K., Yang, B. and Goldhaber-Gordon, D. (2007). *Phys. Rev. Lett.*, **98**, 236803.

Huertas-Hernando, D., Guinea, F. and Brataas, A. (2006). *Phys. Rev. B*, **74**, 155426.

Hummers, W. and Offerman, R. (1958). *J. Am. Chem. Soc.*, **80**, 1339.

Hwang, E., Adam, S. and Das Sarma, S. (2007). *Phys. Rev. Lett.*, **98**, 186806.

Ionescu, A. and Riel, H. (2011). *Nature*, **479**, 329.

Ishigami, M., Chen, J., Cullen, W., Fuhrer, J. and Williams, E. (2007). *Nano Lett.*, **7**, 1643.

Ito, A. and Nakamura, H. (2007). http://nifs-repository.nifs.ac.jp/handle/10655/648 [accessed May 29, 2012].

Jain, N., Bansal, T., Durcan, C. and Yu, B. (2012). *IEEE Electron Device Lett.*, **33**, 925.

Jancovici, B. (1967). *Phys. Rev. Lett.*, **19**, 20.

Jedema, F., Heersche, H., Filip, A., Baselmans, J. and van Wees, B. (2002). *Nature*, **416**, 713.

Jeon, I.-Y., Shin, Y.-R., Sohn, G.-J., Choi, H.-U., Bae, S.-Y., Mahmood, J., Jung, S.-M., Seo, J.-M., Kim, M.-J., Chang, D., Dai, L. and Baek, J.-B. (2012). *Proc. National Acad. Sciences*, **109**, 5588.

Jeong, B., Ihm, J. and Lee, G.-D. (2008). *Phys. Rev. B*, **78**, 165403.

Ji, S.-H., Hannon, J., Tromp, R., Perebeinos, V., Tersoff, J. and Ross, F. (2012). *Nature Mater.*, **11**, 114.

Jia, X., Hofmann, M., Meunier, V., Sumpter, B., Campos-Delgado, J., Romo-Herrera, J., Son, H., Hsieh, Y., Reina, A., Kong, J., Terrones, M. and Dresselhaus, M. (2009). *Science*, **323**, 1701.

Jiang, X., Wang, R., Shelby, R., Macfarlane, R., Bank, S., Harris, J. and Parkin, S. (2005). *Phys. Rev. Lett.*, **94**, 056601.

Jiang, Z., Henriksen, E., Tung, L., Wang, Y.-J., Schwartz, M., Han, Y., Kim, P. and Stormer, H. (2007a). *Phys. Rev. Lett.*, **98**, 197403.

Jiang, Z., Zhang, Y., Stormer, H. and Kim, P. (2007). *Phys. Rev. Lett.*, **99**, 106802.

Jiao, L., Wang, X., Diankov, G., Wang, H. and Dai, H. (2010). *Nature Nanotech.*, **5**, 321.

Jiao, L., Zhang, L., Wang, X., Diankov, G. and Dai, H. (2009). *Nature*, **458**, 877.

Jo, G., Choe, M., Lee, S., Park, W., Kahng, Y. and Lee, T. (2012). *Nanotechnology*, **23**, 112001.

John, D., Castro, L. and Pulfrey, D. (2004). *J. Appl. Phys.*, **96**, 5180.

Johnson, M. and Silsbee, R. (1988). *Phys. Rev. B*, **37**, 5312.

Juselius, J., Sundholm, D. and Gauss, J. J. (2004). *Chem. Phys.*, **121**, 3952.

Kane, C. and Mele, E. (1997). *Phys. Rev. Lett.*, **78**, 1932.

Kane, C. and Mele, E. (2005). *Phys. Rev. Lett.*, **95**, 226801.

Kane, E. (1959). *J. Phys. Chem. Solids*, **12**, 181.

Kantor, Y., Kardar, M. and Nelson, D. (1987). *Phys. Rev. A*, **35**, 3056.

Kantor, Y. and Nelson, D. (1987). *Phys. Rev. A*, **36**, 4020.

Kato, Y., Myers, R., Gossard, A. and Awschalom, D. (2004). *Science*, **306**, 1910.

Katsnelson, M. (2012). *Graphene: Carbon in Two Dimensions*. Cambridge University Press, Cambridge.

Katsnelson, M. and Geim, A. (2008). *Philos. Trans. Roy. Soc. London Ser. A*, **366**, 195.

Katsnelson, M. and Novoselov, K. (2007). *Solid. State Commun.*, **143**, 3.

Katsnelson, M., Novoselov, K. and Geim, A. (2006). *Nature Phys.*, **2**, 620.

Keldysh, L. (1964). *Soviet Phys. JETP*, **18**, 253.

Khakifirooz, A. and Antoniadis, D. (2008). *IEEE Trans. Electr. Devices*, **55**, 1391.

Kharitonov, M. (2012). *Phys. Rev. Lett.*, **109**, 046803.

Khrapach, I., Withers, F., Bointon, T., Polyushkin, D., Barnes, W., Russon, S. and Craciun, M. (2012). *Adv. Mat.* [DOI: 10.1002/adma.201200489].

Ki, D.-K. and Morpurgo, A. (2012). *Phys. Rev. Lett.*, **108**, 266601.

Kim, E. and Castro Neto, A. (2008). *Europhysics Lett.*, **84**, 57007.

Kim, K., Choi, J.-Y., Kim, T., Cho, S.-H. and Chung, H.-J. (2011). *Nature*, **479**, 338.

Kim, K., Lee, Z., Regan, W., Kisielowski, C., Crommie, M. and Zettle, A. (2011a). *ACS Nano*, **5**, 2142.

Kim, K., Sussman, A. and Zettl, A. (2010a). *ACS Nano*, **4**, 1362.

Kim, K., Zhao, Y., Jang, H., Lee, S., Kim, J., Kim, K., Ahn, J.-H., Kim, P., Choi, J.-Y. and Hong, B. (2009). *Nature*, **457**, 706.

Kim, M., Brant, J. and Kim, P. (2010). *Phys. Rev. Lett.*, **104**, 056801.

Kim, R.-H., Bae, M.-H., Kim, D., Cheng, H., Kim, B., Kim, D.-H., Li, M., Wu, J., Du, F., Kim, H.-S., Kim, S., Estrada, D., Hong, S., Huang, Y., Pop, E. and Rogers, J. (2011b). *Nano Lett.*, **11**, 3881.

Kim, S., Jo, I., Dillen, D., Ferrer, D., Fallahazad, B., Yao, Z., Banerjee, S. and Tutuc, E. (2012). *Phys. Rev. Lett.*, **108**, 116404.

Kim, S. and Tomanek, D. (1994). *Phys. Rev. Lett.*, **72**, 2418.

Kirilenko, D., Dideykin, A. and VanTendeloo, G. (2011). *Phys. Rev. B*, **84**, 235417.

Kittel, C. (1986). *Introduction to Solid State Physics* (6th edn). John Wiley & Sons, Inc., New York, pp. 228–32.

Klein, O. (1929). *Z. Phys.*, **53**, 157.

Klitzing, K. von (1986). *Rev. Mod. Phys.*, **58**, 519.

Klitzing, K. von, Dorda, G. and Pepper, M. (1980). *Phys. Rev. Lett.*, **45**, 494.

Knoch, J., Mantl., S. and Appenzeller, J. (2007). *Solid State Electronics*, **51**, 572.

Knox, K., Wang, S., Morgante, A., Ovetko, D., Locatelli, A., Menter, T., Nino, M., Kim, P. and Osgood, R. (2008). *Phys. Rev. B*, **78**, 201408.

Kobayashi, K., Tanimura, M., Makai, H., Yshimura, A., Yoshimura, H., Kojima, K. and Tachibana, M. (2007). *J. Appl. Phys.*, **101**, 094306.

Kobayashi, T., Fukui, K.-I., Enoki, T. and Kusakabe, K. (2006). *Phys. Rev. B*, **73**, 125416.

Kobayashi, T., Fukui, K.-I., Enoki, T., Kusakabe, K. and Kaburagi, Y. (2005). *Phys. Rev. B*, **71**, 193406.

Koch, M., Ample, F., Joachim, C. and Grill, L. (2012). *Nature Nanotech.*, **7**, 713.

Koga, J. and Toriumi, A. (1997). *Appl. Phys. Lett.*, **70**, 2138.

Kohn, W. and Sham, L. (1965). *Phys. Rev.*, **140**, A1133.

Komatsu, K. (1958). *J. Phys. Chem. Solids*, **6**, 381.

Konstantatos, G., Badioli, M., Gaudreau, L., Osmond, J., Bernechea, M., Arquer, F., Gatti, F. and Koppens, F. (2012). *Nature Nanotech.*, [DOI: 10.1038/NNANO 2012.60].

Kopylov, S., Tzalenchuk, A., Kubatkin, S. and Fal'ko, V. (2010). *Appl. Phys. Lett.*, **97**, 112109.

Kosynkin, D., Higginbotham, A., Sinitskii, A., Lomedo, J., Dimiev, A., Price, K. and Tour, J. (2009). *Nature*, **458**, 872.

Kotov, V., Uchoa, B., Pereira, V., Guinea, F. and Castro Neto, A. (2012). *Rev. Mod. Phys.*, **84**, 1067.

Kowaki, Y., Harada, A., Shimojo, F. and Hoshino, K. (2007). *J. Phys. Condens. Matter*, **19**, 436224.

Krishnan, A., Dujardin, E., Ebbeson, T., Yianilos, P. and Treacy, M. (1998). *Phys. Rev. B*, **58**, 14013.

Krumhansl, J. and Brooks, H. (1953). *J. Chem. Phys.*, **21**, 1663.

Kumar, S., Hembram, K. and Waghmare, U. (2010). *Phys. Rev. B*, **82**, 115411.

Kurkijarvi, J. (1972). *Phys. Rev. B*, **6**, 832.

Kuzmenko, A., van Heumen, E., van der Marel, D., Lerch, P., Blake, P., Nosoelov, K. and Geim, A. (2009). *Phys. Rev. B*, **79**, 115441.

Kwon, Y.-K., Berber, S. and Tomanek, D. (2004). *Phys. Rev. Lett.* **92**, 15901.

Landau, L. and Lifshitz, E. (1986). *Statistical Physics* (3rd edn Part 1 Course of Theoretical Physics Volume 5). Pergamon, Oxford, pp. 432–438.

Landau, L. D. (1937). *Phys. Z. Sowjet.*, **11**, 26.

Landau, L. and Lifshitz, E. (1959). *Theory of Elasticity*, Vol. 7 of Course of Theoretical Physics. Addison-Wesley Publishing Company, Reading, Massachusetts.

Laughlin, R. (1983). *Phys. Rev. Lett.*, **50**, 1395.

Laughlin, R. (1998). Nobel Lecture, Dec. 8. [Nobel Foundation, 2002].

Le Doussal, P. and Radzihovsky, L. (1992). *Phys. Rev. Lett.*, **69**, 1209.

Lea, J., Fozooni, P., Kristensen, A., Richardson, P., Djerfi, K., Dykman, J., Fang-Yen, C. and Blackburn, A. (1997). *Phys. Rev. B*, **55**, 16280.

Lee, C., Wei, X., Kysar, J. and Hone, J. (2008). *Science*, **321**, 385.

Lee, C., Wei, X., Li, Q., Carpick, R., Kysar, J. and Hone, J. (2009). *Phys. Status Solidi B*, **246**, 2562.

Lee, G.-H., Jeong, D., Choi, J.-H., Doh, Y.-J. and Lee, H.-J. (2011). *Phys. Rev. Lett.*, **107**, 146605.

Lee, J.-Y., Connor, S., Cui, Y. and Peumans, P. (2008). *Nano Lett.*, **8**, 689.

Lee, K.-J., Chandrakasan, A. and Kong, J. (2011). *IEEE Electron Device Lett.*, **32**, 557.

Leenaerts, O., Partoens, B. and Peeters, F. (2008). *Appl. Phys. Lett.*, **93**, 193107.

Lemme, J., Echtermeyer, T., Baus, M. and Kurz, H. (2007). *IEEE Electron Device Lett.*, **28**, 282.

Lemonik, Y., Aleiner, L., Take, C. and Fal'ko, V. (2010). *Phys. Rev. B*, **82**, 201408.

Levendorf, M., Kim, C.-J., Brown, L., Huang, P., Havener, R., Muller, D. and Park, J. (2012). *Nature*, **488**, 627.

Levine, R., Libby, S. and Pruisken, A. (1983). *Phys. Rev. Lett.*, **51**, 1915.

Li, S., Zhu, Y., Cai, W., Borysiak, M., Han, B., Chen, D., Piner, R., Columbo, L. and Ruoff, R. (2009a). *Nano Lett.*, **9**, 4359.

Li, X., Cai, W., An, J., Kim, S., Nah, J., Yang, D., Piner, R., Velamakanni, A., Jung, I., Tutuc, E., Banerjee, S., Colombo, L. and Ruoff, R. (2009a). *Science*, **324**, 1312.

Li, X., Magnuson, C., Venugopal, A., Tromp, R., Hannon, J., Vogel, E., Colombo, L. and Ruoff, R. (2011). *J. Am. Chem. Soc.*, **133**, 2816.

Li, X., Wang, X., Zhang, L., Lee, S. and Dai, H. (2008). *Science*, **319**, 1229.

Li, X., Zhu, Y., Cai, W., Borysiak, M., Han, B., Chen, D., Piner, R., Colombo, L. and Ruoff, R. (2009a). *Nano Lett.*, **9**, 4359.

Li, Z., Henriksen, E., Jiang, Z., Hao, Z., Martin, M., Kim, P., Stormer, H. and Basov, D. (2009). *Phys. Rev. Lett.*, **102**, 037403.

Li, Z., Wu, P., Wang, C., Fan, X., Zhang, W., Zhai, X., Zeng, C., Li, Z., Yang, J. and Hou, J. (2011). *ACS Nano*, **5**, 3385.

Liao, L., Lin, Y.-C., Bao, M., Cheng, R., Bai, J., Liu, Y., Qu, Y., Wang, K., Huang, Y. and Duan, X. (2010). *Nature*, **467**, 305.

Lide, D. R., Ed. (2004). *Handbook of Chemistry and Physics* (85th edn). CRC Press, Boca Raton.

Lin, Y., Dimitrakopoulos, C., Jenkins, K., Farmer, D., Chiu, H.-Y., Grill, A. and Avouris, P. (2010). *Science*, **327**, 662.

Lin, Y., Valdes-Garcia, A., Han, S.-J., Farmer, D., Meric, I., Sun, Y., Wu, Y., Dimitrakopoulos, C., Grill, A., Avouris, P. and Jenkins, K. (2011). *Science*, **332**, 1294.

Lin, T., Huang, F., Liang, J. and Wang, Y. (2011a). *J. Energy Environ. Sci.*, **4**, 862.

Liu, C., Yu, Z., Neff, D., Zhamu, A. and Jang, Z. (2010). *Nano Lett.*, **10**, 4863.

Liu, H., Liu, Y. and Zhu, D. (2011). *J. Mater. Chem.*, **21**, 3335.

Los, J., Ghiringhelli, L., Meijer, E. and Fasolino, A. (2005). *Phys. Rev. B*, **72**, 214102.

Los, J., Katsnelson, M., Yazyev, O., Zakharchenko, K. and Fasolino, A. (2009). *Phys. Rev. B*, **80**, 121405.

Low, T., Perebeinos, V., Tersoff, J. and Avouris, P. (2012). *Phys. Rev. Lett.*, **108**, 096601.

Lu, C.-K. and Herbut, I. (2012). *Phys. Rev. Lett.*, **108**, 266402.

Lu, C.-Y., Hsieh, K.-Y. and Liu, R. (2009). *Microelectron. Eng.*, **86**, 283.

Lu, Q., Arroyo, M. and Huang, R. (2009a). *Journal of Physics D: Appl. Phys. Lett.*, **42**, 102002.

Lu, X., Yu, M., Huang, H. and Ruoff, R. (1999). *Nanotechnology*, **10**, 269.

Lu, Y., Goldsmith, B., Kybert, N. and Johnson, A. (2010). *Appl. Phys. Lett.*, **97**, 083107.

Luican, A., Li, G., Reina, A., Kong, J., Nair, R., Novoselov, K., Geim, A. and Andrei, E. (2011). *Phys. Rev. Lett.*, **106**, 126802.

Luryi, S. (1988). *Appl. Phys. Lett.*, **52**, 501.

Mak, K., Lee, C., Hone, J., Shan, J. and Heinz, T. (2010). *Phys. Rev. Lett.*, **105**, 136805.

Maassen, T., van den Berg, J., Huisman, E., Dijkstra, H., Fromm, F., Seyller, T. and van Wees, H. (2013). *Phys. Rev. Lett.*, **110**, 067209.

McCann, E. (2006). *Phys. Rev. B*, **74**, 161403.

McCann, E. and Fal'ko, V. (2006). *Phys. Rev. Lett.*, **96**, 086805.

McChesney, J., Bostwick, A., Ohta, T., Seyller, T., Horn, K., Gonzalez, J. and Rotenberg, E. (2010). *Phys. Rev. Lett.*, **104**, 136803.

McClure, J. W. (1957). *Phys. Rev.*, **108**, 612.

McClure, J. W. (1956). *Phys. Rev.*, **104**, 666.

McClure, J. W. (1960). *Phys. Rev.*, **119**, 606.

McCreary, K., Swartz, A., Han, W., Fabian, J. and Kawakami, R. (2012). *Phys. Rev. Lett.*, **109**, 186604.

McDonald, S., Konstantatos, G., Zhang, S., Cyr, P., Klem, E., Levina, L. and Sargent, E. (2005). *Nature Mater.*, **4**, 138.

McEuen, P. L., Bockrath, M., Cobden, D. H., Yoon, Y.-L. and Louie, S. L. (1999). *Phys. Rev. Lett.*, **83**, 5098.

McKenzie, D., Muller, D., Pailthorpe, B., Wang, Z., Kravtchinskaia, D., Segal, D., Lukins, P., Martin, P., Amaratunga, G., Gaskell, P. and Saeed, A. (1991). *Diamond and Related Materials*, **1**, 51.

Malard, L., Pimnta, M., Dresselhaus, G. and Dresselhaus, M. (2008). *Phys. Rept.*, **473**, 51.

Mao, Y., Wang, W., Wei, D., Kaxiras, E. and Sodroski, J. (2011). *ACS Nano*, **5**, 1395.

Marconcini, P., Cresti, A., Triozon, F., Fiori, G., Biel, B., Niquet, Y.-M. Macucci, M. and Roche, S. (2012). *ACS Nano*, **6**, 7942.

Marder, M., Deegan, R. and Sharon, E. (2007). *Physics Today*, **60**, issue 2, p. 33.

Marianetti, C. and Yevick, H. (2010). *Phys. Rev. Lett.*, **105**, 245502.

Mariani, E. and von Oppen, F. (2008). *Phys. Rev. Lett.*, **100**, 076801.

Mariani, E. and von Oppen, F. (2010). *Phys. Rev. B*, **82**, 195403.

Marks, N. (2000). *Phys. Rev. B*, **63**, 035401.

Martin, J., Akerman, N., Ulbricht, G., Lohmann, T., Smet, J., von Klitzing, K. and Yacoby, A. (2008). *Nature Phys.*, **4**, 377.

Martin, J., Akerman, N., Ulbricht, G., Lohmann, T., Smet, J., von Klitzing, K. and Yacoby, A. (2009). *Nature Phys.*, **5**, 669.

Martoccia, D., Willmott, P., Brugger, T., Bjorck, M., Gunther, B., Schleputz, C., Cervellino, A., Pauli, B., Patterson, B., Wintterline, J., Moritz, W. and Greber, T. (2008). *Phys. Rev. Lett.* **101**, 126102.

Matsui, T., Kambara, H., Nimi, Y., Tagami, K., Tsukada, M. and Fukuyama, H. (2005). *Phys. Rev. Lett.*, **94**, 226403.

Mayorov, A., Elias, D., Mucha-Kruczynski, M., Gorbachev, R., Tudorovskiy, T., Zhukov, A., Morozov, S., Katsnelson, M., Fal'ko, V., Geim, A. and Novoselov, K. (2011). *Science*, **333**, 860.

Mayorov, A., Gorbachev, R., Morozov, S., Britnell, L., Jalil, R., Ponomarenko, L., Blake, P., Novoselov, K., Watanabe, K., Taniguchi, T. and Geim, A. (2011a). *Nano Lett.*, **11**, 2396.

Mermin, N. D. (1968). *Phys. Rev.*, **176**, 250.

Mermin, N. D. and Wagner, H. (1966). *Phys. Rev. Lett.*, **17**, 1133.

Meyer, J., Geim, A., Katsnelson, J., Novoselov, K., Obergfell, D., Roth, S., Girit, C. and Zettl, A. (2007a). *Solid State Commun.*, **143**, 101.

Meyer, J., Geim, A., Katsnelson, M., Novoselov, K., Booth, T. and Roth, S. (2007). *Nature*, **446**, 60.

Miao, X., Tongay, S., Petterson, M., Berke, K., Rinzler, A., Appleton, B. and Hebard, A. (2012). *Nano Lett.*, **12**, 2745.

Milaninia, K., Baldo, M., Reina, A. and Kong, J. (2009). *Appl. Phys. Lett.*, **95**, 183105.

Miller, D., Kubista, K., Rutter, G., Ruan, M., de Heer, W., First, P. and Stroscio, J. (2009). *Science*, **324**, 924.

Millman, R. and Parker, G. (1977). *Elements of Differential Geometry*. Prentice-Hall, Englewood Cliffs, N.J.

Mohr, M., Maultzsch, J., Dobardzic, E., Reich, S., Milosevic, I., Damnjanovic, M., Bosak, A., Krisch, J. and Thomsen, C. (2007). *Phys. Rev. B*, **76**, 035439.

Montambaux, G., Piechon, F., Fuchs, J.-N. and Goerbig, M. (2009). *Phys. Rev. B*, **80**, 153412.

Morozov, S., Novoselov, K., Katsnelson, M., Schedin, F., Elias, D., Jaszczak, J. and Geim, A. (2008). *Phys. Rev. Lett.*, **100**, 016602.

Morozov, S., Novoselov, K., Katsnelson, M., Schedin, F., Ponomarenko, L., Jiang, D. and Geim, A. (2006). *Phys. Rev. Lett.*, **97**, 216801.

Moser, J., Barreiro, A. and Bachtold, A. (2007). *Appl. Phys. Lett.*, **91**, 163513.

Mott, N. F. (1949). *Proc. Phys. Soc. (London) A*, **62**, 416.

Mott, N. F. (1968). *Revs. Mod. Phys.*, **40**, 677.

Mounet, N. and Marzari, N. (2005). *Phys. Rev. B*, **71**, 205214.

Murali, R. (2012). *Graphene Nanoelectronics: From Materials to Circuits*. Springer Verlag, Berlin.

Na, H., Kim, H., Adachi, K., Kiritani, N., Tanimoto, S., Okushi, H. and Arai, K. (2004). *J. Electron. Mater.*, **33**, 89.

Nagashima, A., Nuka, A., Itoh, H., Ichinokawa, T., Oshima, C. and Otani, S. (1993). *Surf. Sci.*, **291**, 93.

Nair, R., Blake, P., Grigorenko, A., Novoselov, K., Booth, T., Stauber, T., Peres, N. and Geim, A. (2008). *Science*, **320**, 1308.

Nair, R., Ren, W., Jalil, R., Riaz, I., Kravets, V., Britnell, L., Blake, P., Schedin, F., Mayorov, A., Yuan, S., Katsnelson, M., Cheng, H.-N., Strupinski, W., Bulusheva, L., Okotrub, A., Grigorieva, I., Grigorenko, A., Novoselov, K. and Geim, K. (2010). *Small*, **6**, 2877.

Nair, R., Wu, H., Jayaram, P., Grigoriea, I. and Geim, A. (2012). *Science*, **335**, 442.

Nakada, K., Fujita, J., Dresselhaus, G. and Dresselhaus, M. (1996). *Phys. Rev. B*, **54**, 17954.

Nelson, D. and Peliti, L. (1987). *J. Phys. (Paris)*, **48**, 1085.

Nelson, D. (1982). *Phys. Rev. B*, **26**, 269.

Nicklow, R., Wakabayashi, N. and Smith, G. (1972). *Phys. Rev B*, **5**, 4751.

Niimi, Y., Kambara, H., Matsui, T., Yoshioka, D. and Fukuyama, H. (2006). *Phys. Rev. Lett.*, **97**, 236804.

Novoselov, K. (2011). *Rev. Mod. Phys.*, **83**, 837.

Novoselov, K., Fal'ko, V., Colombo, L., Gellert, P., Schwab, M. and Kim, K. (2012). *Nature*, **490**, 192.

Novoselov, K., Geim, A. K., Morozov, S. V., Jiang, D., Katsnelson, M. I., Grigorieva, I. V., Dubonos, S. V. and Firsov, A. A. (2005). *Nature*, **438**, 197.

Novoselov, K., Geim, A. K., Morozov, S. V., Jiang, D., Zhang, Y., Dubonos, S. V., Grigorieva, I. V. and Firsov, A. A. (2004). *Science*, **306**, 666.

Novoselov, K., Jiang, Z., Zhang, Y., Morozov, S., Stormer, H., Zeitler, U., Maan, J., Boebinger, G., Kim, P. and Geim, A. (2007). *Science*, **315**, 1379.

Novoselov, K., McCann, E., Morozov, S., Fal'ko, V., Katsnelson, M., Zeitler, U., Jiang, D., Schedin, F. and Geim, A. (2006). *Nature Phys.*, **2**, 177.

Ohta, T., Bostwick, A., Seyller, T., Horn, K. and Rotenberg, E. (2006). *Science*, **313**, 951.

Onsager, L. (1949). *Phys. Rev.*, **65**, 117.

Oostinga, J., Heersche, H., Liu, X., Morpurgo, A. and Vandersypen, L. (2007). *Nature Mater.*, **7**, 151.

Oostinga, J., Sacepe, B., Craciun, M. and Morpurgo, A. (2010). *Phys. Rev. B*, **81**, 193408.

Oshima, C., Itoh, A., Rokuta, E. and Tanaka, T. (2000). *Solid State Commun.*, **116**, 37.

Ouyang, Y., Campbell, P. and Guo, J. (2008). *Appl. Phys. Lett.*, **92**, 063120.

Ozyilmaz, B., Jarullo-Herrero, P., Efetov, D., Abanin, D., Levitov, L. and Kim, P. (2008). *Phys. Rev. Lett.*, **99**, 166804.

Paczuski, J., Kardar, J. and Nelson, D. (1988). *Phys. Rev. Lett.*, **60**, 2638.

Painter, G. and Ellis, D. (1970). *Phys. Rev. B*, **1**, 4747.

Parashar, V., Kumar, K., Prakash, R., Pandey, S. and Pandey, A. (2011). *J. Mater. Chem.*, **21**, 6506.

Park, S., Kwon, O., Lee, S., Song, H., Park, T. and Jang, J. (2012). *Nano Lett.*, **12**, 5082.

Park, S. and Ruoff, R. (2009). *Nature Nanotech.*, **4**, 217.

Pauling, L. (1960). *The Nature of the Chemical Bond* (3rd edn). Cornell University Press, Ithaca, p. 235.

Pauling, L. and Wilson, E. (1935). *Introduction to Quantum Mechanics*. McGraw-Hill, New York, pp. 326–336.

Peierls, R. E. (1923). *Helv. Phys. Acta.*, **7**, 81.

Peliti, L. and Leibler, S. (1985). *Phys. Rev. Lett.*, **54**, 1690.

Pereira, V., Castro Neto, A., Liang, H. and Mahadevan, L. (2010). *Phys. Rev. Lett.*, **105**, 156603.

Pilar, F. J. (2001). *Elementary Quantum Chemistry*. Dover, Mineola, NY, p. 125.

Pinto, H., Jones, R., Goss, J. and Briddon, P. (2010). *Physica status solidi A*, **207**, 2131.

Pinto, M. (2007). "Nanomanufacturing Technology: Exa-units at Nano-Dollars", pp. 154–178, in *Future Trends in Microelectronics: Up the Nano Creek* (Eds. S. Luryi, J. Xu and A. Zaslavsky). Wiley, Hoboken.

Polya, G. (1921). *Math. Ann.*, **84**, 149.

Ponomarenko, L., Geim, A., Zhukov, A., Jalil, R., Morozov, S., Novoselov, K., Grigorieva, I., Hill, E., Cheianov, V., Fal'ko, F., Watanabe, K., Taniguchi, T. and Gorbachev, R. (2011). *Nature Phys.*, **7**, 958.

Ponomarenko, L., Yang, R., Gorbachev, R., Blake, P., Mayorov, S., Novoselov, K., Katsnelson, M. M. and Geim, A. (2010). *Phys. Rev. Lett.*, **105**, 136801.

Pop, E., Mann, D., Wang, Q., Goodson, K. and Dai, H. (2006). *Nano Lett.*, **6**, 96.

Preobrajenski, A., Vinogradov, A. and Martensson, N. (2005). *Surf. Sci.*, **582**, 21.

Qiao, Z., Jung, J., Niu, Q. and MacDonald, A. (2011). *Nano Lett.*, **11**, 3453.

Quinn, J., Kawamoto, G. and McCombe, B. (1978). *Surf. Sci.*, **73**, 190.

Radzihovsky, L. (2004). "Anisotropic and Heterogeneous Polymerized Membranes" Chapter 1 in *Statistical Mechanics of Membranes and Surfaces* (2nd edn). (Eds. D. R. Nelson, T. Piran and S. Weinberg). World Scientific, River Edge, NJ.

Radzihovsky, L. and Toner, J. (1998). *Phys. Rev. E*, **57**, 1832.

Ramon, M., Gupta, A., Corbet, C., Ferrer, D., Movva, C., Carpenter, G., Colombo, L., Bourianoff, G., Docy, M., Akinwande, D., Tutuc, E. and Banerjee, S. (2011). *ACS Nano*, **5**, 7198.

Rashba, E. (2000). *Phys. Rev. B*, **62**. R16267.

Raza, H. (Ed.) (2012). *Graphene Nanoelectronics: Metrology, Synthesis, Properties and Applications*. Springer Verlag, Berlin Heidelberg.

Reddick, W. and Amaratunga, A. (1995). *Appl. Phys. Lett.*, **67**, 494.

Reina, A., Jia, X., Ho, J., Nezich, D., Son, H., Bulovic, V., Dresselhaus, M. and Kong, J. (2009). *Nano Lett.*, **9**, 30.

Reina, A., Kong, J., Terrones, M. and Dresselhaus, D. (2009). *Science*, **323**, 1701.

Reina, A., Son, H., Jiao, L., Fan, B., Dresselhaus, M., Liu, Z. and Kong, J. (2008). *J. Phys. Chem. C*, **112**, 17741.

Romero, E. and de los Santos, F. (2009). *Phys. Rev. B*, **80**, 165416.

Rosenberg, K., Edelstein, D., Hu, C.-K. and Rodbell, K. (2000). *Ann. Rev. Mater. Sci.*, **30**, 229.

Rosenstein, B., Lewkowicz, M. and Maniv, T. (2013). *Phys. Rev. Lett.*, **110**, 066603.

Rossi, E., Bardarson, J., Brouwer, P. and Das Sarma, S. (2010). *Phys. Rev. B*, **81**, 121408.

Rossi, E. and Das Sarma, S. (2008). *Phys. Rev. Lett.*, **101**, 166803.

Ruess, G. and Vogt, F. (1948). *Monatsh Chem.*, **78**, 222.

Ruoff, R. (2008). *Nature Nanotech.*, **3**, 10.

Rusin, T. and Zawadski, W. (2007). *Phys. Rev. B*, **76**, 195439.

Rusin, T. and Zawadski, W. (2009). *Phys. Rev. B*, **80**, 045416.

Saito, R., Dresselhaus, G. and Dresselhaus, M. (1998). *Physical Properties of Carbon Nanotubes*. Imperial College Press, London, pp. 170–171.

Saunders, M., Jimenez-Vazquez, H., Cross, R. and Poreda, R. (1993). *Science*, **259**, 1428.

Schedin, F., Geim, A., Morozov, S., Hill, E., Blake, F., Katsnelson, M. and Novoselov, K. (2007). *Nature Mater.*, **6**, 652.

Schelling, P. and Keblinski, P. (2003). *Phys. Rev. B*, **68**, 35425.

Schiffer, P., O'Keefe, M., Osheroff, D. and Fukuyama, H. (1993). *Phys. Rev. Lett.*, **71**, 1403.

Schleyer, P. and Jiao, H. (1996). *Pure and Applied Chemistry*, **68**, 209.

Schneider, J., Piot, B., Sheikin, J. and Maude, D. (2012). *Phys. Rev. Lett.*, **108**, 117401.

Schniepp, H., Li, J.-L., McAllister, M., Sai, H., Herrera-Alonso, M., Adamson, D., Prud'homme, R., Car, R., Saville, D. and Aksay, I. (2006). *J. Phys. Chem. B*, **110**, 8535.

Schwab, M., Narita, A., Hernandez, Y., Balandina, T., Mali, K., De Feyter, S., Feng, X. and Mullen, K. (2012). *J. Am. Chem. Soc.*, **134**, 18169.

Schwierz, F. (2010). *Nature Nanotech.*, **5**, 487.

Sciambi, A., Pelliccione, M., Lilly, M., Bank, S., Gossard, A., Pfeiffer, L., West, K. and Goldhaber-Gordon, D. (2011). *Phys. Rev. B*, **84**, 085301.

Seabaugh, A. and Zhang, Q. (2010). *Proc. IEEE*, **98**, 2095.

Segal, M. (2009). *Nature Nanotech.*, **4**, 612.

Semenoff, G. (1984). *Phys. Rev. Lett.* **53**, 2449.

Sensale-Rodriguez, B., Fang, T., Yan, R., Kelly, M., Jena, D., Liu, L. and Xing, H. (2011). *Appl. Phys. Lett.*, **99**, 113104.

Seol, J., Jo, I., Moore, A., Lindsay, L., Aitken, Z., Pettes, M., Li, X., Yao, Z., Huang, R., Broido, D., Mingo, N., Ruoff, R. and Shi, L. (2010). *Science*, **328**, 213.

Seyller, T. (2012). "Epitaxial Graphene on SiC(0001)", pp. 135–159, in *Graphene Nanoelectronics: Metrology, Synthesis, Properties and Applications* (ed. H. Raza), Springer Verlag, Berlin Heidelberg.

Shao, Y., Wang, J., Wu, H., Liu, J., Aksay, I. and Lin, Y. (2010). *Electroanalysis*, **22**, 1027.

Shioyama, H. (2001). *Journal of Materials Science Letters*, **20**, 499.

Shivaraman, S., Barton, R., Yu, X., Alden, J., Herman, L., Chandrashekhar, M., Park, J., McEuen, P., Parpia, J., Craighead, H. and Spencer, G. (2009). *Nano Lett.*, **9**, 3100.

Shockley, W. (1950). *Phys. Rev.*, **78**, 173.

Shockley, W. (1953). *Electrons and Holes in Semiconductors*. van Nostrand, New York.

Shytov, A., Katsnelson, M. and Levitov, L. (2007). *Phys. Rev. Lett.*, **99**, 246802.

Shytov, A., Rudner, M. and Levitov, L. (2008). *Phys. Rev. Lett.*, **101**, 156804.

Si, C., Duan, W., Liu, Z. and Liu, F. (2012). *Phys. Rev. Lett.*, **109**, 226802.

Sigal, A., Rojas, M. and Leiva, E. (2011). *Phys. Rev. Lett.*, **107**, 158701.

Simmons, J. (1963). *J. Appl. Phys.*, **34**, 1793.

Simmons, J., Blount, M., Moon, J., Lyo, S., Baca, W., Wendt, J., Reno, J. and Hafich, M. (1998). *J. Appl. Phys.*, **84**, 5626.

Simpson, C. D., Brand, J. D., Berresheim, A. J., Pryzbilla, L., Rader, H. J. and Millen, K. (2002). *Chem. Eur. J.*, **8**, 1424.

Snyman, I. and Beenakker, C. (2007). *Phys. Rev. B*, **75**, 045322.

Sojoudi, H., Baltazar, J., Tolbert, L., Henderson, C. and Graham, S. (2012). *ACS Appl. Mater. Interfaces*, **4**, 4781.

Sols, F., Guinea, F. and Castro Neto, A. (2007). *Phys. Rev. Lett.*, **99**, 166803.

Son, Y.-W., Cohen, M. and Louie, S. (2005). *Phys. Rev. Lett.*, **97**, 216803.

Son, Y.-W., Cohen, M. and Louie, S. (2006). *Nature*, **444**, 347.

Sonde, S., Giannazzo, F., Raineri, V., Yakimova, R., Huntzinger, J.-R., Tiberj, A. and Camassel, J. (2009). *Phys. Rev. B*, **80**, 241 406.

Song, J. and Levitov, L. (2012). *Phys. Rev. Lett.*, **109**, 236602.

Sonin, E. (2009). *Phys. Rev. B*, **79**, 195438.

Sprinkle, M., Ruan, M., Hu, Y., Hankinson, J., Rubio-Roy, M., Zhang, B., Wu, X., Berger, C. and de Heer, W. (2010). *Nature Nanotech.*, **5**, 727.

Sprinkle, M., Siegel, D., Hu, Y., Hicks, J., Tejeda, A., Taleb-Ibrahimi, A., Le Fevre, P., Bertran, F., Vizzini, S., Enriquez, H., Chiang, S., Soukiassian, P., Berger, C., de Heer, W., Lanzara, A. and Conrad, E. (2009). *Phys. Rev. Lett.*, **103**, 226803.

Stampler, C., Schurtenberger, E., Molitor, F., Guttinger, J., Ihn, T. and Ensslin, K. (2008). *Nano Lett.*, **8**, 2378.

Stander, N., Huard, B. and Goldhaber-Gordon, D. (2009). *Phys. Rev. Lett.*, **102**, 026807.

Stankovich, S., Dikin, D., Dommett, G., Kohlhaas, K., Zimney, E., Stach, E., Piner, R., Nguyen, S. and Ruoff, R. (2006). *Nature*, **442**, 282.

Stankovich, S., Dikin, D., Piner, R., Kohlhaas, K., Kleinhammes, A., Jia, Y., Wu, Y., Nguyen, S. and Ruoff, R. (2007). *Carbon*, **45**, 1558.

Staudenmaier, L. (1898). *Ber. Dtsch. Chem. Ges.*, **31**, 1481.

Stoller, M., Park, S., Zhu, Y., An, J. and Ruoff, R. (2008). *Nano Lett.*, **8**, 3498.

Subrahmanyam, K., Kumar, P., Maitra, U., Govidaraj, A., Hembram, K., Waghmare, U. and Rao, C. (2011). *Proc. Natl. Acad. Science*, **108**, 2674.

Subrahmanyam, K., Kumar, P., Nag., A. and Rao, C. (2010). *Solid State Commun.*, **150**, 1774.

Subrahmanyam, K., Vivekchand, A., Govindaraj, A. and Rao, C. (2008). *J. Mater. Chem.*, **18**, 1517.

Sutter, P., Flege, P. and Sutter, E. (2008). *Nature Mater.*, **7**, 406.

Szafranek, B., Fiori, G., Schall, D., Neumaier, D. and Kurz, H. (2012). *Nano Lett.*, **12**, 1324.

Tan, Y., Zhang, Y., Bolotin, K., Zhao, Y., Adam, S., Hwang, E., Das Sarma, S., Stormer, H. and Kim, P. (2007). *Phys. Rev. Lett.*, **99**, 246803.

Tani, S., Blanchard, F. and Tanake, K. (2012). *Phys. Rev. Lett.*, **109**, 166603.

Tanner, B. K. (1995). *Introduction to the Physics of Electrons in Solids*. Cambridge University Press, Cambridge.

Tans, S., Devoret, J., Dai, H., Thess, A., Smalley, R., Geerligs, L. and Dekker, C. (1997). *Nature*, **386**, 474.

Tao, L., Lee, J., Chou, H., Holt, M., Ruoff, R. and Akinwande, D. (2012). *ACS Nano*, **6**, 2319.

Terashima, K., Kondoh, M., Takamura, Y., Komaki, H. and Yoshida, T. (1991). *Appl. Phys. Lett.*, **59**, 644.

Teweldebrhan, D., Goyal, V. and Balandin, A. (2010). *Nano Lett.*, **10**, 1209.

Thompson-Flagg, R., Moura, M. and Marder, M. (2009). *Europhysics Lett.*, **85**, 46002.

Titov, M. and Beenakker, C. (2006). *Phys. Rev. B*, **74**, 041401.

Tombros, N., Jozsa, C., Popinciuc, M., Jonkman, H. and Van Wees, P. (2007). *Nature*, **448**, 571.

Topaszto, L., Dobrik, G., Lambin, P. and Biro, L. (2008). *Nature Nanotech.*, **3**, 397.

Topinka, M., Leroy, B., Shaw, S., Heller, E., Westervelt, R., Maranowski, K. and Gossard, A. (2000). *Science*, **289**, 2323.

Tsen, A., Brown, L., Levendorf, M., Ghahari, F., Huang, P., Havener, R., Ruiz-Vargas, C., Mutter, D., Kim, P. and Park, J. (2012). *Science*, **336**, 1143.

Tsui, D. (1999). *Rev. Mod. Phys.*, **71**, 891.

Tsui, D., Stormer, H. and Gossard, A. (1982). *Phys. Rev. Lett.*, **48**, 1559.

Tzalenchuk, A., Lara-Avila, S., Kalaboukhov, A., Paclillo, S., Syvajarvi, M., Yakimova, R., Kazakova, O., Janssen, T., Fal'ko, V. and Kubatkin, S. (2010). *Nature Nanotech.*, **5**, 186.

Ubbelohde, A., Young, D. and Moore, A. (1963). *Nature*, **198**, 1192.

Ugeda, M., Brihuega, I., Guinea, F. and Gomez-Rodriguez, J. (2010). *Phys. Rev. Lett.*, **104**, 096804.

Vafek, O. and Yang, K. (2010). *Phys. Rev. B*, **81**, 041401.

Van Bommel, A., Crombeen, J. and Van Tooren, A. (1975). *Surf. Sci.*, **48**, 463.

Van der Pauw, L. (1958), *Philips Tech. Rev.*, **20**, 220.

Vazquez de Parga, A., Calleje, F., Borca, B., Passeggi, M., Hinarejos, J., Guinea, F. and Miranda, R. (2008). *Phys. Rev. Lett.*, **100**, 56807.

Verberck, B., Partoens, B., Peeters, F. and Trauzettl, F. (2010). *Phys. Rev. B*, **85**, 125403.

Viculis, L., Mack, J. and Kaner, R. (2003), *Science*, **299**, 1361.

Vinogradov, N., Zahkarov, A., Kocevski, V., Rusz, J., Simonov, K., Eriksson, O., Mikkelsen, A., Lundgren, E., Vinogradov, A., Martensson, N. and Preobrajenske, A. (2012). *Phys. Rev. Lett.*, **109**, 026101.

Wakabayashi, H., Ezaki, T., Sakamoto, T., Kawaura, H., Ikarishi, N., Ikezawa, N., Narihiro, M., Ochiai, Y., Ikezawa, T., Takeuchi, K., Yamamoto, T., Hane, M. and Mogami, T. (2006). *IEEE Trans. Electron Dev.*, **53**, 1961.

Wallace, P. R. (1947). *Phys. Rev.*, **71**, 622.

Wang, F., Zhang, Y., Tian, C., Girit, C., Zettl, A., Crommie, M. and Shen, Y. (2008). *Science*, **320**, 206.

Wang, J., Zhu, M., Outlaw, R., Zhao, X., Manos, D. and Holloway, B. (2004). *Appl. Phys. Lett.*, **85**, 1265.

Wang, P., Hilsenbeck, K., Nirschl, T., Oswald, M., Stepper, C., Weis, M., Schmitt-Landsiedel, D. and Hansch, W. (2004a). *Solid State Electronics*, **48**, 2281.

Wang, X., Li, X., Zhang, L., Yoon, Y., Weber, P., Wang, H., Guo, J. and Dai, H. (2009). *Science*, **324**, 768.

Wang, X., Ouyang, Y., Li, X., Wang, H., Guo, J. and Dai, H. (2008a). *Phys. Rev. Lett.*, **100**, 206803.

Wang, X., Zhi, L. and Mullen, K. (2008). *Nano Lett.*, **8**, 323.

Wang, Y., Wong, D., Shytov, A., Brar, V., Choi, S., Wu, Q., Tsai, H.-Z. and Crommie, M. (2013). *Science*, March 2013. [DOI: 10.1126/science.1234320].

Wang, Y.-X., Yang, Z. and Xiong, S.-J. (2010). *Europhysics Letters*, **89**, 17007.

Wehling, T., Novoselov, K., Morozov, S., Vdovin, E., Katsnelson, M., Geim, A. and Lichtenstein, A. (2008). *Nano Lett.*, **8**, 173.

Weitz, R., Allen, M., Feldman, R., Martin, J. and Yacoby, A. (2011). *Science*, **330**, 812.

Wen, X., Garland, C., Hwa, T., Kardar, M., Kokufuta, E., Li, Y., Orkisz, M. and Tanaka, T. (1992). *Nature*, **355**, 426.

Wirtz, L. and Rubio, A. (2004). *Solid State Commun.*, **131**, 141.

Witten, T. (2007) *Rev. Mod. Phys.*, **79**, 643.

Wolf, E. (1985). *Principles of Electron Tunneling Spectroscopy*. Oxford University Press, Oxford. (See also 2nd edn, 2012.)

Wolf, E. (2009). *Quantum Nanoelectronics*. Wiley-VCH, Weinheim.

Wolf, E. (2010). *IEEE Instrumentation and Measurements Magazine*, **13**, 26.

Wolf, E. (2012). *Principles of Electron Tunneling Spectroscopy: Second Edition*. Oxford University Press, Oxford.

Wolf, E. (2012a). *Nanophysics of Solar and Renewable Energy*. Wiley-VCH, Weinheim.

Wu, J., Agrawal, M., Becerril, H., Bao, Z., Liu, Z., Chen, Y. and Peumans, P. (2010). *ACS Nano*, **4**, 43.

Wu, J., Becerril, H., Bao, Z., Liu, Z., Chen, Y. and Peumanns, P. (2008). *Appl. Phys. Lett.*, **92**, 263302.

Wu, Y., Lin, Y.-M., Bol, A., Jenkins, K., Xia, F., Farmer, D., Zhu, Y. and Avouris, P. (2011). *Nature*, **472**, 74.

Xia, F., Mueller, T., Golizadeh-Mojarad, R., Freitag, M., Lin, Y.-M., Tsang, J., Perebeinos, V. and Avouris, P. (2008). *Nano Lett.*, **9**, 1039.

Xia, F., Mueller, T., Lin, Y.-M., Valdes-Garcia, A. and Avouris, P. (2009). *Nature Nanotech.*, **4**, 839.

Xia, J., Chen, F., Li, J. and Tao, H. (2009a). *Nature Nanotech.*, **4**, 505.

Xiao, D., Liu, G.-B., Feng, W., Xu, X. and Yao, W. (2012). *Phys. Rev. Lett.*, **108**, 196802.

Xu, C., Li, H. and Banerjee, K. (2009a). *IEEE Trans. Electr. Devices*, **56**, 1567.

Xu, K., Cao, P. and Heath, J. (2009). *Nano Lett.*, **9**, 4446.

Xu, X., Gabor, N., Alden, J., van der Zande, A. and McEuen, P. (2010). *Nano Lett.*, **10**, 562.

Xu, S., Irle, S., Musaev, D. and Lin, M. (2007). *J. Phys. Chem. C*, **111**, 1355.

Xue, J., Sanchez-Yamagishi, J., Bulmash, D., Jacquod, P., Deshpande, A., Watanabe, K., Taniguchi, T., Jarillo-Herrero, P. and LeRoy, B. (2011). *Nature Mat.*, **10**, 282.

Yacoby, A., Hess, H., Fulton, T., Pfeiffer, L. and West, K. (1999). *Solid State Commun.*, **111**, 1.

Yang, H., Heo, J., Park, S., Song, H., Seo, D., Byun, K., Kim, P., Yoo, I., Chung, H.-J. and Kim, K. (2012). *Science*, **336**, 1140.

Yang, L., Park, C.-H., Son, Y.-W., Cohen, M. and Louie, S. (2007). *Phys. Rev. Lett.*, **99**, 186801.

Yang, S., Feng, X., Ivanovici, K. and Mullen, K. (2010). *Angew. Chem. Int. Edn.*, **49**, 8408.

Yao, X. and Belyanin, A. (2012). *Phys. Rev. Lett.*, **108**, 255503.

Yoo, M., Fulton, T., Hess, H., Willett, R., Dunkleberger, L., Chichester, R., Pfeiffer, L. and West, K. (1997). *Science*, **276**, 579.

Yoo, E., Kim, J., Hosono, E., Zhou, H.-S., Kudo, T. and Honma, I. (2008). *Nano Lett.*, **8**, 2277.

Yoon, D., Son, Y.-W. and Cheong, H. (2011). *Nano Lett.*, **11**, 3227.

Young, A. and Kim, P. (2009). *Nature Phys.*, **5**, 222.

Yu, P. and Cardona, M. (2010). *Fundamentals of Semiconductors, Physics and Materials Properties*. Springer-Verlag, Berlin, pp. 68–96.

Yu, Y.-J., Zhao, Y., Ryu, S., Brus, L., Kim, K. and Kim, P. (2009). *Nano Lett.*, **9**, 3430.

Zabel, H. (2001). *J. Phys.: Condens. Matter*, **13**, 7679.

Zacharia, R., Ulbricht, H. and Hertel, T. (2004). *Phys. Rev. B*, **69**, 155406.

Zakharchenko, K., Fasolino, A. Los, J. and Katsnelson, M. (2011). *J. Phys. Condens. Matter*, **23**, 202202.

Zakharchenko, K., Katsnelson, M. and Fasolino, A. (2009). *Phys. Rev. Lett.*, **102**, 046808.

Zakharchenko, K., Roldan, R., Fasolino, A. and Katsnelson, M. (2010). *Phys. Rev. B*, **82**, 125435.

Zaslavsky, A., Aydin, C., Luryi, S., Christoloveanu, S., Mariolle, D., Fabroulet, D., Deleonibus, S. (2003). *Appl. Phys. Lett.*, **83**, 1653.

Zener, C. (1934). *Proc. Roy. Soc.*, **A145**, 523.

Zhang, F. and MacDonald, A. (2012). *Phys. Rev. Lett.*, **108**, 186804.

Zhang, L. and Fogler, M. (2008). *Phys. Rev. Lett.*, **100**, 116804.

Zhang, Y., Brar, V., Girit, C., Zettl, A. and Crommie, M. (2009). *Nature Phys.*, **5**, 722.

Zhang, Y., Jiang, Z., Small, J., Purewal, M., Tan, Y., Faziollahi, M., Chudow, J., Jaszczak, J., Stormer, H., Kim, P. (2006) *Phys. Rev. Lett.*, **96**, 136806.

Zhang, Y., Tan, Y.-T., Stormer, H. L. and Kim, P. (2005). *Nature*, **438**, 201.

Zhang, Y., Tang, T.-T., Girit, C., Hao, Z., Martin, M., Zettl, A., Crommie, M., Shen, Y. and Wang, F. (2009a). *Nature*, **459**, 820.

Zhao, L., He, R., Rim, K., Schiros, T., Kim, K., Zhou, H., Gutierez, C., Chockalingam, S., Arguello, C., Pálová, L., Nordlund, D., Hybertson, M., Reichman, D., Heinz, T., Kim, P., Flynn, G. and Pasupathy, A. (2011). *Science*, **333**, 999.

Zhao, X., Outlaw, R., Wang, J., Zhu, M., Smith, D. and Holloway, B. (2006). *J. Chem. Phys.*, **124**, 194704.

Zheng, I., O'Connell, M., Doorn, S., Liao, X., Zhao, Y., Akhadov, E., Hoffbauer, M., Roop, B., Jia, Q., Dye, R., Peterson, D., Huang, S., Liu, J. and Zhu, Y. (2004). *Nature Mater.*, **3**, 673.

Zheng, Y., Ni, G.-X., Toh, C.-T., Tan, C.-Y., Yao, K. and Ozyilmaz, B. (2010). *Phys. Rev. Lett.*, **105**, 166602.

Zheng, Y., Ni, Q.-X., Bae, S., Cong, C., Katiya, O., Toh, C.-T., Kim, H., Im, D., Yu, T., Ahn, J., Hong, B. and Ozyilmaz, B. (2011). *Europhysics Lett.*, **93**, 17002.

Zhou, S., Gweon, G.-H., Fedorov, A., First, P., de Heer, W., Lee, D.-H., Guinea, F., Castro Neto, A. and Lanzara, A. (2007a). *Nature Mater.*, **6**, 770.

Zhou, S., Siegel, D., Fedorov, A. and Lanzara, A. (2008). *Phys. Rev. Lett.*, **101**, 086402.

Zhou, W., Kapetanakis, M., Prange, M., Pantelides, S., Pennycock, S. and Idrobo, J.-C. (2012). *Phys. Rev. Lett.*, **109**, 206803.

Zhou, X., Cuk, T., Devereaux, T., Nagaosa, N. and Shen, Z.-X. (2007). In *Handbook of High-Temperature Superconductivity: Theory and Experiment*. (Eds. J. R. Schrieffer and J. S. Brooks), p. 88. Springer Science + Business Media, New York.

Zhu, J., Bhandary, S., Sanyal, B. and Ottosson, H. (2011). *J. Phys. Chem. C*, **115**, 10264.

Zhu, Y., Li, L., Zhang, C., Casillas, G., Sun, Z., Yan, Z., Ruan, G., Peng, Z., Raji, A.-R., Kittrell, C., Hauge, R. and Tour, J. (2012). *Nature* Commun., **3**:1225 [DOI: 10.1038/2234].

Zhu, Y., Murali, S., Stoller, M., Ganesh, J., Cai, W., Ferreira, P., Pirkle, A., Wallace, R., Cychosz, K., Thommes, M., Su, D., Stach, E. and Ruoff, R. (2011). *Science*, **332**, 1537.

Ziegler, K. (2006). *Phys. Rev. Lett.*, **97**, 266802.

Author Index

Subject Index

Page numbers in italics refer to tables and figures.